The New Frontiers
in
Plant Biochemistry

The New Frontiers in Plant Biochemistry

Edited by
T. Akazawa, T. Asahi and H. Imaseki

SPRINGER-SCIENCE+BUSINESS MEDIA, B.V.

© Springer Science+Business Media Dordrecht
Originally published by Japan Scientific Societies Press in 1983
Softcover reprint of the hardcover 1st edition 1983

Supported in part by the Ministry of Education, Science and Culture under the Publication Grant-in-Aid.

Listed in the Library of Congress Cataloging in Publication Data

ISBN 978-94-009-6856-1   ISBN 978-94-009-6854-7 (eBook)
DOI 10.1007/978-94-009-6854-7

To Professor Ikuzo Uritani
on the occasion of his retirement from Nagoya University

PREFACE

A symposium entitled "The New Frontiers and Future Perspectives of Plant Bio-chemistry" was held in Nagoya, Japan, September 1–3, 1981 in honor of Professor Ikuzo Uritani. Recognizing his planned retirement from Nagoya University in March, 1982, the meeting was organized by Professor Uritani's Japanese colleagues and included a number of foreign scientists, many whom were his close friends. This volume is a compilation of the principal papers contributed for the occasion and is dedicated to Professor Uritani as an expression of the high esteem in which he is held for his outstanding achievements in the field of plant biochemistry and plant disease physiology, as well as to convey our warm personal affection and friendship.

The subjects covered in the volume are diverse, reflecting the honoree's broad research interests, but at the same time articles written by experts in each field provide a clear picture of the current frontiers and perspectives of plant biochemistry research. The continuing development of new experimental strategies has spurred rapid and broad research advances in this field of science, and the many interesting concepts now at hand hold promise of further unique progress in the years ahead. It is hoped that this volume will serve as a stimulating text for scientists in this field.

The editors are grateful to the Japan Scientific Societies Press (Tokyo) and the Martinus Nijhoff/Dr. W. Junk Publishers (The Hague, Boston and London) for having graciously agreed to publish the book. Those responsible for organizing the symposium wish to express their hearty thanks to the Ishida Foundation (Nagoya), Nissan Science Foundation (Tokyo), and Yamada Science Foundation (Osaka). Without their financial support the symposium and the subsequent publication of this dedicatory volume could not have been accomplished.

The Editors

CONTRIBUTORS

Akazawa, T.	Research Institute for Biochemical Regulation, Faculty of Agriculture, Nagoya University, Chikusa, Nagoya 464, Japan
Asahi, T.	Laboratory of Biochemistry, Faculty of Agriculture, Nagoya University, Chikusa, Nagoya 464, Japan
Chong, J.	Research Station, Agriculture Canada, 195 Dafoe Road, Winnipeg, Manitoba, Canada R3T 2M9
Comai, L.	Department of Plant Pathology, University of California, Davis, California 95616, U.S.A.
Conn, E.E.	Department of Biochemistry and Biophysics, University of California, Davis, California 95616, U.S.A.
Fukui, T.	Institute of Scientific and Industrial Research, Osaka University, Suita, Osaka 565, Japan
Graham, D.	Plant Physiology Group, CSIRO Division of Food Research and School of Biological Sciences, Macquarie University, North Ryde, Sydney, N.S.W. 2113, Australia
Higuchi, T.	Wood Research Institute, Kyoto University, Uji, Kyoto 611, Japan
Howes, N.K.	Research Station, Agriculture Canada, 195 Dafoe Road, Winnipeg, Manitoba, Canada R3T 2M9
Imaseki, H.	Research Institute for Biochemical Regulation, Faculty of Agriculture, Nagoya University, Chikusa, Nagoya 464, Japan
Kahl, G.	Department of Biology, Johann Wolfgang Goethe-Universität, Frankfurt am Main, Federal Republic of Germany
Kelly, G.J.	Department of Biochemistry and Nutrition, University of New England, Armidale, N.S.W. 2351, Australia
Kim, W.K.	Research Station, Agriculture Canada, 195 Dafoe Road, Winnipeg, Manitoba, Canada R3T 2M9
Kosuge, T.	Department of Plant Pathology, University of California, Davis, California 95616, U.S.A.

Latzko, E.   Botanisches Institut der Westfälische-Wilhelms Universität,
                  Schlossgarten 3, D-4400 Münster, Federal Republic of Germany
Maclachlan, G.   Biology Department, McGill University, 1205 Avenue Docteur
                  Penfield, Montreal, Que., Canada H3A 1B1
Maeshima, M.   Laboratory of Biochemistry, Faculty of Agriculture, Nagoya
                  University, Chikusa, Nagoya 464, Japan
Miflin, B.J.   Biochemistry Department, Rothamsted Experimental Station,
                  Harpenden, Herts, AL5 2JQ, U.K.
Miller, M.E.   Waksman Institute of Microbiology, Rutgers University,
                  Piscataway, New Jersey 08854, U.S.A.
Price, C.A.   Waksman Institute of Microbiology, Rutgers University,
                  Piscataway, New Jersey 08854, U.S.A.
Reardon, E.M.   Waksman Institute of Microbiology, Rutgers University,
                  Piscataway, New Jersey 08854, U.S.A.
Rohringer, R.   Research Station, Agriculture Canada, 195 Dafoe Road, Win-
                  nipeg, Manitoba, Canada R3T 2M9
Shannon, L.M.   Department of Biochemistry, University of California, River-
                  side, California 92521, U.S.A.
Smith, B.N.   Department of Botany and Range Science, Brigham Young
                  University, Provo, Utah 84602, U.S.A.
Stahmann, M.A.   Department of Biochemistry, College of Agricultural and
                  Life Sciences, University of Wisconsin-Madison, Madison,
                  Wisconsin 53706, U.S.A.
Yang, S.F.   Department of Vegetable Crops, University of California,
                  Davis, California 95616, U.S.A.

CONTENTS

Structural Aspects of Ribulose-1,5-bisphosphate Carboxylase (RuBisCO)*
—— A Personal Recollection —

Takashi Akazawa

The first paper on fraction 1 protein (F1P) was published more than 30 years ago in 1947 in *Archives of Biochemistry* under the joint authorship of Sam Wildman and James Bonner (1). To Japanese it seems remarkable that this monumental work in the contemporary history of plant biology research was performed shortly after World War II, because in Japan at that time basic scientific research was severely hampered by postwar conditions. I was very young then and had certainly never thought that I would be involved in this specific research area some years later.

From 1955–1957 I had the opportunity of studying in the USA, first at Purdue University with Professor H. Beevers, and later at the University of California, Berkeley with Professor Eric Conn. I met Professor Sam Wildman at the time of the 1957 annual meeting of the American Society of Plant Physiologists which was held at Stanford University. He presided over the session on "General Biochemistry", and his group presented two papers: (I) "The distribution of F1P in the plant kingdom" and (II) "The soluble protein composition of expanding and aging tobacco leaves".

---

* Fraction 1 protein and ribulose-1,5-bisphosphate carboxylase are synonymous, the two terms sometimes being used interchangeably. Dr. Wildman's group has used the former name continuously, whereas the abbreviated name of RuBisCO (Ribulose-1,5-bisphosphate carboxylase/oxygenase) was proposed by Professor David Eisenberg, UCLA at the Wildman Symposium that was held in honor of the retirement of Professor Wildman in 1979, coined from Nabisco. While Nabisco may be important to the profits of an American biscuit company, RuBisCO or fraction 1 protein is certainly more important to human welfare, with its abundance throughout the world. In fact, Dr. Wildman has sought potential uses of the crystalline form of fraction 1 protein from tobacco. Apart from its possible commercial uses, RuBisCO is one of the most intriguing enzymes in the field of plant biochemistry.

1

These reports became Papers VII and VIII, respectively, of the series of work entitled "The proteins of green leaves" (2,3).

A few months after the Stanford meeting I dropped in at UCLA on my way back to Japan and visited Wildman's laboratory. It was the first time I saw his greenhouse laboratory which, over the years, became my favorite place to visit whenever I passed by the Los Angeles area. Sam Wildman officially retired from UCLA in 1979 and at that time he summarized the research accomplishments on F1P which he and his group had achieved, particularly in relation to the evolutionary aspect of the protein (4).

The isolation of plant proteins free of polyphenols and other hazardous substances present in leaf tissues and their characterization in a pure form are difficult problems, and this was particularly true 20–30 years ago. A major research project initiated by Professor Uritani at Nagoya University around 1952 involved the enzymology of diseased sweet potato roots, and their experimental trials were hampered by the harmful effects of polyphenolic substances. Establishing a method of removing polyphenols seemed to be imperative for Wildman's group as was a method to isolate the leaf protein fraction. In their studies they developed a device for homogenizing tobacco leaf tissues in a chamber filled with N_2-gas (5). I was interested in seeing this super-mini sky-lab in his greenhouse in 1957, because our research work on sweet potato enzyme was always handicapped by an enormously high content of polyphenolics. In subsequent years, many experimental strategies to get rid of polyphenols have been developed, such as the use of Sephadex gel filtration and the inclusion of DTT in the grinding medium.

Dr. Uritani also had occasion to visit the Biochemistry Department of the University of Wisconsin under the support of the Rockefeller Foundation, 1958–1960. During that time he was engaged in studies on the biochemistry of metabolic disorder of black rot sweet potato. From my personal view, his achievements with Dr. Stahmann on the characterization of anomalous protein components in diseased sweet potato root tissues, employing new techniques such as analytical ultracentrifugation, starch gel electrophoresis, immunoelectrophoresis and Ouchterlony double immunodiffusion is one of the landmarks of plant disease biochemistry (6). Although technical progress in the field of plant protein biochemistry is very fast nowadays, the efforts made years ago by a pioneer in introducing a new method must be highly praised. At the same time, I would like to take this opportunity to express my deep appreciation, to which I am sure Professor Uritani will join me, to the American science community and its scientists for the support they have given to us, without which the current level of Japanese biological sciences would not have been reached. Almost all of us avant et après la guerre Japanese students have had the great priviledge of studying at universities or scientific institutions in the USA.

Some years after my return to Nagoya, I had the chance to work at the International Rice Research Institute (IRRI) in the Philippines (1962–1964) through the recommendation of Dr. Uritani. This was established by the Rockefeller Foundation and I initiated my own research work on F1P and starch biochemistry. I was very fortunate to be permitted to carry out some fundamental biochemistry research and I must express my sincere gratitude to many advisors and colleagues who assisted me during those years. Among them I am particularly grateful to an excellent collaborator, Leticia Mendiola, now Dr. L.M. Morgenthaler, in Switzerland. Also, it is my fond memory to recall the warm encouragement given to me by the late Dr. Sterling Hendricks of the U.S. Department of Agriculture in Beltsville, Maryland. He once kindly sent us a 10 g sample package of Sephadex G-200 for our experimental use, since Sephadex of that specific grade was not then commercially available in the Philippines. We repeatedly generated that precious sample of Sephadex to purify our rice leaf F1P, and were finally able to publish the first of a series of papers on F1P in 1964 (7). Hendricks was known not only as an eminent scientist of our time but also an expert mountain climber. I had the special honor of climbing with him, not Mt. McKinley (6,195m) but a small tropical volcanic mountain, Mt. Makiling (300m!), battling a vicious attack by countless small leeches en route.

In the early 1960s, the polyacrylamide gel electrophoresis was still not widely used for characterizing protein molecules and most people were using hydrolyzed starch gel electrophoresis as a routine method (cf.6). Indeed, the rice leaf F1P which we isolated was tested for its homogeneity by the latter method under the guidance of Dr. Uritani.

It should be recognized that although some basic features of F1P had already been uncovered at an early stage of the investigations, the real indication around 1956 (8) concerning its possible identity as ribulose-1,5-bisphosphate carboxylase (RuBisCO) (then commonly called carboxydismutase by Horecker's group at NIH) became a turning point for later intensive research on this important enzyme in photosynthetic carbon metabolism. In our first report in 1964 (7), we presented experimental evidence showing the CO_2 fixation activity (RuBisCO) associated with the rice leaf F1P, and years later the synonymous nature of F1P and RuBisCO was firmly established by Trown (9) and Paulsen and Lane (10). Based on the abundance of F1P in leaf tissues as well as its large molecular size we proposed its possible multifunctional property, and this view appeared to be shared by Wildman and Van Noort who employed immunological techniques to test this theory (11). When Dr. Horecker visited us at IRRI in 1964, I discussed the problem with him and he strongly encouraged us to pursue the possibility quoting the classical work on crystalline aldolase from rabbit muscle carried out by Baranowski (12). In spite of the fact that the multi-enzyme complex hypothesis of F1P has been denied by both Trown and Paulsen-Lane group, I believe that the structure-function relationship of F1P (RuBisCO) in situ still remains as an intriguing research problem to be tackled by future investigations (cf.13). As will be discussed later, not only is the crucial function of subunit B not yet known, but there also exists a possibility that the thylakoid membrane-bound form of RuBisCO may have a role distinguishable from that in the chloroplast stroma (14).

Since my return to Japan, we have almost exclusively shifted to spinach leaf for our RuBisCO studies, which is of course the international favor of plant science research. In parallel, however, we are also studying the enzyme from other autotrophic organisms, e.g. green algae, blue-green algae and photosynthetic bacteria, originally from the comparative biochemistry of the enzyme structure (15), and

TABLE I
Historial Records on Quaternary Structural Models Proposed to RuBisCO

Structural models	Sources	Authors	Year
24 Identical subunits	Chinese cabbage	Haselkorn et al.	1965
"	Oat	Steer et al.	1968
A_8B_6	Tobacco	Kawashima and Wildman	1970
A_8B_{8-10}	Spinach	Rutner	1970
A_8B_8	Spinach	Wishnick et al.	1970
		Siegel and Lane	1972
A_8B_8	Spinach	Sugiyama and Akazawa	1970
	Chlorella	Nishimura et al.	1973
A_8B_8	Tobacco	Baker et al.	1975

more recently from the standpoint of the molecular evolution of RuBisCO (*13,16*).

I think that the advancement of our knowledge on the structure of F1P (RuBis-CO) is an interesting and rather informative one (Table I). Haselkorn, Fernandez-Moran and associates (*17*) proposed an F1P structure isolated from Chinese cabbage containing 24 identical subunits, which was apparently supported by Steer and others (*21*) using protein molecules isolated from the oat plant. These studies were based purely on electron microscopic ultrastructural investigations without chemical analyses. As time went on, however, the quaternary structure of F1P (RuBisCO) was established by Rutner and Lane in. 1967 (*19*), who demonstrated the structural make-up of the spinach enzyme to be comprised of two different subunits (A and B*). This basic structural organization has since been confirmed by many investigators and it is now believed that the enzymes from a wide variety of different photosynthetic organisms have a symmetrical structure containing an octamer each of the large and small subunits, A_8B_8 (*16,20*).

The development of the structural studies on RuBisCO is reminiscent of the research history on the structure of bacterial aspartate transcarbamylase (ATCase). ATCase is the best studied allosteric enzyme containing two different subunits, catalytic (C) and regulatory (R) chains. The enzyme was first discovered by a group in Berkeley, and for some years they held the view that it had a tetrametric structure, C_4R_4 (*21,22*). In 1968, however, Weber at Harvard reported that the enzyme molecule had, in fact, a hexameric structure, C_6R_6, and this was supported by Lipscomb's X-ray crystallographic analysis (*23,24*) (Table II). This outstanding finding was quite sensationally reported in *Nature* using the phrase "Evergreen Enzyme ... " (*25*). I am inclined to believe that F1P/RuBisCO can be categorized as another excellent example of an "Evergreen Enzyme" and, indeed, the constant increase in annual research publications on this protein can be taken as a proof of this notion. The

* In most literature dealing with RuBisCO, abbreviated terms of L and S are commonly used signifying the large and small subunit of the enzyme molecule. However, we have continuously used the symbols, A and B, which were initially designated by A. Rutner and M.D. Lane in their first paper (ref.*19*). It would seem most advantageous to standardize the terminology among the scientists concerned and for future reporting.

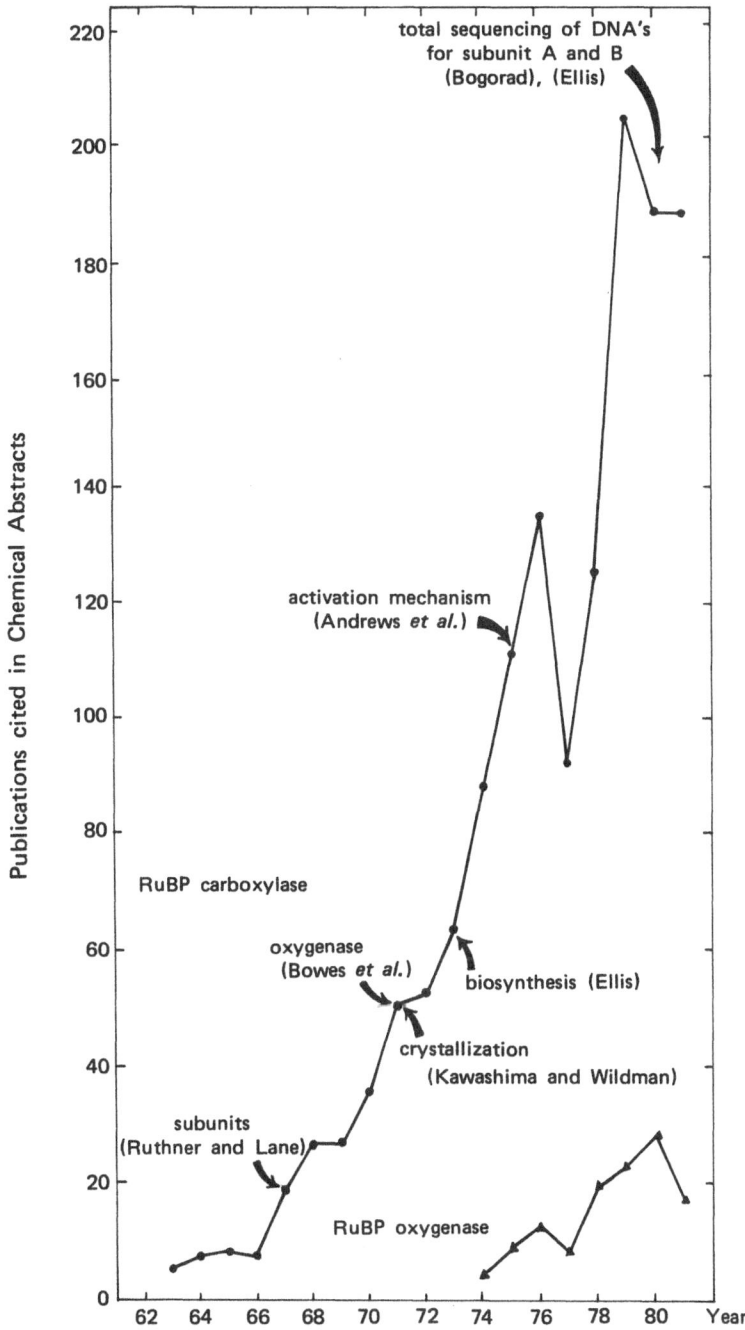

Fig. 1. Numbers of research publications concerning RuBisCO cited in the *Chemical Abstracts*.

yearly number of research papers on RuBisCO cited in *Chemical Abstracts* is illustrated in Fig. 1; for example, 200 papers in 1979 are quite a few more than were written on many other plant enzymes.

It appears that building models of complex molecules is a traditional habit of many chemists; for instance, one can see a picture of a "benzene" ring model in 1886 (Fig.2). Similar attempts made by contemporaries for enzyme molecules seem

TABLE II
Studies on Quaternary Structure of *Escherichia coli* Aspartate Transcarbamylase

Structural models	Authors	Year
C_4R_4	Gerhart and Schachman	1965
	Changeux *et al.*	1968
	Herve and Stark	1967
	Weber	1968
C_6R_6	Weber	1968
	Wiley and Lipscomb	1968
	Meighen *et al.*	1970
	Rosenbusch and Weber	1971
	Richards and Williams	1972
	Cohlberg *et al.*	1972
C_6R_4 (variant)	Jacobson and Stark	1973
	Yang *et al.*	1974
	Evans *et al.*	1974

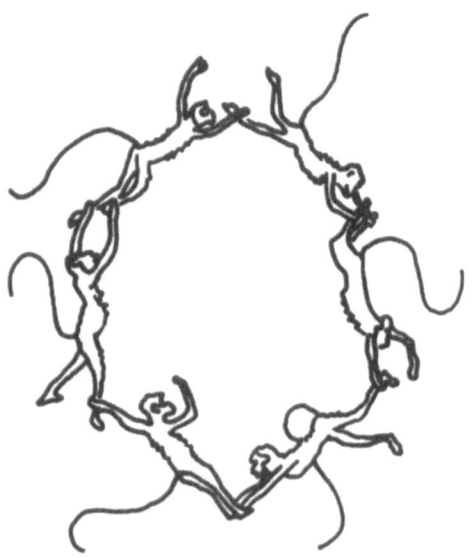

Fig. 2. Structure of benzene ring (1886)

Fig. 3. Research strategies for structure-function relationship of enzyme proteins. Redrawn from Neurath *et al.* (1968) by anonym.

sometimes to be quite constructive for better understanding of the protein structure of enzyme catalysis and regulatory mechanisms (Fig. 3). However, it should be realized that the final goal of understanding the structure-function relationship can be attained by X-ray crystallographic analysis. Wildman's group first succeeded in crystallizing F1P from the tobacco leaf in 1971 (*26*), and recently a more refined method of obtaining crystalline F1P molecules in large quantity from other plant sources usable for X-ray analysis has been developed by Bourque (*27*). However, generally speaking, plant enzyme research along this line has been very backward compared to that of animal and microbial enzymes, and no substantial progress has been made concerning the three dimensional structure of RuBisCO based on the X-ray diffraction method since 1977 (*28*).

There is now strong and convincing experimental evidence that the enzyme catalytic site is located in subunit A of RuBisCO (*20,29*). Such studies indicate that the dissociable nature of RuBisCO under suitable conditions has facilitated solution of the problem. The splitting of the spinach RuBisCO molecule into two subunit species by mercurial (PCMB) treatment at alkaline pH is basically analogous to that employed for the bacterial ATCase used by Gerhart and Schachman (*22,29*). On the other hand, it was found that RuBisCO from the photosynthetic bacterium, *Chromatium vinosum,* can readily be dissociable into two subunits upon mere exposure to an alkaline pH (30); this has been most advantageous in studying the precise nature of this plant-type RuBisCO from the prokaryotic origin ($31,32$).

One of the most prominent research accomplishments made concerning the mechanistic aspect of RuBisCO was the CO_2-Mg^{2+} activation phenomena discovered by Lorimer and his colleagues (33), and through this unique property of the protein molecules many of the hitherto enigmatic features of RuBisCO in chloroplasts relating to photosynthetic CO_2 assimilation have been unveiled (34). The detailed studies by Lorimer and associates ($35,36$) have clearly elucidated the molecular picture of the CO_2-combining site in subunit A, allowing us to understand the molecular mechanism(s) of the enzyme activation.

The only exception of the RuBisCO molecule deviating from the $A_8 B_8$ structure, whose quaternary structure has been studied in depth, is the one from *Rhodospirillum rubrum* which contains only subunit A with a molecular structure of A_2 (37). It has been found that *rubrum* RuBisCO can be equally activated by incubation with CO_2 and Mg^{2+}, providing proof of the presence of the activation site in subunit A ($38-40$). It is known that K_m (CO_2) of the bacterial enzyme as well as the catalytic core moiety (A_8) of the spinach enzyme which is devoid of subunit B is substantially larger than that of the plant-type holoenzyme ($A_8 B_8$). However, the association of subunit B to the larger subunit A was found to decrease the K_m (CO_2) value, and it can be speculated that the role of subunit B lies in making a conformational stabilization of the enzyme molecule, accompanied by an increase in the affinity of the RuBisCO molecule to CO_2 (38).

Our recent experiments concerning the dissociation/association phenomena of RuBisCO isolated from the halophilic blue-green alga, *Aphanothece halophytica*, which can be readily manipulated by the salinity level in the reaction medium (41), as well as an analogous investigation by Andrews and associates at the Australian Institute for Marine Sciences using the enzyme from the marine blue-green alga, *Synechococcus* sp. (42), have demonstrated that the oligomeric form of subunit A is enzymatically inactive, whereas the addition of subunit B results in the reconstitution of the active holoenzyme molecule. In other words, *Aphanothece* RuBisCO requires the quaternary structural organization of $A_8 B_8$ for the expression of enzyme activities. Our previous experiments have shown that a partial reconstitution of spinach leaf RuBisCO from the two separated subunits (A and B) can be attained, but the magnitude of enzyme activity enhancement exerted by the subunit B addition was not large ($29,43$). It now appears likely that the mechanism operating in the blue-green algal RuBisCO is one in which subunit B influences a conformational change at the catalytic site of the enzyme molecule other than that engaged in the CO_2-Mg^{2+} activation localized in subunit A. Taken together we are inclined to speculate that, although the mode of action of subunit B in the RuBisCO molecules may differ kinetically in each individual enzyme of specific origin, its intrinsic role is probably related to enzyme catalysis in conjunction with its association with subunit A. The blue-green algae (cyanobacteria) belonging to the prokaryotic photoautotroph are an interesting class of microorganisms from the standpoint of the evolutionary development of their photosynthetic apparatus (chloroplasts) and many important investigations on them have been carried out. Our recent work may provide us an excellent chance of revealing the essential role of subunit B in the molecule of RuBisCO, hitherto mysterious so far as the enzymes of eukaryotic origin are concerned.

Compared to the time of Wildman in 1947 and ours in 1964, the technological advancement in plant protein biochemistry today is amazingly fast and versatile. Yet I believe that there still exist many important undiscovered features in the structure-function of RuBisCO (F1P), which require intensive research effort in the future.

REFERENCES

1. Wildman, S.G. and Bonner, J. : 1947. *Arch. Biochem.* **14**, 381.
2. Dorner, R.W., Kahn A. and Wildman, S.G. : 1957. *J. Biochem.* **229**, 945.
3. Dorner, R.W., Kahn A. and Wildman, S.G. : 1958. *Biochim. Biophys. Acta* **29**, 240.
4. Wildman, S.G. : 1979. *Arch. Biochem. Biophys.* **196**, 598.
5. Gohn, M., Ginoza, W., Dorner, R.W., Hudson, W.R. and Wildman, S.G. : 1956, *Science,* **124**, 1081.
6. Uritani, I. and Stahmann, M.A. : 1961. *Plant Physiol.* **36**, 770.
7. Mendiola, L. and Akazawa, T. : 1964. *Biochemistry* **3**, 174.
8. Weissbach, A., Horecker, B.L. and Hurwitz, J. : 1956. *J. Biol. Chem.* **218**, 795.
9. Trown, P.W. : 1965. *Biochemistry* **4**, 908.
10. Paulsen, J.M. and Lane, M.D. : 1966. *Biochemistry* **5**, 2350.
11. Wildman, S.G., Van Noort, G. and Hudson, W. : 1961. *Plant Physiol.* **36**, suppl. xix.
12. Baranowski, T. : 1939. *Z. Physiol. Chem,* **260**, 43.
13. Akazawa, T. : 1970. *Progress in Phytochemistry* **2**, 107–141.
14. Akazawa, T. : 1977. Proc. Fourth Int. Cong. Photosynthesis (Reading, England), pp.447–456.
15. Takabe, T. and Akazawa, T. : 1975. *Plant Cell Physiol.* **16**, 1049.
16. Akazawa, T., Asami, S. and Kobayashi, H. : 1982. Abstracts 12th Int. Cong. Biochem. (Perth, Australia), p.74.
17. Haselkorn, R., Fernandez-Moran, H., Kieras, F.J. and van Bruggen, E.F.J. : 1965. *Science* **150**, 1598.
18. Steer, M.W., Gunning, B.E.S., Graham, T.A. and Carr, D.J. : 1968. *Planta* **79**, 254.
19. Rutner, A. and Lane, M.D. : 1967. *Biochem. Biophys. Res. Commun.* **28**, 531.
20. Akazawa, T. : 1979. Encyclopedia of Plant Physiology, Photosynthesis II (Gibbs, M. and Latzko, E., eds.), pp.208–229, Springer-Verlag, Berlin, Heidelberg and New York.
21. Changeux, J.-P. : 1964. *Brookhaven Symp. Biol.* **17**, 232.
22. Gerhart, J.C. and Schachman, H.K. : 1965. *Biochemistry* **4**, 1054.
23. Weber, K. : 1968. *Nature* **218**, 1116.
24. Wiley, D.C. and Lipscomb, W.N. : 1968. *Nature* **218**, 1119.
25. Anonym : 1968. "News and Views. Evergreen Enzymes", *Nature* **218**, 1202.
26. Kawashima, N. and Wildman, S.G. : 1970. *Annu. Rev. Plant Physiol.* **21**, 325.
27. Johal, S. and Bourque, D.P. : 1979. *Science* **204**, 76.
28. Baker, T.S., Suh, S.W. and Eisenberg, D. : 1977. *Proc. Natl. Acad. Sci. U.S.A.* **74**, 1037.
29. Nishimura, M. and Akazawa, T. : 1973. *Biochem. Biophys. Res. Commun.* **54**, 842.
30. Takabe, T. and Akazawa, T. : 1973. *Arch. Biochem. Biophys.* **157**, 303.
31. Takabe, T. and Akazawa, T. : 1975. *Arch. Biochem. Biophys.* **169**, 686.
32. Takabe, T. and Akazawa, T. : 1975. *Biochemistry* **14**, 46.
33. Lorimer, G.H., Badger, M.R. and Andrews, T.J. : 1976. *Biochemistry* **15**, 529.

34. Lorimer, G.H. : 1981. *Annu. Rev. Plant Physiol.* **32**, 349.
35. Lorimer, G.H. and Miziorko, H.M. : 1980. *Biochemistry* **19**, 5321.
36. Lorimer, G.H. and Miziorko, H.M. : 1983. *Annu. Rev. Biochem.* in press.
37. Tabita, F.R. and McFadden, B.A. : 1974. *J. Biol. Chem.* **249**, 3459.
38. Kobayashi, H., Takabe, T., Nishimura, M. and Akazawa, T. : 1979. *J. Biochem.* **85**, 923.
39. O'Leary, M.H., Jaworski, R.J. and Hartman, F.C. : 1979. *Proc. Natl. Acad. Sci. U.S.A.* **76**, 673.
40. Christeller, J.T. and Laing, W.A. : 1978. *Biochem. J.* **173**, 467.
41. Asami, S., Takabe, T. Akazawa, T. and Codd, G.A. : 1983. *Arch. Biochem. Biophys.* in press.
42. Andrews, J.T. and Abel. K.M. : 1981. *J. Biol. Chem.* **256**, 8445.
43. Nishimura, M. and Akazawa, T. : 1974. *J. Biochem.* **76**, 169.

Cyanogenic Glycosides: A Possible Model for the Biosynthesis of Natural Products

Eric E. Conn

INTRODUCTION

The production of hydrogen cyanide (HCN) by living organisms is known as *cyanogenesis*. This phenomenon is exhibited by approximately 2,000 species of higher plants, some ferns, fungi and bacteria, and by a few animals (certain moths and millipedes) (*1,2*). The HCN is released from a bound form, usually a cyanogenic glycoside, as the result of enzymatic action initiated by the disruption of tissues of the organism. Cyanogenesis is considered to be a defense mechanism and, in the case of cyanogenic plants, data exist supporting the thesis that cyanogenic glycosides and cyanogenesis have provided survival value to cyanogenic species during the course of evolution.

Several aspects of cyanogenesis have received renewed attention in recent years : some laboratories have examined the biosynthesis of cyanogenic compounds (glycosides and lipids) in plants (see *3,4* for review). Three other laboratories have studied the production of free HCN by bacteria, a process which may resemble in some aspects the synthesis of plant glycosides although details are few (*5*). With insects, Davis and Nahrstedt (*6*) have shown that moths indeed synthesize their cyanogenic compounds instead of simply sequestering them from cyanogenic plants on which they may feed. All of these studies were initiated because of the fundamentally interesting biochemistry involved ; the work with plants in particular has disclosed a unique biosynthetic pathway involving intermediates not normally encountered in the conventional amino acid metabolism of animals and microorganisms (*3*).

Cyanogenesis also has been studied because some of the most important food

11

crops for man and his animals are cyanogenic species. Sorghum and cassava are the main source of calories for nearly 750 million people ; sorghum is also an important food for livestock. While sorghum grain fortunately is not cyanogenic, the foliage of young plants is highly toxic and animals must be fenced out of sorghum fields. Similarly, sorghum ensilage must be allowed to ferment extensively in order to reduce its HCN content. In the case of cassava, the highly cyanogenic tuber must be processed to remove its HCN and residual cyanogenic glycosides before the starch can be made into flour or other products and consumed. Even so, there is a body of literature suggesting that certain African populations which consume large amounts of cassava may suffer from chronic cyanide poisoning (7).

Cyanogenesis is of interest as a taxonomic character ; Hegnauer, the main worker in this field, has reviewed the subject (8). Cyanogenesis as a plant defense mechanism is recognized by chemical ecologists and evolutionists who study the interactions between plants and herbivores (9).

CYANOGENIC COMPOUNDS

When the cyanogenic glycosides were reviewed in 1970, a total of 18 compounds were described (10). A recent listing of these compounds puts the number now known to exist as 55 (11). Such numbers indicate the renewed research activity in this subject during the past decade, a research effort which has disclosed structural variations unknown in the earlier literature (see 4 for review).

Although the number of cyanogenic glycosides has tripled in the past decade, it is interesting and perplexing that, with one exception (12), all of the newly described compounds are biosynthetically related to compounds known for more than a half-century. In other words, all of the known cyanogenic glycosides, with the exception noted, have their biosynthetic origin in either one of five protein amino acids or are (presumably) biosynthesized from the non-protein, cyclopentenoid amino acid, cyclopentenyl glycine. This precursor-product relationship for the five protein amino acids is shown in Fig. 1.

The relationship between two cyanogenic glycosides whose aglycones are both derived from L-phenylalanine is shown in Fig. 2. Amygdalin, recognized as a cyanogenic substance for 175 years (13), is the D-gentiobioside of (R)-mandelonitrile ; it differs from prunasin, the β-D-glucoside of (R)-mandelonitrile, by containing an additional hexose in its structure. Both compounds are found in the Rosaceae with amygdalin occurring in seeds and prunasin mainly in vegetative tissue. Other glycosides derived from phenylalanine are the epimer of prunasin, sambunigrin [(S)-mandelonitrile-β-D-glucoside), vicianin, whose sugar moiety is the disaccharide vicianose (14) ; and lucumin, the primaveroside of (R)-mandelonitrile (15)]. Recent reviews discussing the physical and chemical properties of these and many other cyanogenic glycosides are available (16,17).

Two epimeric glucosides derived from L-tyrosine are known (Fig. 3) ; the (S)-epimer known as dhurrin was first isolated form *Sorghum bicolor* (18) while taxiphyllin was discovered in *Taxus* species (19). These compounds and triglochinin,

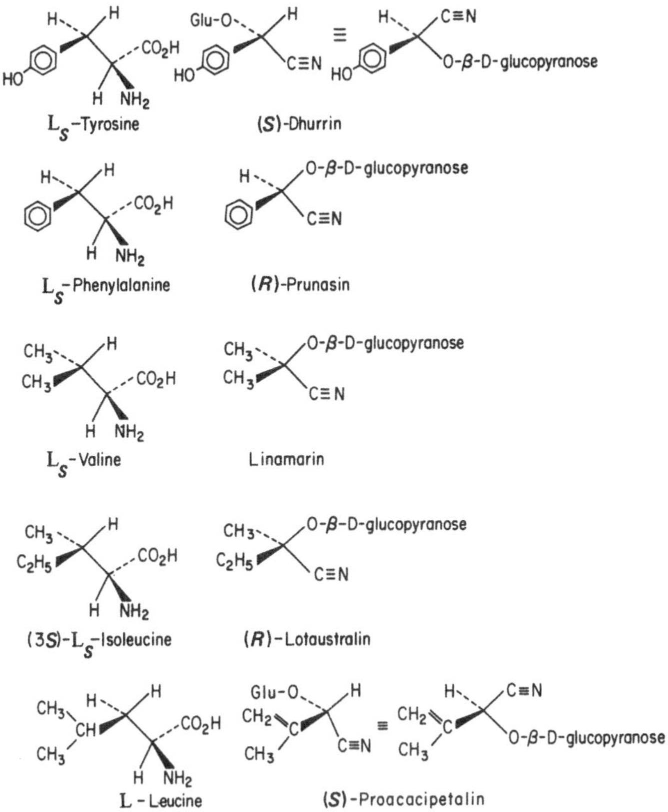

Fig. 1. The precursor-product relationships for five cyanogenic glucosides (Reprinted with permission from *Fed. Proc.* 41(10), 25–34, 1982).

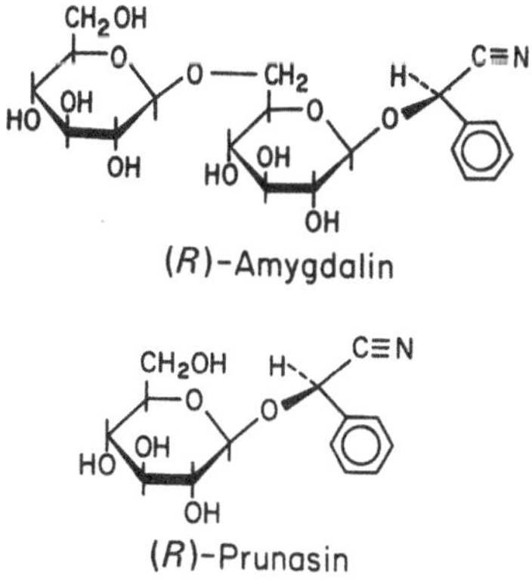

Fig. 2. The structures of amygdalin and prunasin.

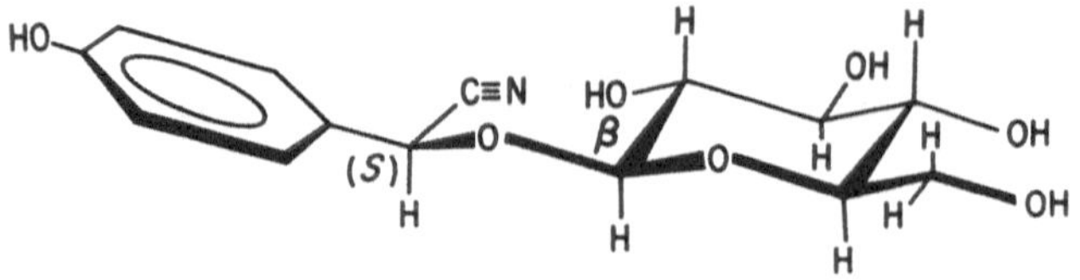

Dhurrin

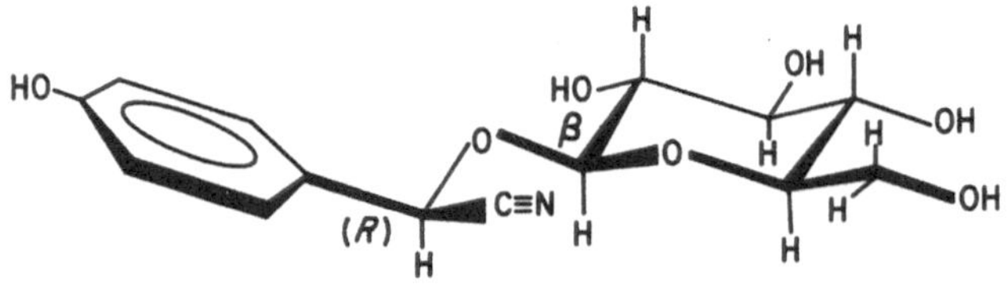

Taxiphyllin

Fig. 3. The structures of dhurrin and taxiphyllin.

which may be formed by oxidative ring cleavage of the aglycone of dhurrin or taxi-phyllin, have special significance as chemosystematic markers in a number of plant families (11). A structural isomer, p-glucosyloxy-mandelonitrile, constitutes the single example of such isomerism in this class of compounds (16).

The structures of linamarin, lotaustralin and proacacipetalin are shown in Fig. 1; their aglycones are biosynthesized from L-valine, L-isoleucine, and L-leucine, res-pectively (4). Both lotaustralin and proacacipetalin are chiral at the carbinol carbon where the sugar is attached and epimers of these two glycosides can be expected to exist in nature. To date, however, only the epimer of proacacipetalin has been re-ported (20). Gentiobioside forms of linamarin, lotaustralin, and proacacipetalin have recently been described (21,22). The aglycone of proacacipetalin has been modified in nature both by reduction of the double-bond and by hydroxylation of the methyl group to yield additional glycosides ; the hydroxyl group in turn can exist as a sulfate or p-hydroxybenzoate derivative (1,11).

Five cyanogenic glycosides having a cyclopentenoid aglycone are known (1), but the configuration of only one, gynocardin, is established (Fig. 4). These compounds obviously cannot be derived directly from a protein amino acid as are the other gly-cosides. The presumed precursor is cyclopentenyl glycine, a non-protein amino acid whose biosynthesis has been studied (23). These cyclopentanoid glycosides are pre-

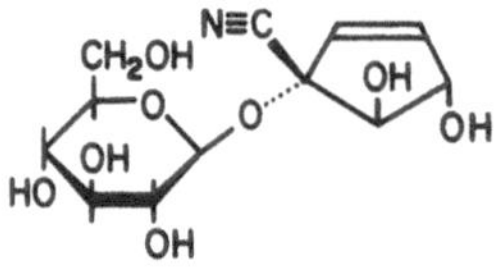

Fig. 4. The structure of gynocardin.

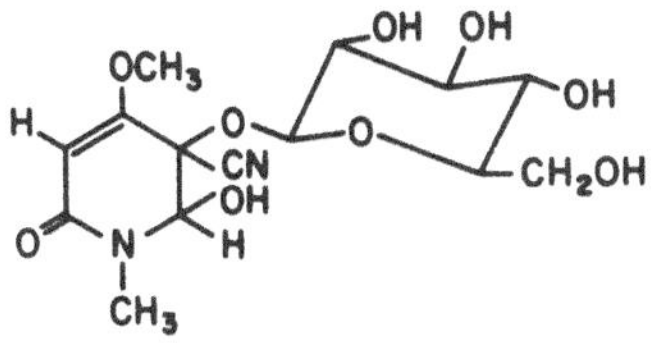

Acalyphin

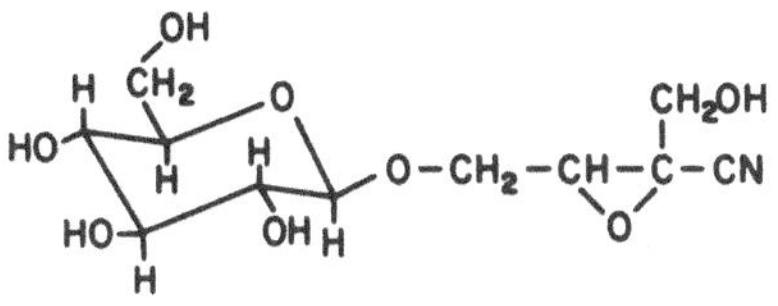

Sarmentosin epoxide

Fig. 5. The structure of sarmentosin epoxide and acalyphin.

sently known only in three plant families (Flacourtiaceae, Turneraceae, Passiflora-ceae). As such, they are valuable markers (11).

Recently, Nahrstedt and coworkers have described two cyanogenic glycosides whose interesting structures deserve comment. Sarmentosin epoxide (Fig. 5), isolated from *Sedum cepaea* (24), is unusual in that the release of HCN is not dependent on the hydrolytic cleavage of the glucosidic bond. A labile cyanohydrin will be formed instead when the oxiran ring is hydrolyzed (by an epoxidase?) and can then liberate HCN. The aglycone of the naturally occurring glycoside is presumed to be bio-synthesized from isoleucine.

Acalyphin (Fig. 5) is of even greater interest because it appears to be the first cyanogenic glycoside discovered in the past 40 years which is derived from a pre-cursor other than the six amino acids listed above. Nahrstedt *et al.* (25), in comment-ing on its formula, have noted the structural similarity of acalyphin to the cyanopyri-dones ricinine, ricinidine, nudiflorine, and mallorepine.

It is not our intent to provide the structures of all of the cyanogenic glycosides. The examples given permit the following generalizations regarding this group of compounds. They possess aglycones which are aromatic, aliphatic, or alicyclic cyano-hydrins (α-hydroxynitriles) of relatively simple structure. The carbinol carbon atoms are usually chiral, thereby providing the possibility of epimeric glycosides when the aglycones are glycosylated. In every case, the sugar linked to the α-hydroxynitrile is β-D-glucose ; in six instances, a second monosaccharide is attached to the first to yield a disaccharide as the cyanogenic glycoside. In the case of cyanogens derived from leucine, the aglycone may be further modified (hydroxylated, reduced or iso-merized) to yield closely related compounds. This group of cyanogenic glycosides is also related to a group of cyanolipids in which the sugar group is replaced with C_{20} long-chain fatty acids.

CYANOGENESIS

The phenomenon of cyanogenesis was discovered at the same time as the cyanogenic glycosides because cyanophoric plant tissues usually contain the enzymes responsible for the release of HCN from the cyanogenic precursors. This process is shown for amygdalin in Fig. 6 and consists of the stepwise hydrolysis of (R)-amygdalin to form (R)-prunasin followed by its hydrolysis to yield (R)-mandelonitrile, a cyanohydrin. While the cyanohydrin can break down non-enzymatically, its rapid dissociation is catalyzed by a stereospecific hydroxynitrile lyase (oxynitrilase).

Several points can be made regarding this process of cyanogenesis. With regard to higher plants, the term cyanogenesis refers to the release, often rapid, of HCN when cyanophoric plant tissues are crushed. Since it can be shown that the plant, prior to crushing, contains intact glycoside, it was early assumed that the cyanogenic substrate was located in one compartment while its catabolic enzymes were in another. The physical disruption of the plant tissue serves to destroy this compartmentation and the substrate is hydrolyzed when it mixes with its catabolic enzymes. This process of hydrolysis of a plant glycoside by a plant glycosidase is not unique for cyanogenic glycosides ; many other glycosides occur in intact plant tissues, only to be hydrolyzed when the tissue is disrupted.

The compartmentation of dhurrin away from its catabolic enzymes has been examined by Kojima et al. (26). Working with purified preparations of protoplasts from the mesophyll and epidermal tissues of young green sorghum leaves, these workers showed that most of the dhurrin was contained in the epidermis while the

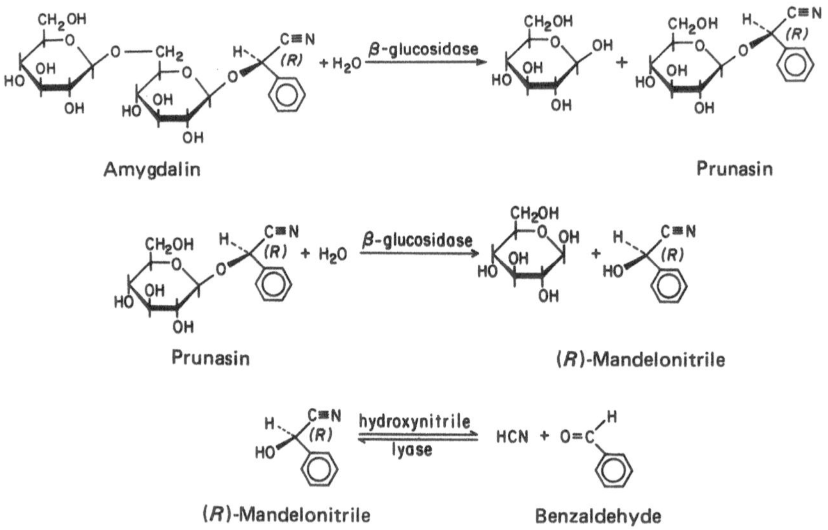

Fig. 6. The enzymatic hydrolysis of amygdalin (Reproduced with permission from *Int. Rev. Biochem.* 27, 21–43, 1979).

catabolic enzymes were found in the mesophyll cells. The third major tissue in such leaves, the bundle-sheath strands, contained neither substrate nor catabolic enzymes. In earlier work, Saunders and Conn (27) had shown that dhurrin was located in vacuoles which could be isolated from sorghum protoplasts ; in view of the later studies (26), these vacuoles had to arise almost exclusively from the epidermal protoplasts.

The compartmentalization of the glucoside of o-coumaric acid and its catabolic glucosidase in leaves of *Melilotus alba* has been studied by Oba *et al.* (28). In this case, the vacuoles of mesophyll cell contained the glucoside, but the glucosidase was absent from those cells. Since the glucosidase activity present in buffered homogenates of *M. alba* leaves could not be accounted for in the protoplasts isolated from the leaves, the authors concluded that the glucosidase was located either on the cell walls, external to the plasma membrane or in the apoplasmic space. Similar findings had been made earlier by Marcinowski and Grisebach (29) who were studying the compartmentation of glycosidic precursors of lignin and their associated glucosidases.

The enzymes responsible for cyanogenesis in cyanophoric plants, *i.e.*, the β-glucosidase and hydroxynitrile lyases, have been the subject of recent investigations. The most important general finding of such studies is that these enzymes are highly specific for the substrates involved. While this is not surprising from the standpoint of enzymology, it is a fact not always appreciated by people working with plant β-glycosidases.

Hösel and Conn (30) have recently summarized the literature on those plant β-glycosidases which show specificity for naturally occurring aglycones. They note that the opinion is generally held, erroneously, that this group of plant enzymes lack specificity for their aglycones. This is due mainly because almond β-glucosidases ("emulsin"), which are unusual because of their comparatively low specificity, have been extensively used in studying the action of plant β-glycosidases. Another cause of this erroneous impression is the widespread and often exclusive use of artificial substrates such as benzyl-, nitrophenyl- or 4-methyl umbellifery-glycosides. Although these substrates have the advantage of being easily assayed, the enzymes which bring about their hydrolysis appear to have little, if any, relationship to the proteins which hydrolyze the glycosidic bonds found in natural products. The hydroxynitrile lyases similarly exhibit specificity for the cyanohydrins, not only for the general nature (aromatic *vs.* aliphatic) of their substrate but also for the stereochemistry (*R vs. S* configuration) of the aglycone.

The specificity of these catabolic enzymes for their substrates implies a physiological role for these proteins and for the process of cyanogenesis itself. As noted elsewhere (9), there is evidence for cyanogenesis and cyanogenic compounds having provided an advantage to cyanophoric plants during evolution. The action of HCN as a respiratory inhibitor is presumably of prime importance in this process, acting to poison herbivores which are exerting selectionary pressure on the species. However, it may not be the HCN in every case which is the primary agent since studies have shown that p-hydroxybenzaldehyde, produced on hydrolysis of dhurrin, can act as a feeding deterrent (31). Other highly volatile compounds, such as benzaldehyde or acetone, could also serve to ward off predatory insects. One might question the role of hydroxynitrile lyases since their cyanohydrin substrates are labile and can dissociate in a non-enzymatic reaction. However, the pH of plant tissues is usually on the acid side of 7, and cyanohydrins are more stable under these conditions. Therefore, the lyase is essential for the rapid release of HCN which occurs when plant tissues are homogenized.

Cyanogenesis is not necessarily exhibited by every tissue of a cyanophoric plant, nor does it appear that a cyanogenic tissue necessarily retains its cyanide-generating potential throughout the life of a plant. For example, the seed of *S. bicolor* contains only traces of cyanogenic glycoside, but after 5 days of germination in the dark, the young seedling contains about 5% (d.wt.) dhurrin. As growth continues, there is a decrease in concentration of cyanogenic glycoside of whole seedling, mainly because the cyanogen is being diluted out by the mass of the plant as it enlarges. However, synthesis must be occurring in some tissue (*e.g.*, meristematic tissue). In one of the few studies on this subject, Loyd and Gray showed that intact sorghum plants generally increased in their content of cyanogenic glycoside over a growth period of 4 months (*32*). In another study, Clegg *et al.* (*33*) provided indirect evidence that linamarin present in high concentration in wild lima beans is transferred intact from the cotyledonary tissue into the newly formed seedling as germination occurs. In this case, the catabolic enzymes presumably have no access to the cyanogen as it is translocated into the seedling. On the other hand, if cyanogenic glycosides undergo metabolic turnover in the intact plant, the catabolic glycosidase and hydroxynitrile lyase would presumably be involved in this process.

BIOSYNTHESIS

Studies in several laboratories during the past two decades have been concerned with the biosynthesis of cyanogenic glycosides. Early work, reviewed in detail elsewhere (*3,4,34*), not only established the precursor-product relationship shown in Fig. 1 at an early date but permitted a biosynthetic pathway to be proposed in 1968 (*35*). That version of the pathway is nearly identical to a revised one proposed in 1979 (*3*) ; the more recent one (Fig. 7) contains only one additional intermediate, the *N*-hydroxyamino acid.

The biosynthetic pathway which was postulated in 1968 (*35*) and contained an aldoxime, nitrile, and cyanohydrin was supported by two types of evidence. The first involved experiments in which the precursor amino acids and postulated intermediates were labeled with ^{14}C, ^{15}N or ^{3}H and administered to intact seedling or seedling parts (*e.g.*, excised shoots). In such "feeding experiments", the incorporation of radioactivity from labeled amino acids and/or intermediates into the product cyanogenic glycosides was quantitated. Depending on the effectiveness of incorporation, the compound was concluded to be or excluded from being a precursor and/or intermediate in biosynthesis.

In these early studies, a second type of experiment was also involved in establishing the pathway. These were "trapping experiments" in which the precursor amino

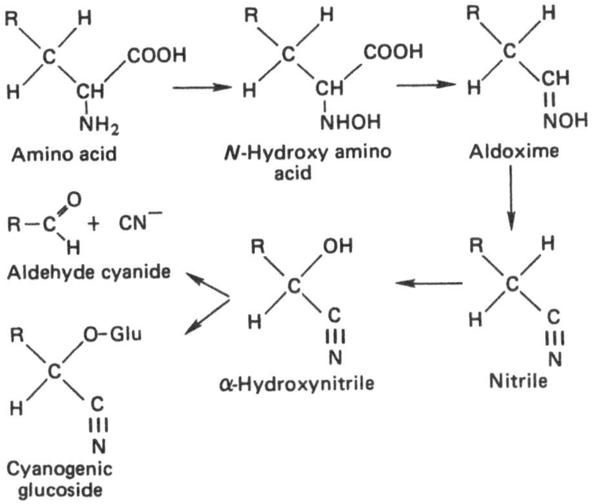

Fig. 7. The biosynthetic pathway for cyanogenic glycosides (Reproduced with permission from Ref. *3*).

acid labeled with isotope, usually ^{14}C, was administered to a plant or excised shoot simultaneously with unlabeled, postulated intermediate. Then after a period of metabolism, the intermediate was reisolated from the plant and examined for isotope. If, during its metabolism in the plant the proposed intermediate had become labeled with isotope, it was said to have trapped isotope. This presumably occurred because the originally labeled precursor had been metabolically converted into the intermediate by biosynthetic enzymes functioning in the intact plant.

Feeding and trapping experiments while useful are limited in interpretation because the act of administering these compounds to plant tissues may induce artefactual reactions which are unrelated to the biosynthesis being studied. For this reason, enzyme studies which can identify the reaction involved are greatly preferred. Such studies which involved the last biosynthetic step in the process had been carried out in flax (*36*) and sorghum (*37*) by 1970 and 1974, respectively. These were studies showing that the cyanogenic glycosides in those plants were synthesized by UDP-glucosyl transferases which glucosylated the cyanohydrins of acetone and *p*-hydroxy-[*S*]-mandelonitrile, respectively. Such work established the nature of the last step in biosynthetic pathway.

All of the earlier steps of the biosynthetic pathway remained a mystery until I.J. McFarlane showed that a particulate (microsomal) fraction from dark-grown sorghum seedlings would convert L-tyrosine to *p*-hydroxy-[*S*]-mandelonitrile (*38*) in the presence of NADPH and O_2. This remarkable conversion corresponded to a series of four reactions catalyzing all but the last step in the biosynthetic pathway (Fig. 7) and it soon became clear that the particulate enzyme system could be used in clarifying the details of the sequence. Specifically, studies with the particulate enzyme system eventually established that an N-hydroxyamino acid was an intermediate in the formation of the oxime from the precursor amino acid (39). Whether the N-hydroxyamino acid is the only intermediate in that conversion of the amino acid to the aldoxime is not yet clear ; since the overall conversion involves a 4-electron oxidative decarboxylation of the protein amino acid, one might expect more than one intermediate in the process. Indeed, 3-(p-hydroxyphenyl)-2-nitrosopropionic acid has been proposed as an intermediate in the 2-electron oxidative decarboxylation of N-hydroxytyrosine by sorghum microsomes (39).

To date, cell-free biosynthetic particulate systems have been characterized from seedlings of S. bicolor (39), Triglochin maritima (40), and Linum usitatissimum (41). While these three systems share a number of common properties (42), the most striking feature is that the four step conversion of amino acid to cyanohydrin (Fig. 7) proceeds without accumulation of any of the pathway intermediates. This observation cannot be explained on the basis of observed kinetic parameters of the individual enzyme reactions. In each of the three plant systems, the concentrations of intermediates which do accumulate are well below those expected if a steady state had been established between independent, soluble enzymes. Also, the measured V_{max} values do not increase consistently with each intermediate in the order of their utilization on the pathway. This finding is perplexing since it implies that there should be rate limiting steps in each sequence which would lead to a buildup of intermediates.

When the properties of the sorghum microsomal system were examined in detail (43) it became apparent that the four enzymes of the pathway were arranged in such a way that they constituted a highly efficient system for channeling the flow of carbon atoms from tyrosine into dhurrin, the cyanogenic glycoside of that species. Similar conclusions have been reached regarding the particulate systems obtainable from T. maritima and flax (L. usitatissimum). An increasing number of examples of metabolic channeling are appearing and indeed have been reported earlier in the literature ; these include the fatty acid synthetase of yeast (44) ; the membrane-bound association of phenylalanine ammonia-lyase and cinnamic hydroxylases of cucumber cotyledons (45), potato tubers (46), and Dunaliella marina (47) ; the tryptophan synthetase of Neurospora (48) ; pyrimidine biosynthesis (49) and degradation of cyclic AMP in beef adrenal cortex (50). The much studied "arom complex" of Neurospora, which originally was thought to be metabolically channeled and exhibit catalytic facilitation (51), may instead owe some of its interesting kinetic properties to the coordinate activation of the multifunctional protein (enzyme cluster) which catalyzes the five-step reaction sequence of the complex (52,53).

The most plausible explanation of channeling in both soluble and membrane-bound systems is that intermediates remain associated with an enzyme either by covalent bonding (as in the fatty acid synthetase) or by being released into a microcompartment in the protein from which, and into which, diffusion is limited. This microcompartment also obviously must contain the active site for the next enzyme in the pathway.

Finally, it is important to consider what advantages channeling provides to a cell. One obvious advantage is that the metabolism of intermediates common to more than one metabolic pathway can be more easily regulated if those intermediates are retained in separate compartments. Where intermediates are chemically labile, a short transit time between sequential catalytic sites provides less opportunity for decomposition to occur. Gaertner has also pointed out that channeling may be a technique for reducing the number of freely-diffusing low molecular weight compounds in the cytoplasm (53). This could be an advantage if such compounds tax the solvent capacity of the cell. Finally, it is possible that a channeled system may exhibit "catalytic facilitation", i.e., may carry out the overall sequential reaction sequence at a faster rate than a reaction initiated later in the sequence with established intermediates. Although the channeled "arom complex" reaction system exhibits catalytic facilitation, it does not appear to be due to the channeling process (see 53 for detailed discussion).

In 1974, Stafford (54) pointed out that multifunctional enzyme complexes might be involved in the formation of plant phenols and lignin. Although channeling was not specifically mentioned, catalytic facilitation was discussed and its advantages noted. Since that time Kindl and coworkers (45–47) have observed several examples of channeled bifunctional systems involved in the early steps of phenylpropanoid metabolism. Wink et al. (55) have also postulated that a channeled process is involved in the biosynthesis of quinolizidine alkaloids in lupin chloroplasts. It is therefore tempting to speculate that metabolic channeling may be involved in the biosynthesis of many secondary compounds. Since the primary metabolites (e.g., amino acids, acetic acid, glucose) which serve as precursors of these secondary metabolites are extremely active in intermediary metabolism, channeling could be a means whereby the synthesis of the secondary metabolite is insured.

REFERENCES

1. Conn, E.E. : 1981. *In* Biochemistry of Plants, Vol. 7 (Stumpf, P.K. and Conn, E.E., eds.), pp. 479–500, Academic Press, New York.
2. Conn, E.E. : 1980. *Annu. Rev. Plant Physiol.* 31, 433–451.
3. Conn, E.E. : 1979. *Naturwissenschaften* 66, 28–34.
4. Conn, E.E. : 1980. Encyl. Plant Physiol., New Series, Vol. 7 (Bell, E.A. and Charlwood, B.V., eds.), pp.461–492, Springer-Verlag, Berlin.
5. Knowles, C.J. : 1976. *Bacteriol. Rev.* 40, 652–680.
6. Davis, R.H. and Nahrstedt, A. : 1979. *Comp. Biochem. Physiol.* 64B, 395–397.
7. Nestel, B. and MacIntyre, R. : 1973. *In* Chronic Cassava Toxicity, Int. Dev. Res. Center Monograph IDRC-010e, Ottawa, Ont.
8. Hegnauer, R. : 1977. *Plant Syst. Evol.* (Suppl.) 1, 191–209.
9. Jones, D.A. : 1981. *In* Cyanide in Biology (Vennesland, B., ed.), pp. 509–516, Academic Press, London.
10. Eyjolfsson, R. : 1970. *Fortsch. Chem. Organ. Naturstoff.* 28, 74–108.
11. Seigler, D.S. : 1981. *In* Cyanide in Biology (Vennesland, B., ed.), pp.132–143, Academic Press, London.
12. Nahrstedt, A. : 1982. *Phytochemistry* 21, 101–105.
13. Schrader, J.C.C. : 1803. *Anal. Phys.* (Leipzig) [1] 13, 503–504.
14. Kofod, H. and Eyjolfsson, R. : 1969. *Phytochemistry* 8, 1509–1511.
15. Eyjolfsson, R. : 1971. *Acta Chem. Scand.* 25, 1898–1900.
16. Seigler, D.S. : 1977. *Prog. Phytochem.* 4, 83–100.

17. Nahrstedt, A. : 1981. *In* Cyanide in Biology (Vennesland, B., ed.),pp.145–181, Academic Press, London.
18. Dunstan, W.R. and Henry, T.A. : 1902. *Phil. Trans. Roy. Soc.* A199, 399–410.
19. Towers, G.H.N., McInnes, A.G. and Neish, A.C. : 1964. *Tetrahedron* 20, 71–77.
20. Seigler, D.S., Dunn, J.E., Conn, E.E. and Pereira, J. : 1982. *Biochem. Syst. Ecol.* in press.
21. Smith, C.R., Jr., Weisleder, D., Miller, R.W., Palmer, I.S. and Olsen, O.E. : 1980. *J. Org. Chem.* 45, 507–510.
22. Nartey, R., Brimer, L. and Christensen, S.B. : 1981. *Phytochemistry* 20, 1311–1314.
23. Cramer, U., Rehfeldt, A.G. and Spener, F. : 1980. *Biochemistry* 19, 3074–3080.
24. Nahrstedt, A., Walther, A. and Wray, V. : 1982. *Phytochemistry* 21, 107–110.
25. Nahrstedt, A., Kant, J.D. and Wray, V. : 1982. *Phytochemistry* 21, 101–105.
26. Kojima, M., Poulton, J.E., Thayer, S.S. and Conn, E.E. : 1979. *Plant Physiol.* 63, 1022–1028.
27. Saunders, J.A. and Conn, E.E. : 1978. *Plant Physiol.* 61, 154–157.
28. Oba, K., Conn, E.E., Canut, H. and Boudet, A.M. : 1981. *Plant Physiol.* 68, 1359–1363.
29. Marcinowski, S. and Grisebach, H. : 1978. *Eur. J. Biochem.* 87, 37–44.
30. Hösel, W. and Conn, E.E. : 1982. *Trends Biochem. Sci.* 7, 219–221.
31. Woodhead, S., Galeffi, C. and Bettolo, G.B.M. : 1982. *Phytochemistry* 21, 455–456.
32. Loyd, R.C. and Gray, E. : 1970. *Agron. J.* 62, 394–397.
33. Clegg, D.O., Conn, E.E. and Janzen, D.H. : 1979. *Nature* 278, 343–344.
34. Conn, E.E. : 1973. *Biochem. Soc. Symp.* 38, 277–302.
35. Hahlbrock, K., Tapper, B.A., Butler, G.W. and Conn, E.E. : 1968. *Arch. Biochem. Biophys.* 125, 1013–1016.
36. Hahlbrock, K. and Conn, E.E. : 1970. *J. Biol. Chem.* 245, 917–922.
37. Reay, P.F. and Conn, E.E. : 1974. *J. Biol. Chem.* 249, 5826–5830.
38. McFarlane, I.J., Lees, E.M. and Conn, E.E. : 1975. *J. Biol. Chem.* 250, 4708–4713.
39. Møller, B.K. and Conn, E.E. : 1979. *J. Biol. Chem.* 249, 5826–5830.
40. Cutler, A.J., Hosel, W., Sternberg, M. and Conn, E.E. : 1981. *J. Biol. Chem.* 256, 4253–4258.
41. Cutler, A.J. and Conn, E.E. : 1981. *Arch. Biochem. Biophys.* 212, 468–474.
42. Cutler, A.J. and Conn, E.E. : 1982. *Recent Adv. Phytochem.* 14, 249–271.
43. Møller, B.L. and Conn, E.E. : 1980. *J. Biol. Chem.* 255, 3049–3056.
44. Stoops, J.K. and Wakil, S.J. : 1981. *J. Biol. Chem.* 256, 5128–5133.
45. Czichi, U. and Kindl, H. : 1977. *Planta* 134, 133–143.
46. Czichi, U. and Kindl, H. : 1975. *Planta* 125, 115–125.
47. Czichi, U. and Kindl, H. : 1975. *Hoppe-Seyler's Z. Physiol. Chem.* 356, 475–485.
48. Matchett, W.H. : 1975. *J. Biol. Chem.* 249, 4041–4049.
49. Mally, M.I., Grayson, D.R. and Evans, R.E. : 1980. *J. Biol. Chem.* 255, 11372–11380.
50. Wombacher, H. : 1980. *Arch. Biochem. Biophys.* 201, 8–19.
51. Gaertner, F.H., Ericson, J.C. and DeMoss, J.A. : 1980. *J. Biol. Chem.* 545, 595–600.
52. Welch, G.R. and Gaertner, F.H. : 1976. *Arch. Biochem. Biophys.* 172, 476–489.
53. Gaertner, F.H. : 1978. *Trends Biochem. Sci.* 3, 63–65.
54. Stafford, H.A. : 1974. *Recent Adv. Phytochem.* 8, 53–79.
55. Wink, M., Hartmann, T. and Witte, L. : 1980. *Z. Naturforsch.* 35c, 93–97.

Biosynthesis and Microbial Degradation of Lignin

Takayoshi Higuchi

INTRODUCTION

Lignins are a characteristic component of woody tissues found only in the true vascular plants, ferns, and plants higher than the fern, and are absent from the algae, mosses, fungi, and bacteria (*1*). The roles of lignins, which are aromatic polymers, in xylem tissues are to impart rigidity to the cell walls and enable terrestrial plants to develop large upright forms resistant to various stresses, to assist in the smooth transportation of solutes by decreasing the permeability of cell walls in the conductive xylem tissues, and to resist attack by microorganisms. Chemical and biochemical investigations showed that lignins are classified into three major groups, guaiacyl lignin (ferns and gymnosperms), guaiacyl-syringyl lignin (angiosperms) and guaiacyl-syringyl-*p*-hydroxyphenyl lignin (grasses) based on their structural monomer units.

In the first section of this paper, differences in biosynthesis of guaiacyl and syringyl lignins in gymnosperms and angiosperms are explained in terms of the different functions of *O*-methyltransferases, *p*-hydroxycinnamate:CoA ligases and *p*-hydroxycinnamyl alcohol oxidoreductases which participate in the formation of monolignols. In addition, the role of quinonemethide intermediates in the formation of lignin-carbohydrate complexes (L.C.C.) is discussed based on the analytical results of the reaction products of the quinonemethide of guaiacylglycerol-β-guaiacyl ether with sugars, and DHP-polysaccharide complexes. In the second section, microbial degradation pathways of major substructures of lignins, such as arylglycerol-β-aryl ether, phenylcoumaran, diarylpropane-1,3-diol and resinol are proposed and discussed in relation to lignin biodegradation.

23

BIOSYNTHESIS

The biosynthetic pathway of lignin has been mostly elucidated by feeding experiments (2) with labeled precursors and enzyme studies (3–5) involved in the respective reactions in the pathway. The pathway of lignin biosynthesis established is illustrated in Fig. 1. Attention has recently been focused on the mechanisms of the development of the enzyme system of lignification during tissue differentiation, and the

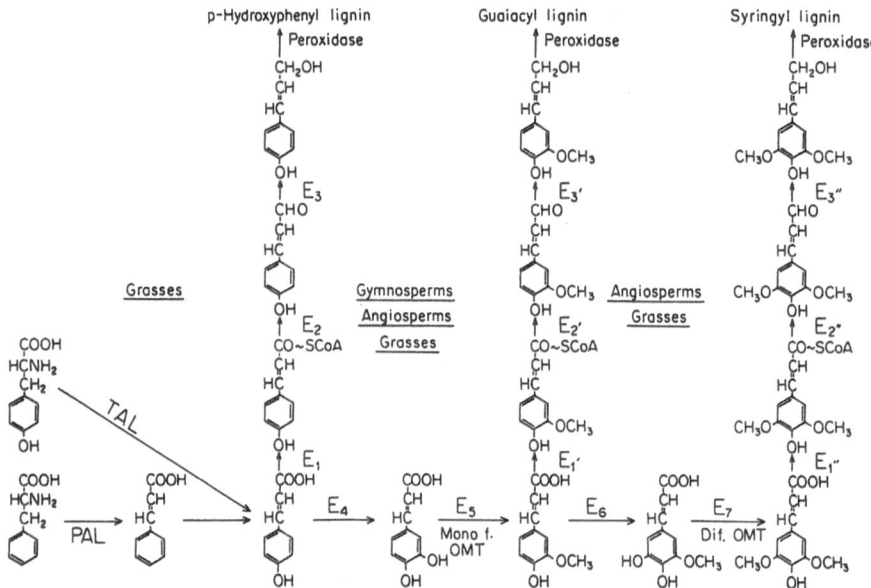

Fig. 1. Biosynthetic pathways of p-hydroxyphenyl, guaiacyl and syringyl lignins. Enzymes ; (E1), (E1'), (E1'') Hydroxycinnamate : CoA ligase, (E2), (E2'), (E2'') Hydroxycinnamoyl-CoA reductase, (E3), (E3'), (E3'') Cinnamyl alcohol dehydrogenase, (E4) p-Coumarate-3-hydroxylase, (E5) Hydroxycinnamate-O-methyltransferase, (E6) Ferulate-5-hydroxylase, (E7) Hydroxycinnamate-O-methyltransferase.

regulation of lignin biosynthesis in plant cell (6,7). The enzymes mediating the following reactions were characterized : deamination of L-phenylalanine and L-tyrosine to *trans*-cinnamic and *trans-p*-coumaric acids, successive hydroxylations of cinnamic acid to p-coumaric and caffeic acids, methylation of caffeic and 5-hydroxy-ferulic acids to ferulic and sinapic acids, reduction of these p-hydroxycinnamic acids to the corresponding alcohols (monolignols), and dehydrogenative polymerization of these monolignols to lignin polymers.

Only ferulate-5-hydroxylase, which is a key enzyme in the differentiation of lignin biosynthesis into guaiacyl- and syringyl lignin pathways in angiosperms and which could be involved in the conversion of ferulate to 5-hydroxyferulate, remains uncharacterized among the enzymes in lignin biosynthesis.

5-Hydroxyferulate was found for the first time to be incorporated efficiently into both sinapate and syringyl lignin in wheat (8), and was postulated by the author to be an intermediate between ferulate and sinapate. Enzyme studies in our laboratory showed that 5-hydroxyferulate is the best substrate for angiosperm O-methyltransferases (OMTs) (9) to give sinapate which is converted to syringyl lignin via sinapyl alcohol. It is thus expected that ferulate-5-hydroxylase can be characterized by further investigations.

Since 1967 we have been investigating metabolic differences in the formation of guaiacyl, syringyl and p-hydroxyphenyl lignins in conifers, dicotyledons and monocotyledons (grasses). We have found that the synthetic pathways of lignins in gymnosperms and angiosperms are separated at the stages of hydroxylation of ferulate to 5-hydroxyferulate, the subsequent methylation of 5-hydroxyferulate to sinapate, and the reduction of p-hydroxycinnamates to the corresponding alcohols. In gymnosperms ferulate-5-hydroxylase is usually absent, the affinity of the OMT towards 5-hydroxyferulate is poor and the reducing enzymes of sinapate are lacking. On the other hand, an excess of p-coumarate formed by the mediation of tyrosine ammonialyase which is only found in grasses is converted to p-hydroxyphenyl lignin and the esterified p-coumaric acid which are characteristic of grass lignin (2).

In the present paper the roles of OMTs (EC 2.1.1−), p-hydroxycinnamate : CoA ligases (EC 6.2.1.12) and p-hydroxycinnamyl alcohol oxidoreductases (EC 1.1.1−) are further discussed in relation to the regulation of guaiacyl and syringyl lignin biosynthesis.

Distribution and Role of OMT in Lignin Biosynthesis

The methoxyl content has been recognized as an essential criterion in the characterization of lignins : conifer lignin which is mainly composed of guaiacyl units contains 15% methoxyl, hardwood lignin composed of guaiacyl and syringyl units contains 20%, and grass lignin composed of guaiacyl, syringyl and p-hydroxyphenyl units contains an average of 15%.

We found that the OMT is one of the key enzymes in determining the fate of lignins synthesized in plants, i.e., guaiacyl lignin in gymnosperms and guaiacyl-syringyl lignin in angiosperms ; the substrate specificities of purified OMTs are completely different between gymnosperm (i.e., Japanese black pine) and angiosperm (i.e., poplar). Angiosperm OMTs well catalyze the conversion of caffeate and 5-hydroxyferulate to ferulate (FA) and sinapate (SA) which are finally converted to guaiacyl and syringyl lignins, respectively, whereas gymnosperm OMTs only catalyze the conversion of caffeate to ferulate but not of 5-hydroxyferulate to SA (10). We further found that the two methylating activities for FA and SA by angiosperm OMTs are involved in a single enzyme protein and that plants possessing higher SA/FA ratios such as angiosperms give greater S/V (syringaldehyde/vanillin) ratios on nitrobenzene oxidation of lignin and a positive Mäule reaction. Plants with lower SA/FA ratios such as conifers and ferns give lesser S/V ratios and negative

TABLE I

Relationship between the SA/FA Ratio, the S/V Ratio and Mäule Reaction of Lignins*

Plant species	SA/FA (OMT)	S/V (lignin)	Mäule reaction
Pteridophyta			
Psilotum nudum	0.2	0	–
Angiopteris lygodifolia	0.3	–	–
Gymnospermae			
Ginkgo biloba	0.1	0	–
Taxus cuspidata	0.1	–	–
Pinus densiflora	0.1	0	–
Pinus thunbergii	0.1	0	–
Pinus taeda	0.3	0	–
Pinus strobus	0.4	0	–
Thuja orientalis	0.83	0	–
Thuja standishii	0.03	0	–
Podocarpus macrophylla	0.6	0	–
Angiospermae (Dicotyledoneae)			
Magnolia grandiflora	3.0	2.2	+
Liriodendron tulipifera	2.5	2.7	+
Trocodendron aralioides	1.6	2.6	+
Cercidiphyllum japonicum	3.2	2.9	+
Euptelea polyandra	3.6	–	+
Nuphar japonicum	2.3	–	+
Ranunculus acris	3.6	–	+
Poplus euramericana	3.2	2.1	+
Betula nigra	3.1	2.5	+
Quercus myrsinaefolia	2.5	–	+
Ulmus americana	3.2	2.6	+
Viscum album	2.2	1.7	+
Erythrina japonica	2.3	0.1	–
Tilia japonica	2.3	–	+
Paulownia tomentosa	2.9	1.4	+
Pueraria thunbergiana	2.5	–	+
Angiospermae (Monocotyledoneae)			
Alöe arborescens	1.1	–	+
Tradescantia virgiana	1.6	–	+
Oryza sativa	0.9	1.0	+
Triticum aestivum	1.0	1.0	+
Zizania latifolia	1.0	–	+
Phyllostachys pubescens	1.3	1.1	+
Sparganium stoloniferum	1.5	–	+
Scirpus triqueter	2.2	–	+

The enzyme assay is based on the transfer of $^{14}CH_3$ group of S-adenosyl-L-methionine to caffeate or 5-hydroxyferulate forming ferulate (FA) or sinapate (SA)-O-$^{14}CH_3$, respectively at 30°C for 30–60 min. The incubation mixture contained enzyme solution 0.2ml, phenolic substrates 0.5 μmole, K-phosphate (pH 7.5) 20 μmole, NaN_3 10 μmole, $MgCl_2$ 1 μmole, cysteine 10 μmole, 2-mercaptoethanol 10 μmole and isoascorbate 10 μmole.

* SA,FA , activity against sinapate and ferulate, respectively. S , syringaldehyde; V , vanillin.

Mäule reaction (11).

The distribution and SA/FA ratio of OMTs in various vascular plants are shown in Table I (12). The SA/FA ratios listed are only of the enzyme activities which represent more than 10^3 cpm/hr of the products by the present assay method. In general, the various plant ratios showed a sharp contrast between gymnosperm and angiosperm. As postulated by the authors, gymnosperm OMTs primarily catalyzed only FA-formation, while those of the angiosperm catalyzed both FA- and SA-formations. The results well explain why gymnosperm lignin is almost entirely composed of guaiacyl units and angiosperm lignin is composed of both guaiacyl and syringyl units.

Gymnosperm OMTs

Most of the gymnosperm OMTs, as represented by the Pinaceae enzyme, showed low SA/FA ratios. *Thuja*, *Podocarpus* and *Ephedra* OMTs were exceptions which showed rather high ratios; seedlings of *Thuja orientalis* also gave a high ratio, 0.83 (13). Most of the species in Cupressaceae contain only guaiacyl lignin except *Tetraclinis* which contains guaiacyl-syringyl lignin (14). The high SA/FA ratio in *Thuja* seems to indicate that the lignin biosynthetic pathway in Cupressaceae might be tinged with characteristics of the angiosperm type : it is known that a lignan of *Thuja orientalis*, dihydroxythujaplicatin methyl ether, contains a syringyl unit (13). The ratio, SA/FA probably reflects genetic codes in the biosynthesis of guaiacyl and syringyl lignins, and the higher ratios seem to reflect a genetic variation in this family. This situation is probably similar in Podocarpaceae and Cycadaceae because some species in these families are known to contain both guaiacyl and guaiacyl-syringyl lignins (15). These groups are conceivably in transition stages between gymnosperm and angiosperm in the point of lignin biosynthesis. The lignin biosynthetic pathway in Gnetales which contains guaiacyl-syringyl lignin is presumed to be completely transformed into an angiosperm type.

Angiosperm OMTs

Angiosperm OMTs generally showed a normal ratio (2.2–2.8). *Trochodendron* OMT showed rather a low SA/FA ratio. This plant is known to lack vessel elements but contains guaiacyl-syringyl lignin (16). However, crude OMT of mistletoe (*Viscum album*), the lignin of which was characterized to be a normal angiosperm type, apparently showed a very high ratio, although the ratio by the purified enzyme was found to be normal (SA/FA=2.2). The result suggested that the crude enzyme solution of mistletoe contains some inhibitors for FA activity (17). Another interesting example is *Erythrina*, the SA/FA ratio of which is normal angiosperm type, although the secondary xylem of this wood is comprised almost entirely of guaiacyl lignin (18,19). Poulton et al. found (20) that one of the OMTs in soybean (*Glycine max*) suspension cells was specific for flavonoid and the SA/FA ratio was ca. 1.4, while another lignin specific OMT in the same cells showed a ratio of ca. 1.9. If the lignin specific OMT in the tissues is present at a lower level than the flavonoid specific OMT, the ratio could be affected by the level of the latter. A reported low SA activity of swede root (*Brassica napo-brassica*) (*21*), the lignin content of which is usually very low, might be ascribed to such an effect.

Gramineae OMTs

The present investigation showed that Gramineae OMTs gave an SA/FA ratio around 1.0, whereas other monocotyledon ones gave rather low ratios compared with normal angiosperm OMTs, the bamboo, *Phyllostachys pubescens*, OMT is specific for lignin, and the SA/FA ratio remained constant during purification (*11, 22*). No isoenzymes were detected in the crude preparation, the crude bamboo OMT showed a net ratio of lignin specific enzyme and was unaffected by isoenzymes, as is found in herbaceous angiosperms. Grass lignin differs from other gymnosperm and angiosperm lignins in that it contains *p*-hydroxyphenyl units. Since Gramineae is thought to be more evolved phylogenically than gymnosperms and dicotyledons, it is recognized that the SA/FA ratio (Gramineae, *ca.* 1.0) does not simply increase with phylogenical evolution. These results led to the conclusion that OMTs are roughly classified into three groups, *i.e.*, gymnosperm, angiosperm and grass types which are related to the lignin evolution, although unusual ratios were found in some cases.

Role of p-Hydroxycinnamate : CoA Ligase in Lignin Biosynthesis

p-Hydroxycinnamate:CoA ligase has been found to be widely distributed in many taxonomically different plants using ferulate as substrate (*4*). However, no investigations have yet been carried out on the distribution of the CoA ligase specific for sinapate in woody plants. In search of enzymes involved in the regulation of the preferential synthesis of guaiacyl and syringyl lignins in woods, we investigated the CoA ligase which is active for sinapate as well as ferulate (*23*) (Fig. 2).

Table II shows the substrate specificity of CoA ligases which were extracted from the xylem of 8 angiosperms, 4 gymnosperms and one monocotyledon. *p*-Hydroxycinnamate was found to be the best among the substrates tested. Ferulate was converted efficiently to feruloyl-CoA by all the enzymes. On the other hand, sinapate was converted to sinapoyl-CoA only by the enzymes of *Erythrina*, acacia and a bamboo, and the ratio of activity to that with ferulate was 0.40 on the average. The enzymes from other plants were completely inactive with sinapate.

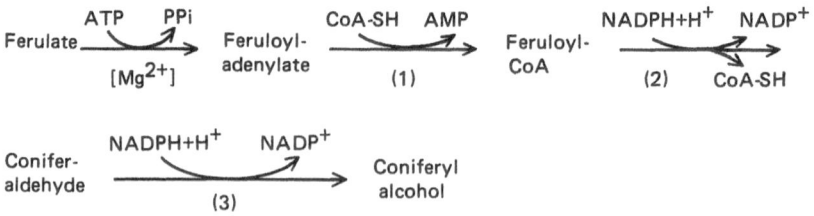

Fig. 2. Enzymic reduction of ferulate to coniferyl alcohol. (1) Hydroxycinnamate: CoA ligase, (2) Cinnamoyl-CoA reductase, (3) Cinnamyl alcohol oxidoreductase.

TABLE II
Substrate Specificities of p-Hydroxycinnamate : CoA Ligases of Gymnosperms and Angiosperms*

Plant species	Activity (pKAT/mg)				
	PA	FA	SA	CA	5-HFA
Gymnospermae					
Chamaecyparis psifera	–	73	0	–	–
Juniperus chinensis	–	51	0	–	–
Thuja orientalis	206	213	0	144	100
Metasequoia glyptostroboides	337	289	0	179	91
Angiospermae					
Erythrina crista-galli	128	70	30	55	15
Robinia pseudoacacia	301	143	57	112	49
Sophora japonica	–	44	0	–	–
Populus euramericana	155	137	0	95	51
Paulownia tomentosa	190	141	0	110	48
Liriodendron tulipifera	–	63	0	–	–
Acer buergeriana	–	260	0	–	–
Prunus yedoensis	–	43	0	–	–
Angiospermae (Monocotyledoneae)					
Phyllostachys bambusoides	27	34	13	–	–

The incubation mixture contained 500 nmole cinnamates, 2 μmole ATP and $MgSO_4$, 50 nmole CoASH and 10–100 μg enzymes in 2.74 ml of Tris buffer (pH 7.3) 100 mM. The reaction mixture was incubated at 30°C for 5 min in UV cuvettes, and the increase in absorbance of λ_{max} values of respective cinnamoyl-CoA esters was measured.
PA, p-hydroxycinnamate; FA, ferulate; SA, sinapate; CA, caffeate; 5-HFA, 5-hydroxyferulate.

Sinapate:CoA ligase is presumed to be indispensable for biosynthesis of syringyl lignin in angiosperms. The CoA ligases active with sinapate have been reported in swede root (24), soybean (25), petunia (26), pea, runner bean, leek (27), and poplar (28), while no activities were observed in Forsythia, Acer, Taxus (4), tomato (29), or carrot (30). Poplar, which was reported to contain sinapate : CoA ligase by Grant and Ranjeva (28), showed no activity for sinapate under any conditions in our investigation : using enzymes prepared from xylem of poplar shoots harvested from May to July at 4°C under liquid nitrogen or under ultrasonic, the addition of ethylene gylcol, mercaptoethanol and polyclar AT in the standard extraction medium, or the effects of other additives, enzyme stabilizers, extraction reagents for enzymes in membraneous systems, etc.

Since the occurrence of sinapate:CoA ligase in angiosperms does not seem to be common, we presume the possible participation of an additional pathway in biosynthesis of syringyl lignin : sinapyl alcohol is formed via successive reactions of 5-hydroxyferulate, 5-hydroxyferuloyl-CoA, 5-hydroxyconiferaldehyde, sinapaldehyde and sinapyl alcohol as illustrated in Fig. 3. It was found that 5-hydroxyferulate was well converted to its CoA ester by both angiosperm and gymnosperm ligases and that 5-hydroxyconiferaldehyde added to the reaction medium methylated efficiently to sinapaldehyde by angiosperm OMTs (23). However, the involvement of sinapate:

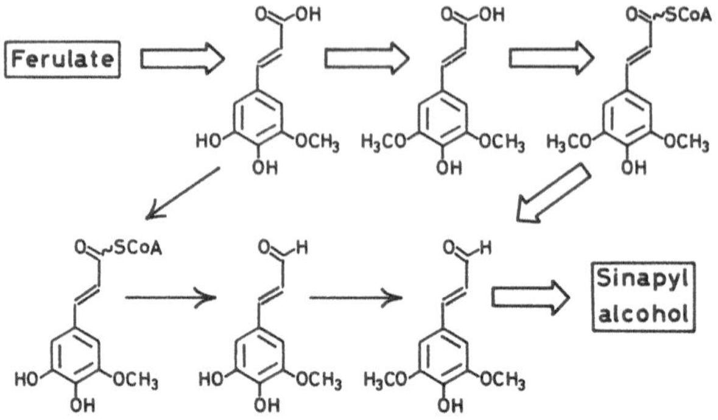

Fig. 3. Biosynthetic pathways of sinapyl alcohol in angiosperms. ⟹, main pathway; ⟶, additional pathway.

CoA ligase in syringyl lignin biosynthesis is not ruled out, especially in grasses which generally contain both free- and esterified sinapate as well as active sinapate:CoA ligase. Sinapate:CoA ligase would be too labile to be extracted from some angiosperms which have shown no activity for sinapate.

Activity for ferulate but not for sinapate of gymnosperm enzymes found in the present investigation is in accord with preferential formation of guaiacyl lignin in gymnosperms.

Since the CoA ligase of *Erythrina crista-galli* was found to be active with sinapate, the enzyme was partly characterized (*23*). No multiple forms of the ligase were observed by chromatography on DEAE-Sephadex, hydroxyapatite and Sepharose 4B. The enzyme activity for both sinapate and ferulate was eluted as one peak from the calibrated column of the Sephacryl S-200, and the molecular weight was estimated as about 40,000. The enzyme reaction was specific for ATP. *p*-Coumarate was the best substrate followed by ferulate and isoferulate, and the activity for sinapate

TABLE III
Substrate Specificity of *Erythrina* Ligase

Substrate	Activity (nKAT/mg)	Relative activity (%, FA = 100%)
p-Coumarate	3.5	140
Ferulate	2.5	100
Isoferulate	2.5	100
3,4-Dimethoxycinnamate	2.1	84
p-Methoxycinnamate	1.7	68
Sinapate	1.2	48
Caffeate	1.2	48
5-Hydroxyferulate	0.4	17
Cinnamate	0	0

The enzyme assay was carried out under the same conditions as described in Table II.

was 48% that of ferulate (Table III). The remarkably high affinity for sinapate of the *Erythrina* enzyme was found from its K_m value of 1 μM in comparison with 11 μM (soybean isoenzyme) (25), and 170 μM (petunia) (26). Characteristics of purified *Erythrina* enzyme were almost the same as the sinapate : CoA ligases reported earlier.

Erythrina is unique in that it contains guaiacyl-rich lignin although it belongs to the Leguminosae family, whereas the plant contains angiosperm type OMT and *p*-hydroxycinnamyl alcohol oxidoreductase which reduces sinapaldehyde to sinapyl alcohol. We therefore concluded that the lack of syringyl units in *Erythrina* lignin could be ascribed to the low activity of ferulate-5-hydroxylase as suggested by a tracer experiment (19).

Regulatory Role of p-Hydroxycinnamyl Alcohol Oxidoreductase in Lignin Biosynthesis

Substrate specificities of *O*-methyltransferase (OMT) give a reasonable explanation for most cases of the biosynthesis of guaiacyl lignin in gymnosperms and guaiacyl-syringyl lignin in angiosperms. However, in some cases OMT was found not to be a controlling enzyme in the formation of guaiacyl and syringyl lignins : the OMTs of tissue cultures of angiosperms, and of xylem of *Erythrina crista-galli* showed high SA activities, although their lignins were almost totally lacking in syringyl units (18,19). A similar phenomenon was observed for some exceptional gymnosperms such as *Thuja standishii* and *T. orientalis* (13) and bamboos in different lignification stages. The content of syringyl units in the lignins of bamboo shoots was found to increase greatly as growth progressed (32), although the SA/FA ratio of the OMT remained constant (11,22).

We recently found that *p*-hydroxycinnamyl alcohol oxidoreductases (CAD) which catalyze the last step of the formation of lignin precursors (monolignols) are another possible enzyme for regulation of the biosynthesis of guaiacyl and syringyl lignins in gymnosperms and angiosperms. It was found that substrate specificities of enzymes of the seedlings of Japanese black pine (gymnosperm) and *Zelkova serata* (angiosperm) were remarkably different (33) : sinapaldehyde was a good substrate for angiosperm CAD to give sinapyl alcohol but was a poor substrate for gymnosperm CAD. On the other hand, both coniferaldehyde and *p*-coumaraldehyde were well reduced to the corresponding alcohols by both gymnosperm and angiosperm enzymes. Further investigations on the substrate specificities of gymnosperm and angiosperm CADs (16 species) showed that gymnosperm CADs are remarkably more active with coniferaldehyde than with sinapaldehyde, and the ratio of sinapyl alcohol to coniferyl alcohol formed (Salc/Calc) ranged from 0.05 to 0.55 (average 0.22), whereas in angiosperm CADs Salc activity was almost equal to Calc activity and the ratios were in the range of 0.65 to 1.53 (average 1.09) (Table IV).

In our further investigations Japanese black pine CAD was purified and characterized in relation to the guaiacyl specific nature of gymnosperm CAD. It was found that the ratio of Salc/Calc remained constantly low (average 0.07) through all purification steps, and that both Salc and Calc activities were eluted as a single peak from

TABLE IV
Specific Activities of p-Hydroxycinnamyl Alcohol Oxidoreductases of Gymnosperms and Angiosperms*

Plant species	Activity (pKAT/mg)		
	Calc	Salc	Salc/Calc
Gymnospermae			
Pinus thunbergii	330	37	0.11
Pinus densiflora	–	–	0.16
Larix leptolepis	–	–	0.22
Ginkgo biloba	300	15	0.05
Chamaecyparis obtusa	–	–	0.07
Thuja orientalis	1500	500	0.33
Cryptomeria japonica	680	290	0.43
Metasequoia glyptostroboides	3300	1800	0.55
Angiospermae			
Populus euramericana	29	43	1.50
Liriodendron tulipifera	460	380	0.83
Robinia pseudoacacia	1600	1900	1.20
Erythrina crista-galli	200	130	0.65
Prunus yedoensis	860	1320	1.53
Prunus persica	1300	840	0.65
Zelkova serrata	–	–	1.20
Angiospermae (Monocotyledoneae)			
Phyllostachys bambusoides	–	–	1.69

The incubation mixture contained 150 nmole cinnamaldehydes, 300 nmole NADPH, 100 μl enzyme solution in 2.7 ml, 200 mM K-phosphate buffer (pH 6.5). Incubations were carried out at 30°C for 5 min in UV cuvettes, and the decrease in A_{340} was measured continuously.
*Calc, coniferyl alcohol; Salc, sinapyl alcohol.

TABLE V
Substrate Specificity of p-Hydroxycinnamyl Alcohol Oxidoreductase of Japanese Black Pine

Substrate	K_m (μmole)	V_{max} (nKAT/mg)	V_{max}/K_m (nKAT/mole/mg)
NADPH	6.8	3.9	0.57
Coniferaldehyde	9.1	3.3	0.36
Cinnamaldehyde	14	7.6	0.53
p-Coumaraldehyde	30	5.7	0.19
3,4-Dimethoxycinnamaldehyde	43	0.46	0.01
Sinapaldehyde	–*	0.07	–

The enzyme assay was carried out under the same conditions as described in Table IV.
* Activity was too low to calculate K_m.

DEAE-cellulose, Sephadex G-100 and hydroxyapatite columns. These results indicated that the CAD of Japanese black pine is considerably more specific to coniferaldehyde than sinapaldehyde and that no multiple forms are present. The enzyme reaction was completely dependent on NADPH. Coniferaldehyde, p-coumaraldehyde and cinnamaldehyde gave higher V_{max} values, but V_{max} of sinapaldehyde was extremely low (Table V) : the ratio of Salc/Calc on the V_{max} value was 0.02. The molecular weight of the CAD estimated by column chromatography on a calibrated Sephadex G-100 was about 67,000, which was close to that of isoenzyme 2 of soybean CAD (69,000). However, the ratio of Salc/Calc to that of isoenzyme 2 was 0.84 (*34*) indicating that the enzyme was remarkably more active with sinapaldehyde than coniferaldehyde. It was thus concluded that Japanese black pine CAD possibly regulates the lignin synthesis towards preferential formation of guaiacyl lignin in cooperation with OMT.

As mentioned earlier, OMT of the *Thuja* species gave relatively high SA/FA ratios, although the lignin was composed entirely of guaiacyl units. This suggests that some regulations are involved after OMT reaction in the formation of guaiacyl lignin in this species. Such regulations seem to be determined partly by the CAD which reduces coniferaldehyde more specifically than sinapaldehyde (Salc/Calc=0.33).

The ratio of guaiacyl to syringyl units in angiosperm lignins calculated on the basis of the methoxyl content is about 1.0, which is in good accordance with the Salc/Calc ratios in angiosperm CADs (average 1.09). Conclusively, we presume that syringyl lignin formation in angiosperms is roughly regulated first by OMT and then more precisely by cinnamoyl-CoA reductase and CAD (Fig. 4).

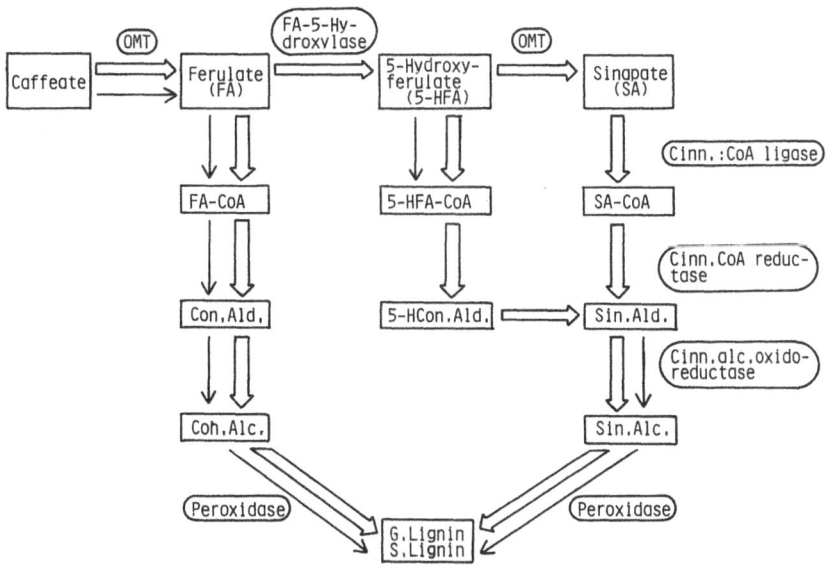

Fig. 4. Regulatory enzymes involved in the biosynthesis of guaiacyl and syringyl lignins. ——→, pathway in gymnosperm; ⟹, pathway in angiosperm.

As mentioned earlier, *Erythrina crista-galli* (Leguminosae) contains guaiacyl-rich lignin, although the OMT is of the angiosperm type. Since the Salc/Calc ratio of the CAD was found to be relatively low among angiosperms tested, the enzyme must be partly involved in the regulation of lignin formation, but a low activity of ferulate-5-hydroxylase should be mostly responsible for guaiacyl lignin formation in this plant (*19*).

Formation of Lignin-Carbohydrate Complexes (L.C.C.) in Cell Walls

In plant cell walls lignin occurs in close association with cellulose and hemi-celluloses. Lignin-carbohydrate complexes (L.C.C.) were isolated from milled woods, and the mode of chemical linkages has been investigated by several researchers. Various linkages have been proposed between lignin and hemicelluloses, although they are still inconclusive (*35*).

We recently found (*36,37*) that both the carboxyl group of D-glucuronic acid and the C_6 hydroxyl group of D-glucose preferentially attack the C_α of the quinonemethide to give ester and ether, respectively, when these sugars are added to a solution of the quinonemethide of guaiacylglycerol-β-guaiacyl ether. Analyses of the reaction products proved that the carboxyl group of the uronic acid and the C_6 hydroxyl group of the sugar are predominantly connected to C_α of the quinonemethide *via* ester and ether linkages, respectively, and that no glycosidic linkages are formed. Further investigations showed that the quinonemethide reacts with glucuronic acid remarkably well, and with glucose to a small extent even in the presence of water. Based on these results the authors postulated the formation of L.C.C. by such reactions during lignification in plant cell walls. In order to prove L.C.C. formation by this reaction coniferyl alcohol was dehydrogenatively polymerized by peroxidase and H_2O_2 in concentrate solutions of either pectin, mannan or xylan in dialysis tubes (*38*). In addition, the quinonemethide of isoeugenol and dehydrodiisoeugenol were reacted with pectin in DMSO solution as a model. In both cases the reaction products were separated and lyophilized.

Analysis of the thermal softening point and dielectric relaxation process of the products were adopted as a tool to characterize DHP/polysaccharide complexes (*38*). The DHP of coniferyl alcohol and mannan alone were softened at 140°C and 228°C, and their physical mixture also softened at the same temperatures. However, when coniferyl alcohol was polymerized in a mannan solution by peroxidase and H_2O_2, the thermal softening temperature (T_s) of DHP/mannan complexes shifted to 245°C, which is 17° higher than of mannan itself, and no T_s peaks corresponding to that of DHP or mannan were found (Fig. 5). We therefore presume that chemical bonding was formed between the DHP and mannan, and that essentially all of the DHP was present as L.C.C. When DHP/mannan complexes were hydrolyzed by mild acidolysis the peak at 245°C disappeared indicating that the DHP was linked to mannan by a benzyl ether bond.

The T_s of the DHP/pectin complexes (water-insoluble fraction) shifted to a 15° higher temperature (213°C), and no peak was found at 140°C due to DHP itself. By mild acidolysis of the complexes the peak at 213°C disappeared and new peaks

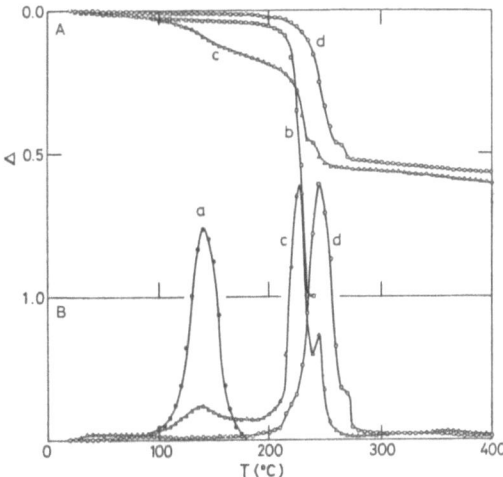

Fig. 5. Thermal softening processes of mannan, DHP and DHP/mannan complexes. A , relative deformation curves; B , differential curves ; (a) DHP, (b) mannan, (c) a mixture of DHP and mannan, (d) DHP/mannan complexes.

appeared at $136°C$ (DHP) and $199°C$ (pectin), indicating that the peak at $213°C$ was due to the L.C.C. linked by a benzyl ester bond between the DHP and pectin.

The T_s of DHP/xylan complexes, on the other hand, were around $150°C$ and $245°C$, the same temperature as DHP and xylan, which suggests that no L.C.C. was formed in the dehydrogenative polymerization in this condition. The result seems in accord with the theory that xylose residues in a xylan molecule take pyranose form without having a primary hydroxyl group, and the attack of the quinonemethides by secondary hydroxyl groups at C_2 and C_3 is remarkably slower than that by the primary hydroxyl group $(36,37)$.

T_s of the model L.C.C. obtained by the addition of pectin to the DMSO solution of quinonemethides of isoeugenol and dehydrodiisoeugenol were $218°C$ and $226°C$, respectively, which were again $20°C$ and $28°C$ higher than pectin itself. The quinonemethide/pectin complexes (model L.C.C.) showed a maximum absorption in the range between 10 kHz and 100 kHz in the range of $25°C$ to $75°C$, which is ascribable to the relaxation of the benzyl ester linkages between carboxyl groups of pectin and C_α of the C_6-C_3 units. In DHP/pectin complexes, on the other hand, the maximum absorption appeared in the range between 330 Hz and 1 kHz $(25°-75°C)$ which is due to benzyl ester linkages between the pectin and DHP.

Based on these results, we presumed that L.C.C. were formed by nucleophilic addition of carboxyl and primary hydroxyl groups of glycosyl residues to quinonemethide intermediates when coniferyl alcohol was polymerized by peroxidase and H_2O_2 in the concentrated solution of the carbohydrates. It is conceivable that in plant cell walls and middle lamella a considerable part of the lignin molecules are linked by C_α to these primary alcohol and carboxyl groups in the hemicelluloses and pectins. Eriksson et al. (39) recently showed that in black spruce L.C.C. lignin is linked to the 4-*O*-methylglucuronic acid units by ester linkages, probably benzyl ester and to the other sugar units in the hemicelluloses by ether bonds, probably benzyl ether bond. These observations are well in accord with our results.

MICROBIAL DEGRADATION

Analytical comparisons (*40*) of sound and white-rotted lignins have shown that the degradative process of lignin is heavily oxidative : degradation of lignin polymer occurs in the side chains, which are oxidized with formation of α-carbonyl and α-carboxyl groups, and in the aromatic nuclei, which are oxidatively cleaved following demethylation and introduction of hydroxyl groups in phenolic units to give 2,3- and 3,4-dihydroxyphenyl moieties. It is suggested that various lignin structural units are degraded by the mediation of extracellular and/or membrane-bound enzymes which attack both low molecular weight and polymeric substrates *via* many intermediate products along several different pathways.

Degradation of Dilignols

Since lignin has no easily hydrolyzable repeating units as do polysaccharides, proteins and nucleic acids, the use of model compounds containing major structural units in lignin is indispensable to elucidate the degradation mechanism of various linkages in the lignin macromolecule. For this purpose we synthesized several oligolignols and used as substrate for lignin degrading fungi. *Fusarium solani* M-13-1 which was isolated from soil by an enrichment technique using DHP as sole carbon source and *Phanerochaete chrysosporium* have been used for these degradation studies.
1) Arylglycerol-β-aryl Ether (β-*O*-4) Model Compounds

Mycelial suspensions of *F. solani* M-13-1 were incubated (*41*) with shaking with guaiacylglycerol-β-coniferyl ether [1]*, which is the most common inter-phenylpropane linkage in lignin (30–60%), and the degradation of the compound was followed *via* UV spectral analysis of the culture filtrate. From the acid-free fraction of the ethyl acetate extract of the culture filtrate after 40 hrs of incubation guaiacylglycerol-β-coniferaldehyde ether [2] was isolated and identified by NMR and MS spectrometry ; the acid fraction of the extract after 76 hrs of incubation contained guaiacylglycerol-β-ferulic acid ether [3]. From the filtrate of similar cultures in which the β-ferulic acid ether and the β-vanillin ether [4] were used as substrate, β-vanillin and β-vanillic acid ether [5], respectively were identified. The guaiacylglycerol-β-vanillic acid ether was degraded to glycerol-2-vanillic acid ether [6] which was converted to vanillic acid [8] with other several compounds. 2-Methoxy-*p*-benzoquinone [7], which is expected as another cleavage product of guaiacylglycerol-β-vanillic acid ether, could not be identified because of comparatively rapid degradation of the quinone by the fungus. However, when syringylglycerol-β-vanillic acid ether [5'] was used as substrate 2,6-dimethoxy-*p*-benzoquinone [7'] was isolated

---

* The number in [ ] indicates the compounds used and/or identified in biodegradation of dilignols as shown in Figs. 6–15.

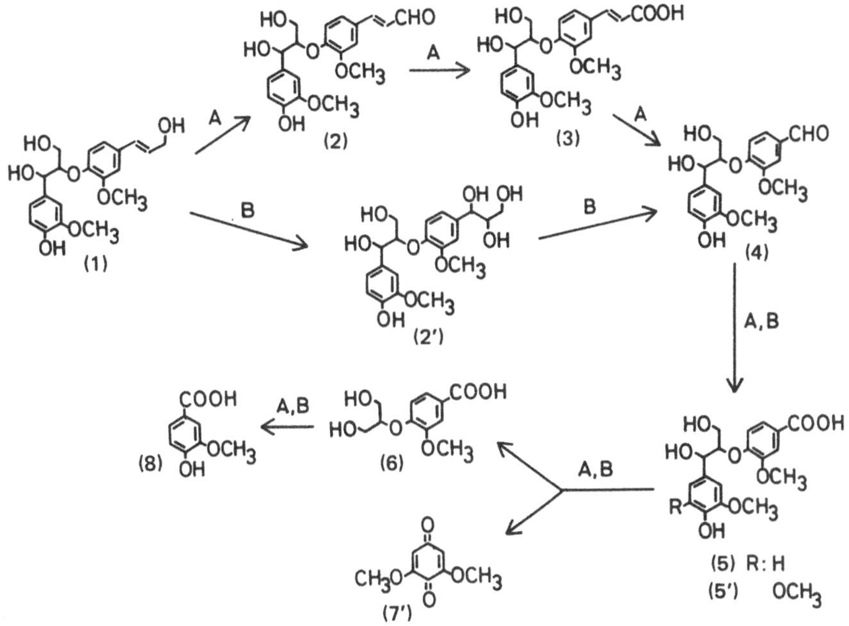

Fig. 6. Degradation pathways of guaiacylglycerol-β-coniferyl ether by *Fusarium solani* M-13-1 and *Phanerochaete chrysosporium*. A, pathway by *Fusarium*; B, pathway by *Phanerochaete*.

and identified, indicating that the β-vanillic acid ethers were degraded *via* alkayl-phenyl cleavage, possibly by a phenol oxidizing enzyme as shown in Fig. 6 (*42*).

Veratrylglycerol-β-vanillin ether was oxidized to the vanillic acid ether, but the latter was not degraded further under similar cultural conditions, showing that the phenolic hydroxyl group is necessary for degradation of the arylglycerol moiety by this fungus.

It is thus evident that the degradation of arylglycerol-β-aryl ether by the fungus proceeds at least in part *via* oxidative cleavage of the carbon-to-carbon linkage between the aromatic ring and the propyl side chain. Enoki *et al.* (*43*) recently found that *P. chrysosporium* also gave glycerol-2-guaiacyl ether from guaiacylglycerol-β-quaiacyl ether in agreement with our results.

In our alternative investigation with *P. chrysosporium* we (*44*) found that veratrylglycerol-β-coniferyl ether [9] was converted to veratrylglycerol-β-guaia-cylglycerol ether [10] which was converted to the vanillic acid ether [12] *via* the vanillin ether [11]. The vanillic acid ether [12] was converted to 4-ethoxy-3-meth-oxybenzyl alcohol [13] *via* C_α-C_β cleavage instead of alkyl-phenyl cleavage as in the phenolic β-O-4 model (Fig. 7). Recently cleavage of the C_α-C_β and/or alkyl-phenyl linkages in veratrylglycerol-β-guaiacyl ether was also reported in *P. chrysosporium* (*43*) and in bacteria (*45*).

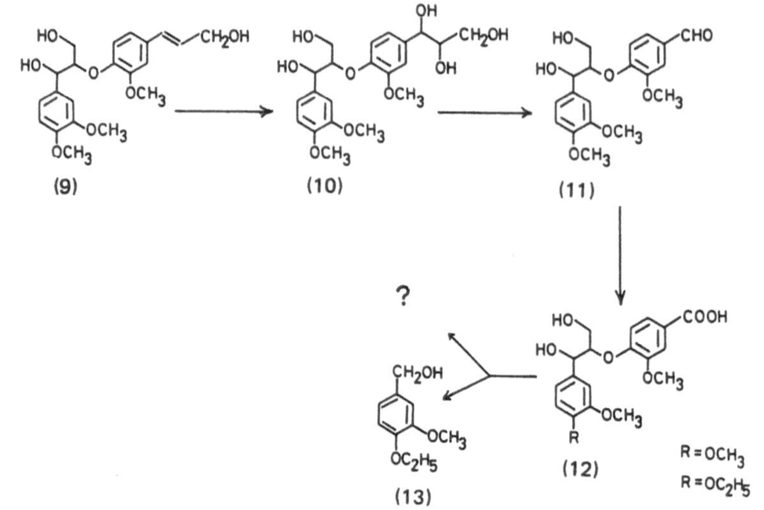

Fig. 7. Degradation pathway of veratrylglycerol-β-coniferyl ether by *Phanerochaete chrysosporium*.

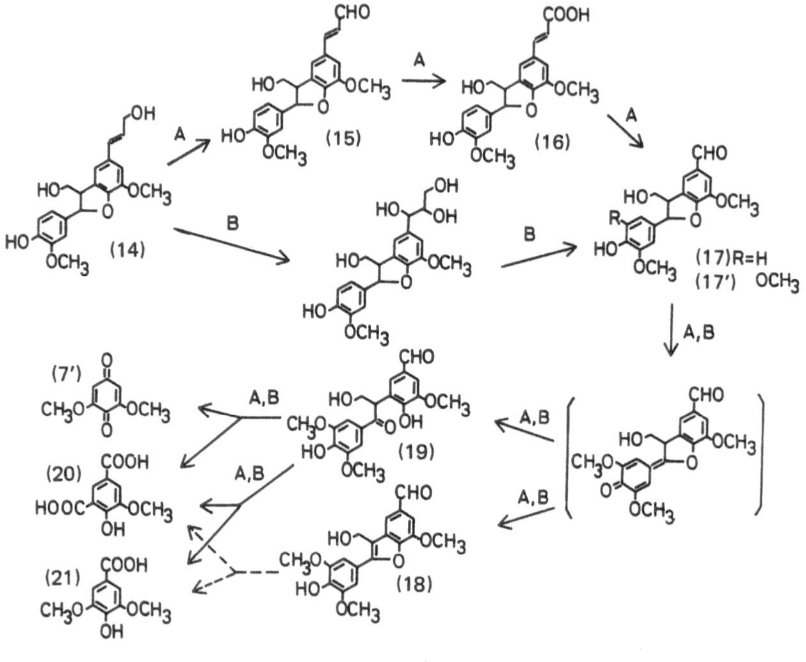

Fig. 8. Degradation pathways of dehydrodiconiferyl alcohol by *Fusarium solani* M-13-1 and *Phanerochaete chrysosporum*. A, pathway by *Fusarium*; B, pathway by *Phanerochaete*.

2) Phenylcoumaran (β-5) Model Compounds

From the culture filtrate of *F. solani* M-13-1 incubated with dehydrodiconiferyl alcohol [14], which is a major dilignol (10–15% in lignin), the following six degradation products were identified by NMR and MS spectrometry (46) : phenylcoumaran-γ'-aldehyde [15], phenylcoumaran-γ'-carboxylic acid [16], phenylcoumaran-α'-aldehyde [17], 5-acetylvanillyl alcohol, and 5-carboxyvanillyl alcohol. Vanillic acid which had been expected as another degradation product could not be detected. However, further investigations using phenylcoumaran-α'-aldehyde with syringyl group [17'] as substrate (47) showed that a part of the substrate was converted either to phenylcoumarones [18] or propiosyringone derivative [19], 5-carboxyvanillic acid [20], syringic acid [21] and 2,6-dimethoxybenzoquinone [7']. The degradation pathway shown in Fig. 8 is indicated by these results.

Iwahara *et al.* (48) recently found that an enzyme which catalyzes oxidation of α, β-unsaturated primary alcohol group in the side chain of dehydrodiconiferyl alcohol and of guaiacylglycerol-β-coniferyl ether to the corresponding aldehyde is secreted into the culture medium of *F. solani* M-13-1 during incubation with DHP of coniferyl alcohol. The constitutive enzyme was found to oxidize α, β-unsaturated primary alcohol groups in a wide range of lignin model compounds as well as lignin macromolecules. Further investigation indicated that the aldehydes formed are oxidized to the corresponding acids which are then oxidized to the C_6-C_1 acids by inducible enzymes. Because the alcohol oxidase is extracellular and nonspecific in its attack on lignin and models, it is thought to be involved in the oxidative degradation of chinnamyl alcohol groups of lignin.

In ligninolytic cultures of *P. chrysosporium* we found recently (49) that 4-O-methyl dehydrodiconiferyl alcohol [22] is first converted to the corresponding glycerol [23], which is then converted to a vanillin derivative [24]. The coumaran

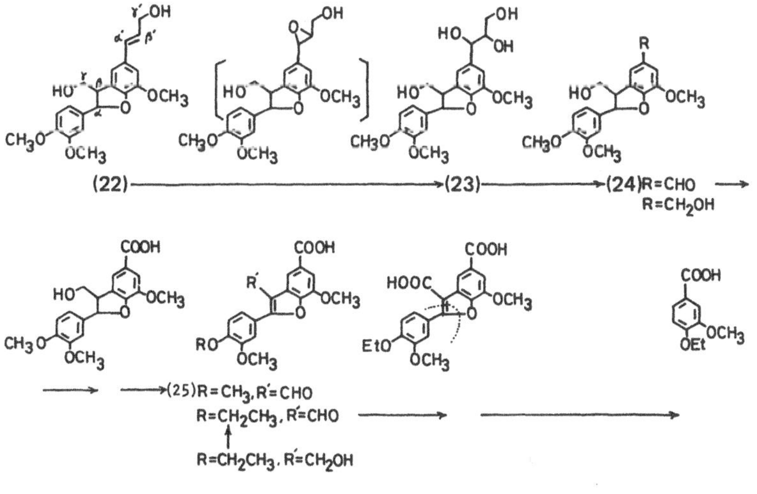

Fig. 9. Degradation pathway of 4-O-methyldehydrodiconiferyl alcohol by *Phanerochaete*.

ring is then converted to coumarone [25] which is converted to C_6-C_1 acid (Fig. 9).

Further investigation using phenylcoumaran-α'-aldehyde with syringyl group [17'] showed the formation of a propiosyringone derivative [19], phenylcoumarones [18], 5-carboxyvanillic acid [20], syringic acid [21], and 2,6-dimethoxybenzoquinone [7']. The results indicated that a very similar degradation pathway is operated in both *Fusarium* and *Phanerochaete* for phenolic phenylcoumaran (Fig. 8).

3) Pinoresinol (β-β') Model Compounds

With *F. solani* M-13-1 syringaresinol [28], which is a major substructure of lignin (5–10% in lignin), was converted to the compounds [29], [30], [31] and 2,6-dimethoxy-*p*-benzoquinone [7']. The formation of these degradation products seems to indicate that compounds [28] was degraded *via* oxidative cleavage of the alkyl-

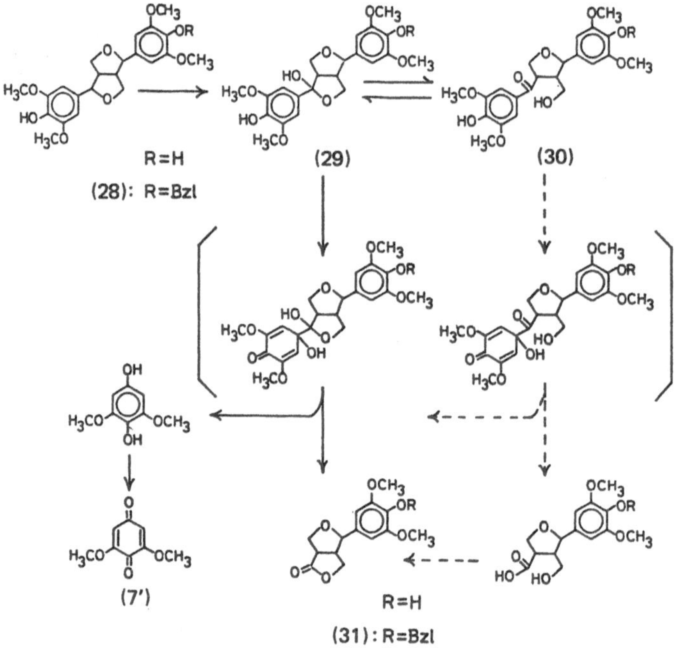

Fig. 10. Degradation pathway of syringaresinol by *Fusarium solani* M-13-1.

phenyl bond as shown in Fig. 10 *(50)*. α-Hydroxysyringaresinol [29] seems to be an intermediate for alkyl-phenyl cleavage by this fungus. Dibenzyl and dimethyl ethers of syringaresinol were not degraded under similar conditions, but the monobenzyl ether was converted to [31], which was not metabolized further by this fungus. The result indicated that both phenolic hydroxyl groups are indispensable for complete degradation of syringaresinol by this fungus.

In recent investigations by Iwahara *et al. (51)* an enzyme which catalyzes oxidative cleavage of syringaresinol and pinoresinol in a similar way was isolated from homogenate of the fungal cells and partly characterized.

4) Diarylpropane-1,3-diol (β-1) Model Compounds

Diarylpropane-1,3-diol is one of the major substructures of lignin (5-15% in lignin). From the culture filtrate of *F. solani* M-13-1 incubated with 1,2-diguaiacyl-propane-1,3-diol [32] 2-guaiacylpropane-1,3-diol [33] and its biphenyl dimer [34] were identified. No 2-methoxy-*p*-benzoquinone was detected but 2,6-dimethoxy-*p*-benzoquinine, [7'], [35] and [36] were identified as degradation products when 1, 2-disyringylpropane-1,3-diol [32'] was used as substrate. Thus, the result again indicated that the cleavage of the alkyl-phenyl bond of [32] occurred as in the cases of β-O-4 and β-β' dilignols as shown in Fig. 11 (52).

On the other hand, with *Phanerochaete chrysosporium* non-phenolic diarylpropane, 1,2-bis(3-methoxy-4-alkoxyphenyl)propane-1,3-diol [37] was found to be degraded to three major products, [38],[39] and [40] by Nakatsubo *et al.* (53) (Fig.12).

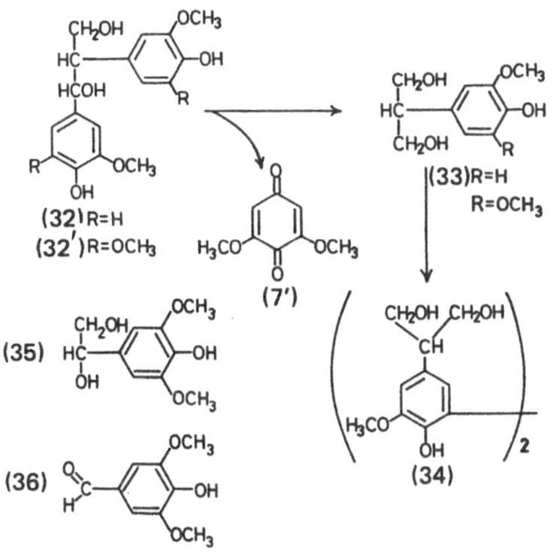

Fig. 11. Degradation pathway of 1,2-diarylpropane-1,3-diols by *Fusarium solani* M-13-1.

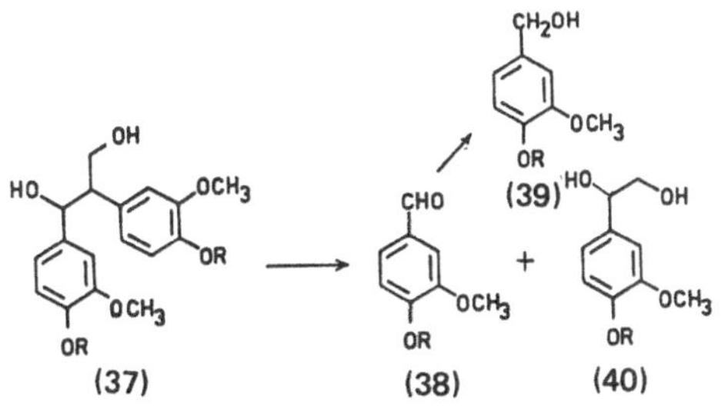

Fig. 12. Degradation pathway of 1,2-bis(3-methoxy-4-alkoxyphenyl)propane-1,3-diol by *Phanerochaete chrysosporium*.

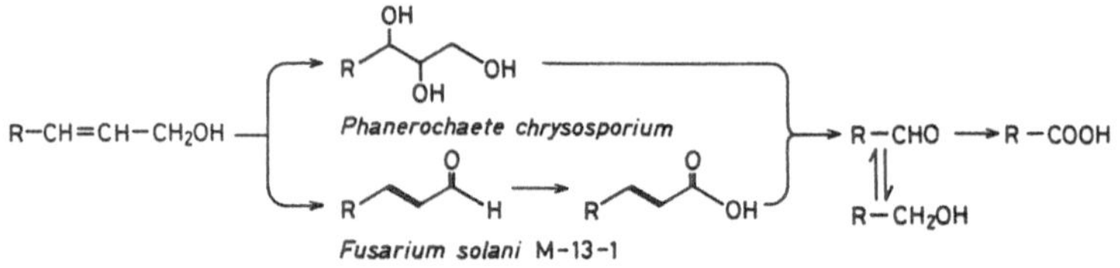

Fig. 13. Oxidative degradation of allyl alcohol side chains of dilignols by *Fusarium* and *Phanerochaete*.

The degradation products of main dilignols by *F. solani* M-13-1 and *Phanerochaete chrysosporium* were the result of the following degradation reactions (54) : 1) Cinnamyl alcohol groups are oxidized to the corresponding cinnamic acid groups *via* cinnamaldehyde groups by *Fusarium*, but are converted to glycerol groups by *Phanerochaete* ; 2) both cinnamic and glycerol groups are converted to the C_6-C_1 acid derivatives *via* C_6-C_1 aldehyde derivatives (Fig. 13) ; 3) α-hydroxydilignols such as arylglycerol-β-aryl ethers and diarylpropane-1,3-diols are degraded *via* cyclohexadienone radical derivatives to the corresponding hydroquinones and glyceraldehydes by alkyl-phenyl cleavage mediated by phenol oxidizing enzymes ; 4) α-ether (alkoxy and phenoxy) dilignols such as pinoresinol and phenylcoumaran are first converted to α-hydroxy ethers *via* quinonemethide intermediates and then α-hydroxy ethers are degraded to the corresponding hydroquinone and lactone or ester derivative by both fungi by the mediation of phenol oxidizing enzymes (Fig. 14) ; 5) non-phenolic dilignols are not degraded by *Fusarium* but cleaved mainly between C_α and C_β by *Phanerochaete*.

Degradation processes of these phenolic dilignols seem similar between *Fusarium* and other white-rot (lignin degrading) fungi including *Phanerochaete*. Even in *Phanerochaete* phenolic dilignols are degraded much faster than non-phenolic dilignols.

CONCLUDING REMARKS

Chemical analysis of degraded lignins and the degradation of dilignols have led to the conclusion that lignin is degraded oxidatively in both aliphatic side chains and in the aromatic nuclei from the surface of the lignin macromolecules, simultaneously. Aromatic moieties with free phenolic hydroxyl groups (20–30% lignin) would be preferentially attacked *via* alkyl-phenyl cleavages by phenol-oxidizing enzymes and produce moieties with newly formed phenolic groups which could again be oxidized by the enzymes. On the other hand, non-phenolic aromatic moieties are mainly cleaved between C_α and C_β giving C_6-C_1 fragments. It is thus conceivable that phenol oxidizing enzymes which attack phenolic units of lignin, and the enzymes which mediate the cleavage of C_α-C_β linkages synergistically cooperate in the degradation of lignin, as in exo- and endo-cellulases in the hydrolysis of cellulose. Aromatic alcohol dehydrogenases and oxygenases would be involved as key enzymes in the oxidative degradation of side chains, and dioxygenases would be indispensable for cleavage of aromatic rings of lignin.

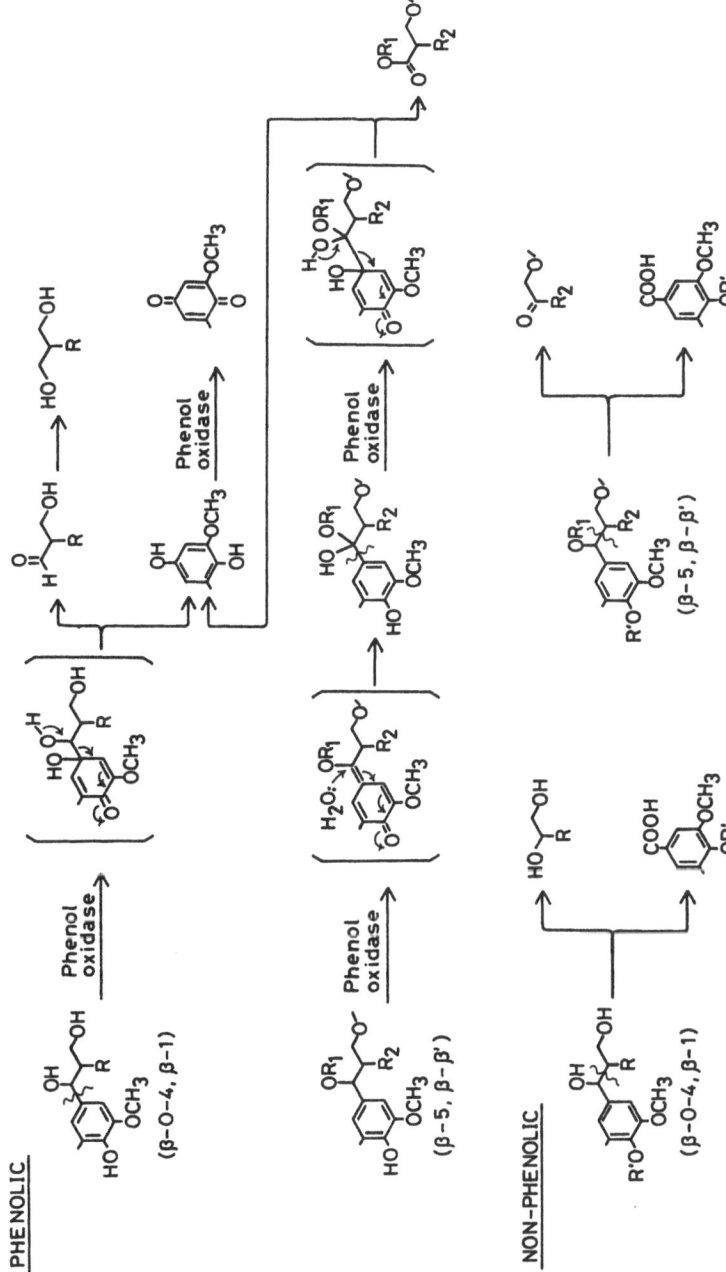

Fig. 14. Side chain cleavage of phenolic and non-phenolic dilignols by *Fusarium* and *Phanerochaete*.

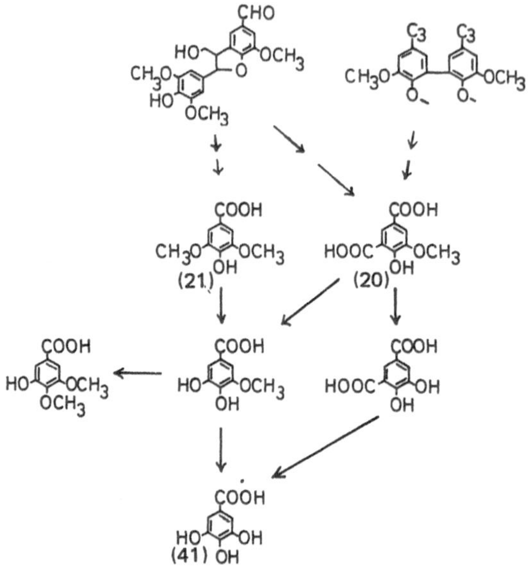

Fig. 15. Metabolic pathway of syringic acid and 5-carboxyvanillic acid by *Phanerochaete chrysosporium.*

Vanillic acid and quinones usually found as degradation products of noncondensed units of lignin would be metabolized *via* protocatechuic acid and/or hydroquinones by either demethylation or decarboxylation (*55,56*). Syringic acid [21] and 5-carboxyvanillic acid [20] as degradation products of syringyl lignin and condensed units of lignins would be metabolized *via* gallic acid [41] by demethylation and oxidative decarboxylation (Fig. 15) (*57*). To fill the missing links of the degradation processes between dilignols and lignin polymers, investigations using trilignols as substrate are in progress.

Acknowledgements

This paper is based on recent investigations related to the biosynthesis and biodegradation of lignin by members of the Research Section of Lignin Chemistry, Wood Research Institute, Kyoto University. The author is indebted to Drs. M. Shimada, F. Nakatsubo, H. Kuroda and H. Kutsuki, and Messrs. M. Tanahashi, T. Katayama, Y. Kamaya and T. Umezawa for their cooperation in these investigations.

This investigation was supported in part by Grant-in-Aid No. 548047 for Scientific Research, and 1980 Grant-in-Aid for Environmental Science (R-33-8), No. 403064, both from the Ministry of Education, Science and Culture of Japan, and a 1980 Weyerhaueuser grant.

REFERENCES

1. Kawamura, I. and Higuchi, T. : 1964. *In* Chimie Biochimie de la Lignine, Cellulose, Hemi-celluloses, pp. 439–456, Les Imperimeries Reunies de Chambery, France.
2. Higuchi, T. : 1971. *Adv. Enzymol.* **34**, 207–283.
3. Higuchi, T., Shimada, M., Nakatsubo, F. and Tanahashi, M. : 1977. *Wood Sci. Technol.* **11**, 153–167.
4. Gross, G.G. : 1977. *Recent Adv. Phytochemistry* **11**, 141–184.
5. Grisebach, H. : 1977. *Naturwissenschaften* **64**, 619–625.
6. Hahlbrock, K. and Grisebach, H. : 1979. *Annu. Rev. Plant Physiol.* **30**, 105–130.
7. Gross, G.G. : 1979. *Recent Adv. in Phytochemistry* **12**, 177–220.
8. Higuchi, T. and Brown, S. : 1963. *Can. J. Biochem. Physiol.* **41**, 613–619.
9. Shimada, M., Ohashi, H. and Higuchi, T. : 1970. *Phytochemistry* **9**, 2463–2470.
10. Shimada, M., Fushiki, H. and Higuchi, T. : 1973. *Mokuzai Gakkaishi* (J. Japan Wood Res. Soc.) **19**, 13–21.
11. Shimada, M., Kuroda, H. and Higuchi, T. : 1973. *Phytochemistry* **12**, 2873–2875.
12. Kuroda, H. and Higuchi, T. : 1982. *Wood Research* **68**, 1–7.
13. Kutsuki, H., Shimada, M. and Higuchi, T. : 1981. *Mokuzai Gakkaishi* **27**, 39–48.
14. Erickson, M. and Miksche, G.E. : 1974. *Holzforschung* **28**, 135–138.
15. Gibbs, R.D. : 1958. *In* The Physiology of Forest Trees (Thimann, K.V., ed.), pp. 269–312, Ronald Press, New York.
16. Kawamura, I. and Higuchi, T. : 1965. *Mokuzai Gakkaishi* **11**,19–22.
17. Kuroda, H. and Higuchi, T. : 1976. *Phytochemistry* **15**, 1511–1514.
18. Kawamura, I., Shinoda, Y., Ai, T. and Tanada, T. : 1977. *Mokuzai Gakkaishi* **23**, 400–404.
19. Kutsuki, H. and Higuchi, T. : 1978. *Mokuzai Gakkaishi* **24**, 625–631.
20. Poulton, J.E., Hahlbrock, K. and Grisebach, H. : 1977. *Arch. Biochem. Biophys.* **180**, 543–549.
21. Rhodes, M.J.C., Hill, A.C.R. and Wooltorton, L.S.C. : 1976. *Phytochemistry* **15**, 707–710.
22. Kuroda, H., Shimada, M. and Higuchi, T. : 1981. *Phytochemistry* **20**, 2635–2639.
23. Kutsuki, H., Shimada, M. and Higuchi, T. : 1982. *Phytochemistry* **21**, 267–271.
24. Rhodes, M.J.C. and Wooltorton, L.S.C. : 1973. *Phytochemistry* **12**, 2381–2387.
25. Knobloch, C.-H. and Hahlbrock, K. : 1975. *Eur. J. Biochem.* **52**, 311–320.
26. Ranjeva, R., Boudet, A.M. and Faggion, R. : 1976. *Biochimie* **58**, 1255–1262.
27. Wallis, P.J. and Rhodes, M.J.C. : 1977. *Phytochemistry* **16**, 1891–1894.
28. Grand, C. and Ranjeva, R. : 1979. *Physiol. Veg.* **17**, 433–444.
29. Rhodes, M.J.C. and Wooltorton, L.S.C. : 1976. *Phytochemistry* **15**, 947–951.
30. Rhodes, M.J.C. and Wooltorton, L.S.C. : 1975. *Phytochemistry* **14**, 2161–2164.
31. Heinzmann, U., Seitz, U. and Seitz, U. : 1977. *Planta* **135**, 313–318.
32. Higuchi, T. : 1957. *Physiol. Plant.* **10**, 633–648.
33. Kutsuki, H., Shimada, M. and Higuchi, T. : 1981. *Phytochemistry* **21**, 19–23.
34. Wyrambik, D. and Grisebach, H. : 1975. *Eur. J. Biochem.* **59**, 9–15.
35. Lai, Y.Z. and Sarkanen, K.V. : 1971. *In* Lignins, Occurrence, Formation, Structure and Re-actions (Sarkanen, K.V. and Ludwig, C.H., eds.), pp. 218–228, Wiley-Interscience, New York.
36. Tanaka, K., Nakatsubo, F. and Higuchi, T. : 1976. *Mokuzai Gakkaishi* **22**, 589–590.
37. Tanaka, K., Nakatsubo, F. and Higuchi, T. : 1979. *Mokuzai Gakkaishi* **25**, 653–659.
38. Tanahashi, M., Aoki, T. and Higuchi, T. : 1981. *Mokuzai Gakkaishi* **27**, 116–124.
39. Eriksson, O., Goring, D.A.I. and Lindgren, O. : 1980. *Wood Sci. Technol.* **14**, 267–279.
40. Chang H.-m., Chen. C.L. and Kirk, T.K. : 1980. *In* Lignin Biodegradation : Microbiology, Chemistry, and Potential Applications I (Kirk, T.K., Higuchi, T. and Chang, H.-m., eds.), pp. 215–230, CRC Press, Boca Raton, Florida.
41. Katayama, T., Nakatsubo, F. and Higuchi, T. : 1980. *Arch. Microbiol.* **126**, 127–132.

42. Katayama, T., Nakatsubo, F. and Higuchi, T. : 1981. *Arch. Microbiol.* **130**, 198–203.
43. Enoki, A., Goldsby, G.P. and Gold, M.H. : 1980. *Arch. Microbiol.* **125**, 227–232.
44. Nakatsubo, F., Umezawa, T. and Higuchi, T. : 1981. The Ekman-Days 1981 (Intern. Symp. Wood Pulping Chemistry) 3, 16–24, Stockholm.
45. Rast, H.G., Engelhardt, G., Ziegler, W. and Wallnofer, P.R. 1980. *FEMS Microbiol. Lett.* 8, 259–263.
46. Ohta, M., Higuchi, T. and Iwahara, S. : 1979. *Arch. Microbiol.* **121**, 23–28.
47. Katayama, T. and Higuchi, T. : 1980. Abst. 25th Symp. Lignin Chemistry, Fukuoka, Japan.
48. Iwahara, S., Nishihira, T., Jomori, T., Kuwahara, N. and Higuchi, T. : 1980. *J. Ferm. Technol.* 58, 183–188.
49. Nakatsubo, F., Kirk, T.K., Shimada, M. and Higuchi, T. : 1980. *Arch. Microbiol.* **128**, 416–420.
50. Kamaya, Y., Nakatsubo, F., Higuchi, T. and Iwahara, S. : 1981. *Arch. Microbiol.* **129**, 305–309.
51. Iwahara, S., Ishiki, K. and Higuchi, T. : 1981. *Nippon Nogeikagaku Kaishi* **55**, 991–995.
52. Namba, H., Nakatsubo, F. and Higuchi, T. : 1980. Abst. 25th Symp. Lignin Chemistry, Fukuoka, Japan.
53. Nakatsubo, F., Reid, I.D. and Kirk, T.K. : 1981. *Biochem. Biophys. Res. Commun.* **102**, 484–491.
54. Higuchi, T. : 1981. *Wood Research* **67**, 47–58.
55. Ander, P., Hatakka, A. and Eriksson, K.-E. : 1980. *Arch. Microbiol.* **125**, 189–202.
56. Kuwahara, M., Takagami, H., Yonehara, M., Sato, T. and Iwahara, S. : 1981. *Mokuzai Gakkaishi,* **27**, 885–892.
57. Umezawa, T., Higuchi, T. and Nakatsubo, F. : 1982. Abst. 32th Annu. Meeting Wood Res. Soc., Fukuoka, Japan.

Legume Lectins: Structural Relationships and Properties

Leland M. Shannon

INTRODUCTION

Lectins are carbohydrate-binding proteins, of unknown physiological function, which have been isolated from many plant and animal sources (*1-3*). Because of their ability to bind carbohydrate receptors on the surfaces of animal cells, lectins provide valuable probes of the cell surface (*4*). Further, lectins elicit a variety of responses in certain plant and animal cells including : redistribution of surface receptors, stimulation of transport, increases in protein and nucleic acid synthesis, changes in cyclic nucleotide metabolism, alteration in microtubule assembly, and stimulation of cell division (*4,5*). In this report the words lectin and hemagglutinin are used interchangeably and as general terms for carbohydrate binding proteins which can agglutinate blood cells.

Lectins participate in cell agglutination by forming bridges between carbohydrate receptors located on adjacent cells. As more and more lectin molecules bind the cell surface, the aggregate grows in size, until finally it gets so large it settles out of solution. The lectin mediated aggregation of cells is reversible. Upon adding a dilute solution of an appropriate sugar, the aggregate dissociates. The added sugar molecules compete with the cell surface carbohydrate for the lectin binding sites. These observations indicate that lectins : 1) are carbohydrate binding proteins ; 2) that they possess at least two carbohydrate binding sites ; and 3) that they cause cell agglutination by binding reversibly to carbohydrate receptors on the cell surface.

Lectins were discovered about 100 years ago (*6*) and were originally called "phytohemagglutinins". The term "lectin" is commonly used today because similar agglutinins have now been isolated from bacteria (*7*), fungi (*8*), eels (*9*), and mammals (*10*). Lectins were considered curiousities until about 1965 when it was shown that the lectin from kidney bean was able to induce mitosis in lymphocytes (*11*). Two years later cancer cells were reported to be more susceptible than normal cells to agglutination by the lectin from wheat germ (*12*). These discoveries stimulated scores of investigators to employ lectins as tools to study cell surfaces and cellular growth, differentiation and development. Since 1965 many plant lectins have been purified. Because lectins frequently represent 1–3% of the seed protein and because the molecules possess multiple carbohydrate binding sites, affinity chromatography provides a simple and rapid one-step purification scheme. A large number of purified lectins have been characterized with respect to their physical and chemical properties (*13*). Most lectins possess a native molecular weight between 130,000–150,000, are tetramers composed of one or two different types of subunits, and possess 2 to 4 carbohydrate binding sites. The K_m for the preferred monosaccharide generally falls within the 2–5 mM range. Several lectins are dependent upon tightly associated manganese and/or calcium for sugar binding. Nearly all lectins are glycoproteins containing 1–5% carbohydrate.

Although many plant lectins have been characterized with respect to their physical and chemical properties, relatively little is known about their synthesis, localization, and function(s) in plants. A number of hypothetical roles for lectins in plants have been advanced, *i.e.*, transport of carbohydrates (*1*), involvement in seed formation and/or seed germination (*13*), plant "immune" system (*14*), specific recognition processes (*15*), and control of cell growth (*16*). To date, however, there is no convincing evidence to support any of these as a general role for plant lectins. Research in my laboratory is directed toward achieving a better understanding of the physiology and biochemistry of plant lectins, with the ultimate goal of elucidating their biological function(s). Progress to date has been both exciting and substantial. The following is a brief summary of recent work from this laboratory.

STRUCTURAL SIMILARITIES AMONG LEGUME LECTINS

Based upon their common source and upon their possessing similar physical, chemical, and biological properties, many of the legume lectins might be homologs (*17*). Immunochemistry provides a powerful method for determining relationships among proteins. Antibodies raised against a given lectin will react with a different lectin only when the two proteins share a common determinant(s). Different proteins can be functionally similar and yet antigenically totally unrelated. However, when proteins are found to be antigenically similar, there is a strong probability of both an evolutionary and a functional relationship. The following experiments were designed to assess the immunological relatedness among legume lectins.

All available lectins were screened by Ouchterlony double diffusion against normal rabbit sera and antisera prepared against pure lectins from six legume species. The tests were done on agar plates containing sugars to prevent any spurious lectin-glycoprotein interactions. Figure 1 shows the results of a typical set of tests using

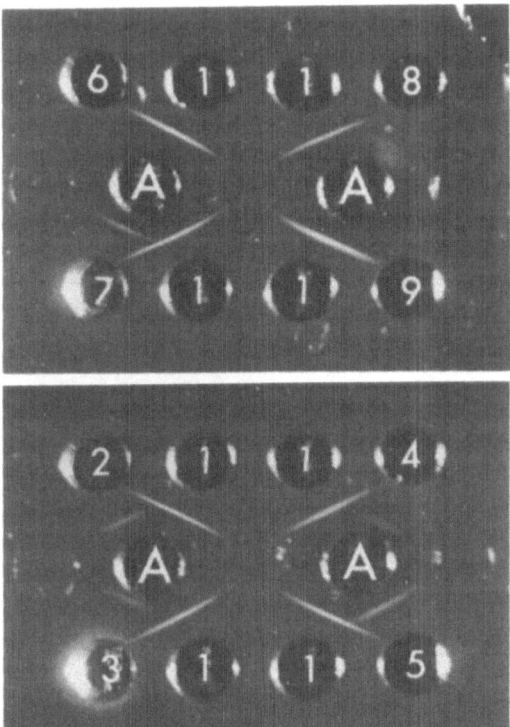

Fig. 1. Typical immunological cross-reactions between *Bandeiraea* antiserum and assorted lectins. A : *Bandeiraea* antiserum, 10 μl ; 1, *Bandeiraea* lectin, 10 μg ; 2, *Bauhinia* lectin, 10 μg ; 3, *Lotus* lectin, 100 μg ; 4, *Ulex* lectin, 20 μg ; 5, *Sophora* lectin, 10 μg ; 6, *P. vulgaris* lectin, 50 μg ; 7, *Ricinus* lectin, 90 μg ; 8, *Vigna* lectin, 10 μg ; 9, soybean lectin, 10 μg ; Photograph taken after 4 hr at 37 °C.

Bandeiraea antisera. Under these conditions no precipitin lines obtained with pre-immune sera. In all cases where a cross-reaction was observed, a single "spur" was present indicating that the cross-reacting lectin was immunologically related, but not identical to the homologous lectin. The results of all of these screenings are summarized in Table I. Note that 11 of the 13 galactose-specific lectins tested cross-reacted with one or more of the antisera.

Since lectins are multivalent carbohydrate binding proteins, it was conceivable that the lectins could bind to a carbohydrate component in the antiserum, thereby causing precipitin bands similar to those obtained in antibody reactions. This possibility was ruled out in the following manner. First, carbohydrate binding was inhibited by saturating the lectin binding sites with sugars. Second, the lectins were tested with purified IgG rather than whole antiserum. These precautions failed to diminish the ability of the antisera to form precipitin bands with the lectins. Third, heat treatment of the lectin abolished its ability to agglutinate red blood cells, but failed to prevent it from forming precipitin bands with antiserum. If the precipitin

TABLE I

Immunological Cross-reactions between Legume Lectins

Lectin	Antisera					
	Bandeiraea simplicifolia	*Bauhinia purpurea alba*	*Glycine max*	*Phaseolus lunatus*	*Sophora japonica*	*Vigna radiata*
1. *Arachis hypogaea* (Gal)[a]	−	−	−	−	−	−
2. *Bandeiraea simplicifolia* (Gal)	⊕	+	−	−	+	+
3. *Bauhinia purpurea alba* (GalNAc)	+	⊕	−	+	+	+
4. *Dolichos biflorus* (GalNAc)	−	+	+	+	+	+
5. *Glycine max* (GalNAc)	−	−	⊕	−	−	−
6. *Lotus tetragonolobus* (fucose)	+	+	−	−	+	+
7. *Phaseolus lunatus* (GalNAc)	+	−	+	⊕	−	−
8. *Phaseolus vulgaris* (Gal)	+	+	−	−	−	−
9. *Ricinus communis* (Gal)	+	+	−	+	+	+
10. *Sophora japonica* (GalNAc)	+	+	−	−	⊕	+
11. *Ulex europeus* (fucose)	+	+	−	−	+	+
12. *Vinga radiata* (Gal)	+	+	−	−	+	⊕
13. *Wisteria floribunda* (GalNAc)	−	+	−	−	+	−
14. *Concanavalia ensiformis* (Glc/Man)	−	−	−	−	−	−
15. *Lens culinaris* (Glc/Man)	−	−	−	−	−	−
16. *Pisum sativum* (Glc/Man)	−	−	−	−	−	−
Total positive CRM	8	9	2	3	8	7

[a] Sugar specificity.

⊕, homologous reaction; +, precipitin band observed; −, no precipitin band observed.

bands resulted from lectin interactions with carbohydrate components, the bands should not appear when lectin agglutination activity is destroyed. Since the precipitin bands were not diminished by heat treatment of the lectins, we again concluded that the precipitin bands resulted from antibody specific reactions.

As a further test of both the strength and specificity of the cross-reactions, we examined the ability of the antisera to inhibit lectin-induced hemagglutination. The hemagglutinating activity of every lectin which cross-reacts with an antiserum, is also specifically inhibited by that serum. The inhibitory potency of each serum was for the most part closely related to the relative strength of the cross-reactions seen by double diffusion. That is, each antiserum most potently inhibited the hemagglutinin activity of those lectins with which it most strongly cross-reacted. In no case where we could make the test did we observe a cross-reaction between serum and lectin without also seeing inhibition of lectin activity by that serum.

All of the legume lectins that have been characterized to date can be placed into two main groups with respect to their carbohydrate-binding specificities : (a) those specific for galactose (and its derivatives such as fucose and N-acetyl galactosamine

(GalNAc)) ; and (b) those specific for glucose or mannose (and their derivatives). The overwhelming majority of the legume lectins that have been studied fall into the galactose-specific group (Table I).

The results presented here indicate that there is extensive immunological relatedness among the lectins in the galactose group. Of 13 galactose-specific lectins tested, only peanut and soybean agglutinins failed to show a cross-reaction with at least one of the sera used. Antisera against soybean lectin did, however, react with two other lectins. We believe that these results coupled with what is already known about these proteins, i.e., they are all lectins, they have similar chemical and physical properties, they are all glycoproteins, they are all from the same plant family, and they are all isolated from seeds, suggest that many of the galactose lectins from legumes are homologs. A similar conclusion has been reached (17–19) from comparative amino acid sequence analysis and, although the sequence studies involved a very limited number of lectins, the similarities in N-terminal amino acid sequences among several galactose lectins was striking.

In addition to the immunological cross-reactions among most legume lectins, we also observed cross-reactivity between certain legume lectins and a lectin from castor bean, a dicotyledonous plant outside the legume family. The galactose-specific castor bean lectin has many properties in common with the legume lectins and is also isolated from seeds. We believe that these results suggest that evolutionary and functional homologs of a unique lectin species occur widely in legumes and perhaps even in dicotyledonous plants in general. We have not observed any immunological relationship between the galactose lectins and lectins displaying other specificities, but evidence consistent with an evolutionary relationship has been presented (19).

The evidence now available suggests that most of the well characterized legume lectins are members of a specific class of evolutionarily related legume proteins. Many of these lectins are likely to be homologs.

LEGUME SPECIES DEVOID OF LECTINS

In view of the above findings and the fact that GalNAc lectins often account for 1% or more of the seed protein, and are in species widely spread within the legume family (15), we wondered if lectins might be ubiquitious legume proteins. A review of the 'literature suggested that this possibility was unlikely, since many legume species, perhaps 50% or more of all species tested, appeared to be totally devoid of hemagglutinin activity of any kind. Furthermore, many of the species which contained hemagglutinins appear to contain proteins with carbohydrate specificities distinct from the GalNAc lectins (13). To confirm the validity of the earlier literature a survey of a number of legume species was performed. Indeed, seed extracts of many legume species displayed very little or no hemagglutinin activity, or contained hemagglutinins which were not inhibited by GalNAc. One cannot rule out the presence of a hemagglutinin activity by the limited assay methods used. However, since all of the GalNAc hemagglutinins which have been described are very easily detected in crude extracts by the assay methods used, it is reasonable to conclude

TABLE II
Immunological Cross-reactions between Crude Seed Extract (Which Are Devoid of Hemagglutinin Activity) and Antisera Raised against Purified GalNAc Specific Legume Lectins

| Seed extract | Antisera | | |
	Bauhinia	Dolichos	Sophora
1. Acacia	+	+	+
2. Amorpha	+	−	+
3. Cercis	±	−	+
4. Colutea	+	−	+
5. Delonix	+	±	+
6. Genista	+	−	+
7. Gleditsia	+	−	+
8. Lupinus	+	−	+
9. Mimosa	+	+	+
10. Parkinsonia	±	−	+
11. Poinciana	+	−	+
12. Pueraria	+	+	+
13. Spartium	+	−	+
14. Thermopsis	+	−	+
15. Lespedeza	+	−	+
Total positive CRM	13	3	15

+, Precipitin line observed; −, no precipitin line observed; ±, very weak reaction, questionable.

that none of these plants contain a detectable quantity of hemagglutinin activity comparable to the GalNAc hemagglutinins.

We decided to examine extracts of a number of GalNAc hemagglutinin "negative" species to determine if they contained proteins immunologically related to any of the GalNAc hemagglutinins (CRM). The results, given in Table II, reveal that every plant tested contained CRM to antisera raised against one or more of the GalNAc hemagglutinins. The antisera used in these studies were from rabbits immunized with very pure hemagglutinin preparations. Although many plants showed CRM with antisera raised against several different lectins, it appeared likely that one major protein was responsible, since adsorption of the extracts by any one of the reactive sera results in a loss of reactivity with all the other sera.

Since the GalNAc hemagglutinins are glycoproteins, the antisera raised against them might recognize glycoproteins in plant extracts which are totally different from the hemagglutinins, but which possess similar carbohydrate determinants. This possibility was ruled out in the following way : The quantity of hemagglutinin (antigen) required to totally inhibit the homologous antisera was determined. About 100 times this quantity of antigen was denatured by boiling and then subjected to several proteolytic treatments. The resulting material retained no hemagglutinin activity, showed no immunological reaction by Ouchterlony double diffusion and most importantly, showed no ability to block the reaction of antisera with CRM containing extracts. All the buffers and protease solutions were tested for the presence of several glycosidases and none were found. If carbohydrate determinants constituted a dominant part of the CRM reaction, then the glycopeptides remaining in the proteolyzed antigen preparation would be expected to have been potent hapten inhibitors of the CRM reaction. Since o inhibition was seen, it was concluded that carbohydrate plays little or no role in determining the immunological similarities between hemagglutinins and their CRMs.

Most plant hemagglutinins can be purified by affinity chromatography (3). Sepharose to which GalNAc has been attached provides an excellent adsorbent for the purification of most of the GalNAc specific hemagglutinins. There are now many examples of proteins which have carbohydrate binding properties, but which are not themselves hemagglutinins, that can be purified by carbohydrate affinity chromatography. It is also known that hemagglutinins may sometimes exist in forms, or can be converted to forms, which retain carbohydrate binding properties but are no longer hemagglutinins. This information leads one to question whether or not the CRMs, even though they do not display hemagglutinin activity, can be adsorbed by affinity matrices. Therefore, a number of CRM containing plant extracts were chromatographed on GalNAc Sepharose columns with the result that very little, if any of the total protein or CRM was retained by the columns. These findings suggest that the CRMs, which do not display classic hemagglutinin activity, also do not possess the carbohydrate binding properties typical of GalNAc lectins.

α-GALACTOSIDASES WITH HEMAGGLUTININ ACTIVITY

The recent discovery that certain legume enzymes possess hemagglutinin properties (20–23) opened an exciting area of research. In this section I will summarize what is currently known about enzymatic-hemagglutinins and how they are related to "classic" legume lectins.

Working with seed extracts of mung bean, a very potent agglutinin of trypsinized rabbit erythrocytes was observed. The mung bean extract displayed two unique characteristics. Firstly, the hemagglutinin activity was inhibited by xylose and inositol, in addition to galactose, and thus possessed a specificity distinct from previously described hemagglutinins. Secondly, the extracts displayed an interesting property referred to as "clot dissolving activity". Apparently, under the appropriate conditions, this protein is capable of enzymically altering those erythrocyte receptors with which it interacts, resulting in a dissolution of cell aggregates. The disaggregated erythrocytes are permanently altered and are no longer agglutinable by the mung bean extract (although they remain agglutinable by many other legume lectins). Since these observations could be explained by an enzymatic alteration of receptors on the erythrocytes, we examined the mung bean extract for various hydrolytic activities and found a very strong α-galactosidase. During purification we followed both hemagglutin and α-galactosidase activities.

Figure 2 depicts the gel filtration elution profile from a Sephacryl S-200 column, the final purification step. Note that both hemagglutinin and α-galactosidase activities elute at coincident positions. Inset A shows the appearance of SDS gels across the

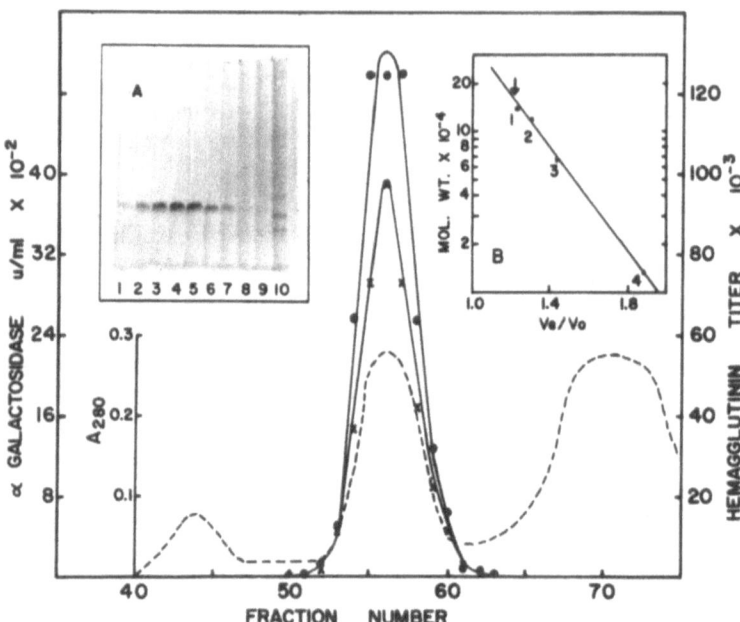

Fig. 2. Sephacryl S-200 elution profile. O, hemagglutinin ; ×, α-galactosidase ; ----, A_{280}. Inset A shows the appearance of SDS gels of 100 μl portions of fractions across the peak. Inset B shows the calibration curve of column.

peak. A single band is noted, the intensity of which coincides with A_{280} and activity peaks. Inset B shows the calibration curve of the column. The extrapolated native molecular weight for α-galactosidase-hemagglutinin was 160,000. The molecule appeared to be a tetramer composed of a single type of subunit (45,000 MW). All attempts to separate the α-galactosidase from hemagglutinin activity were unsuccessful. What was observed instead, was that these two activities exactly co-purified by every separation method tested. Attempts to selectively inactivate one of the activities by the use of heat or PCMB were also unsuccessful. Both activities were equally sensitive to these treatments and were equally protected from inactivation by the hapten sugars galactose, xylose, and inositol (21).

A number of carbohydrates were examined for their ability to inhibit the agglutination and α-galactosidase reactions. The only compounds showing significant inhibitory properties were p-nitro-phenyl α galactoside [pNPG(α)], galactose, xylose, and inositol. All sugars which inhibited agglutination either inhibited the α-galactosidase or functioned as substrate. In every case, the concentration of sugar causing 50% inhibition of enzyme activity was very similar to that causing 50% inhibition of hemagglutination (21). Data presented thus far indicate that hemagglutinin and α-galactosidase activities co-purify to yield a single size class of polypeptide. The fact that both activities are simultaneously bound to erythrocytes and are simultaneously released, are equally heat labile (in presence and absence of protective sugars), possess indistinguishable substrate and inhibitor specificity — these facts, we believe, provide compelling evidence that both the hemagglutinin and α-galactosidase activities reside within a single protein species.

The implications of these findings could be very important. The fact that a phytohemagglutinin may possess hydrolytic activity offers a completely new dimension to our thinking about the function of lectins. For this reason it is important to know if the mung bean protein is some freak enzyme that just happens to possess hemagglutinin activity, or if it is genuinely homologous to other legume seed lectins. In our search for additional α-galactosidase hemagglutinins we examined seed extracts

TABLE III
Kinetic Studies of Hemagglutinin and Non-hemagglutinin Forms of α-Galactosidase Using pNPG (α) as Substrate

	K_m (mM)	K_i^a (mM)		
		Gal	Xyl (mM)	Inos
Non-hemagglutinin α-galactosidases				
Amorpha	1.2	0.5	2.9	12.6
Bandeiraea	1.1	0.6	4.4	14.7
Bauhinia	1.0	0.5	3.1	15.8
Caragana	0.8	0.6	3.8	9.5
Cercis	1.8	0.5	3.4	9.0
Colutea	0.4	0.4	3.2	17.3
Conavalia	1.2	0.2	1.7	9.5
Cytisus	1.0	0.3	1.8	6.1
Dolichos	1.2	0.5	3.6	–
Genista	0.6	0.3	1.6	11.0
Laburnum	1.4	0.5	2.7	14.7
Lathyrus	1.3	0.6	3.5	8.7
Lens	1.2	1.2	3.3	6.9
Lespedeza	0.5	0.3	2.7	11.7
Mimosa	0.7	0.4	2.2	11.6
Phaseolus	0.7	0.5	3.1	12.8
Sophora	0.6	0.3	2.2	9.0
Spartium	1.7	0.4	3.5	11.4
Ulex	0.7	0.4	1.7	10.0
Wisteria	0.7	0.6	3.6	17.8
Hemagglutinin α-galactosidases				
Soybean	0.2	1.1	3.8	12.9
Lima bean	0.4	0.3	2.0	15.0
Lupine	0.2	0.3	1.5	7.5
Thermopsis	0.4	0.4	1.6	8.8
Pueraria	0.4	0.9	2.9	8.8
Mung bean	0.3	0.7	5.2	20.0

^{a} K_i determinations were performed using pNPG(α) as the substrate.

from 26 genera of legumes (24). We detected clot-dissolving-enzymatic-phytohem-agglutinins in five additional species. The kinetic and hemagglutinin properties are virtually identical with each other and with the mung bean protein. They possess practically identical K_m values and each extract is inhibited by the same sugars with virtually identical K_i values from plant to plant. The same three sugars inhibited hemagglutinin activity with virtually identical K_i values from plant to plant. The soybean α-galactosidase-hemagglutinin has been purified to apparent homogeniety and appears to be immunologically identical to the mung bean protein (23). The α-galactosidase-hemagglutinin has also been isolated from soybean leaves indicating that this protein is not limited to seed tissues. The α-galactosidase-hemagglutinin in soybean, and lima, are distinct proteins from the well characterized "classic" lectins present in these species. The two groups of proteins can be separated from each other by affinity chromatography or by immunoprecipitation, they have different sugar specificities, and they can exist simultaneously in the same tissue.

The above survey detected clot-dissolving α-glactosidase-hemagglutinin activity in only six genera. α-Galactosidase activity, however, was present in every legume examined (25). The kinetic properties of hemagglutinin and non-hemagglutinin forms of α-galactosidase are presented in Table III. The α-galactosidase activities are vir-tually identical. They possess the same K_m for pNPG(α) and each extract was in-hibited by the same sugars with virtually identical K_i from plant to plant. Each ex-tract displayed Michaelis-Menten substrate utilization with competitive inhibition by galactose, xylose, and inositol.

We next challenged each of the above extracts with antibody raised against mung bean α-galactosidase-hemagglutinin. Each extract contained materials which cross-reacted with the mung bean antibody, i.e., each extract showed CRM — even though the extract did not contain a clot dissolving α-galactosidase-hemagglutinin. This observation led us to inquire if the α-galactosidase activity present in each extract was responsible for the CRM. To answer this, IgG was purified from mung bean antibody and from normal sera. Each IgG was then coupled to Sepharose and extracts from the different legume species, passed over the columns. In each case virtually 100% of the α-galactosidase activity could be specifically adsorbed from the extract by anti-sera raised against the mung bean protein. No specific binding was seen for β-galacto-sidase, β-glucosidase, or α-mannosidase. These immunoprecipitation studies indicate that the non-hemagglutin α-galactosidases are immunologically closely related to the mung bean α-galactosidase-hemagglutinin and were at least partially responsible for the CRM. This result, coupled with the similarities in their physical and kinetic pro-perties, suggests that both types of α-galactosidases are members of one specific functional class of protein, i.e., they are homologues.

The evidence collected to date indicates that the α-galactosidase-hemagglutinins and the "classic" legume hemagglutinins are members of distinct classes of protein : they can be separated from each other ; they can exist simultaneously in the same tissue ; and they have different sugar specificity. Immunological studies reveal, how-ever, that these two classes of proteins share at least some antigenic determinants (26). Antisera raised against several "classic" hemagglutinins cross-react with mung bean α-galactosidase-hemagglutinin and antisera to the mung bean protein cross-react with several "classic" hemagglutinins. Thus, while the α-galactosidase-hemagglutinins and the "classic" legume hemagglutinins appear to be members of distinct classes of proteins, immunochemical data suggest they may be evolutionarily related.

CONCLUDING REMARKS

Evidence at this time suggests that there are at least two evolutionarily related, but distinct classes of legume proteins which possess hemagglutinin properties : these are the "classic" lectins and the α-galalctosidase-hemagglutinins. All legumes appear to contain a specific α-galactosidase, but only rarely does this enzyme possess hemagglutinin activity. Hence, we suspect the hemagglutinin activity associated with the α-galactosidase-hemagglutinin molecule may not be important with respect to the *general* functioning of this class of protein. It is clear that a great many questions about these interesting and yet elusive proteins remain to be answered. We hope the observations reported here will stimulate additional work and help provide the foundation from which the physiological roles of lectins will ultimately emerge.

Acknowledgements
The author expresses his sincere appreciation to the Japan Society for Promotion of Science for the Short Term Fellowship which helped make participation in this symposium possible. Major portions of this research were supported by National Science Foundation, grant PCM-7910450.

REFERENCES

1. Lis, H. and Sharon, N. : 1973. *Annu. Rev. Biochem.* 42, 541–574.
2. Liener, I.E. : 1976. *Annu. Rev. Plant Physiol.* 27, 291–319.
3. Goldstein, I.J. and Hayes, C.E. : 1978. *Adv. Carb. Chem. Biochem.* 35, 127–340.
4. Nicolson, G.L. : 1974. *Int. Rev. Cytol.* 39, 89–190.
5. Hoffstein, S., Soberman, R., Goldstein, I. and Weissmann, G. : 1976. *J. Cell. Biol.* 68, 781–787.
6. Stillmark, H. : 1888. *Inaug. Diss. Dorpat.*
7. Gilboa-Garber, N. : 1972. *Biochim. Biophys. Acta* 273, 165–173.
8. Fujita, Y., Oishi, K., Suzuki, K. and Imahori, K. : 1975. *Biochemistry* 14, 4465–4470.
9. Desai, P.R. and Springer, G.F. : 1972. *Methods Enzymol.* 28, Part B, 383–388.
10. Kawasaki, T. and Ashwell, G. : 1976. *J. Biol. Chem.* 251, 1296–1302.
11. Nowell, P.C. : 1960. *Cancer Res.* 20, 462–466.
12. Aub, J.C., Tieslau, C. and Lankester, H. : 1963. *Proc. Natl. Acad. Sci. U.S.A.* 50, 613–619.
13. Toms, G.C. : 1971. *In* Chemotaxonomy of Legumes (Harbourne, J.B., Boulter, D. and Turner, B.L., eds.), pp. 367–462, Academic Press, New York.
14. Bird, G.W.G. : 1959. *Br. Med. J.* 15, 165–168.
15. Bohlool, B.B. and Schmidt, E.L. : 1974. *Science* 185, 269–281.
16. Nagl, W. : 1972. *Exp. Cell Res.* 74, 599–602.
17. Foriers, A., Wuilmart, C., Sharon, N. and Strosberg, A.D. : 1977. *Biochem. Biophys. Res. Commun.* 75, 980–986.
18. Etzler, M.G., Talbot, C.F. and Ziaya, P.R. : 1977. *FEBS Lett.* 82, 29–41.

19. Foriers, A., de Neve, R., Kanarek, L. and Strosberg, A.D. : 1978. *Proc. Natl. Acad. Sci. U.S.A.* **75**, 1136–1139.
20. Albersheim, P. and Wolpert, J.S. : 1976. *Plant Physiol.* **57**, S79.
21. Hankins, C.N. and Shannon, L.M. : 1978. *J. Biol. Chem.* **253**, 7791–7797.
22. Paus, E. and Steen, H.B. : 1978. *Nature* **272**, 452–454.
23. del Campillo, E. and Shannon, L.M. : 1982. *Plant Physiol.* **69**, 628–631.
24. Hankins, C.N., Kindinger, J.I. and Shannon, L.M. : 1980. *Plant Physiol.* **65**, 618–622.
25. Hankins, C.N., Kindinger, J.I. and Shannon, L.M. : 1980. *Plant Physiol.* **66**, 375–378.
26. Hankins, C.N., Kindinger, J.I. and Shannon, L.M. : 1979. *Plant Physiol.* **64**, 104–107.

The Frontiers of Influence of Some Plant Enzymes

Erwin Latzko and Grahame J. Kelly

INTRODUCTION

The classic concept of an enzyme is that of a protein which efficiently and specifically speeds up a chemical reaction to an extent (many-million-fold) that makes life possible. The frontier of direct influence of the enzyme can, at this point, be thought of as extending little further than the realm of the reaction being catalysed. However, some (but not all) enzymes have evolved the ability to detect and respond to certain changes in their microenvironments, such as changes in pH and in the concentrations of particular metabolites. More often than not, this has given them the capability of regulating the supply of substrates to other enzymes, and of thereby regulating metabolism. Termed regulatory enzymes, the frontier of direct influence of these enzymes extends beyond the sphere of catalysis and into the sphere of metabolic regulation.

Examples of regulatory enzymes in plants include fructosebisphosphatase, phosphofructokinase, phosphoribulokinase, and pyruvate P_i dikinase. These are considered in this chapter, particular emphasis being given to the enzymes in photosynthetic cells, since enzymes in these cells must regularly experience the considerable microenvironmental changes that accompany light/dark transitions. A fifth enzyme, the newly-discovered ascorbate peroxidase, is also examined as an example of an enzyme with a vital catalytic function but no known regulatory function.

FRUCTOSEBISPHOSPHATASE OF THE SPINACH-LEAF CYTOSOL

$$\text{D-fructose-1,6-P}_2 + \text{H}_2\text{O} \longrightarrow \text{D-fructose-6-P} + \text{P}_i$$

Photosynthetic cells contain two fructosebisphosphatases, one in the cytosol (1) and the other in the chloroplast (2,3). The spinach-leaf cytosolic enzyme purified to homogeneity (4) was found to have properties similar to those of the fructosebis-phosphatases from non-photosynthetic cells such as seeds and animal tissues : it shows a high affinity for fructose-1,6-P$_2$ (K_m 2.5 μM), it is active at all pH values between 7.0 and 8.5, and it can monitor the energy status of the cytosol through an interaction with AMP ; this interaction reduces the enzyme activity. The enzyme normally must be involved with cytosolic gluconeogenesis from triose-P (the product of photosynthesis exported from chloroplasts) to sucrose, and it seems possible that it could control the rate of this gluconeogenesis by slowing down when AMP levels become elevated, thus diverting the triose-P into glycolysis as would be required for the eventual replenishment of the ATP supply. However, the concentration of AMP required to reduce enzyme activity by half was observed to be about 0.5 mM (4) and doubts might be expressed as to whether this sensitivity is sufficient for the enzyme to have a significant regulatory role. The enzyme also responds to accumu-lation of its product (fructose-6-P) with reduced activity (5), but here again the re-latively low sensitivity lessens the likelihood of any physiological significance.

It is worth nothing that any reduction in the activity of cytosolic fructosebis-phosphatase will not only reduce the rate of sucrose biosynthesis, but may also reduce the export of triose-P from chloroplasts because the P$_i$ required to be im-ported by counter-exchange with exported triose-P will no longer be available (6). Metabolism within the chloroplast would consequently alter (e.g., accelerated starch biosynthesis) in response to the accumulated triose-P. In this way the frontier of in-fluence of cytosolic fructosebisphosphatase extends across the chloroplast envelope and into the chloroplast.

SPINACH-CHLOROPLAST FRUCTOSEBISPHOSPHATASE

The other fructosebisphosphatase (the chloroplast enzyme) is one of the enzymes of the Calvin cycle. Its properties differ markedly from those of other fructosebisphos-phatases. As exemplified by the spinach-chloroplast enzyme (3,7,8) it has an affinity for fructose-1,6-P$_2$ which varies widely, depending on environmental conditions such as Mg^{2+} concentration, pH, and the balance between reductive and oxidative conditions. Under mildly oxidative conditions it is inactive below pH 8.0, but under reductive conditions it is quite active at pH values as low as 7.5. Finally, it is insensi-tive to AMP (9).

Chloroplast fructosebisphosphatase has evolved the ability to respond to illumi-nation of the chloroplast. At the physiological pH of the stroma in illuminated chloroplasts (about 8.0), the enzyme activity increases upon illumination from zero to a level sufficient to accommodate observed rates of CO$_2$ fixation (10). This light-

mediated activation almost certainly involves the reduction of one or more disulphide linkages to disulphydryls, since the activation can be mimicked *in vitro* by
incubating purified enzyme with dithiothreitol, and when this is done, the number of
sulphydryl groups on the enzyme molecule is more than doubled (*3*).

Since chloroplasts contain only just sufficient fructosebisphosphatase activity to
support CO_2 fixation, the initial impression of the physiological significance of the
observed light-mediated activation was that it constituted, at least in part, the "light-
on/dark-off" switch of CO_2 fixation. A static situation was envisaged, in which the
enzyme was reductively activated at sunrise (the reducing equivalents presumably
originating from the photosynthetic electron transport chain), and oxidatively deactivated at sunset (the oxidant being unknown). However, with the observation by
Kaiser (*11*) that reductively-activated enzymes could be oxidatively-deactivated by
H_2O_2 (which is a product of pseudocyclic photosynthetic electron transport), a
dynamic picture has emerged, in which light-mediated activation is required throughout the illumination period in order to repair those enzyme molecules that fall prey
to photosynthetically generated oxidants such as H_2O_2 (*12,13*).

This does not mean that the concept of a "light-on/dark-off" switch can be dis-
carded. If light-mediated activation represented nothing more than a mechanism to
protect susceptible enzymes against photosynthetically generated oxidants, then
chloroplast fructosebisphosphatase would be an enigma in that it would be seen as
an enzyme so foolish as to have evolved oxidant-sensitive sulphydryl groups essential
for activity at the pH (8.0) of the chloroplast stroma, yet not essential for activity at
the higher pH of 8.8 (*10*), and, in addition, to have done so when there exists in the
leaf cytosol another fructosebisphosphatase that is presumably far less sensitive to
oxidants at pH 8.0 (*4*). Clearly, there is a strong temptation to believe that the cell
found some advantage in having a chloroplast fructosebisphosphatase so sensitive to
the reducing environment of the illuminated chloroplast at physiological pH. This
advantage may lie in the opportunity for the enzyme to participate in the "light-
on/dark-off" switch of CO_2 fixation, and it might be fortuitous that the reductive
mechanism that constitutes the "light-on" portion of the switch also happens to
protect the activated enzyme against certain oxidants generated by the photochemi-
cal apparatus.

Another, perhaps more subtle advantage of the light-mediated activation is that
the extent of this activation could dictate whether or not a significant proportion of
fixed carbon is diverted to the synthesis of starch in the chloroplast. As pointed out
elsewhere (*14*), the activity of chloroplast fructosebisphosphatase needs to be 50%
greater when the cell synthesizes starch from CO_2 than when it synthesizes sucrose
from CO_2. Hence, under low light the activation of the fructosebisphosphatase may
be insufficient to support starch synthesis, and all fixed CO_2 will instead be chan-
nelled (in the form of triose-P) into the cytosol for the manufacture of sucrose.

GROWTH TEMPERATURE AND SPINACH-LEAF
FRUCTOSEBISPHOSPHATASES

The observation that, in the absence of reducing agents, spinach-chloroplast fructose-bisphosphatase is inactive at pH 7.5 but active at pH 8.8 while the spinach cytosolic fructosebisphosphatase is about equally active at both these pH values (3,4), allows estimations of the activities of both these enzymes in crude extracts from spinach leaves (Fig. 1). A comparison between warm months and cool months showed a remarkable variation in the ratio of the two enzyme activities, with a maximum content of the chloroplast enzyme coinciding with a minimum content of the cytosolic enzyme during the warmest months of the year (Fig. 1). Taken together with the comments in the preceding paragraph (above), the results indicate a tendency to synthesize starch in the warmer months and sucrose in the cooler months.

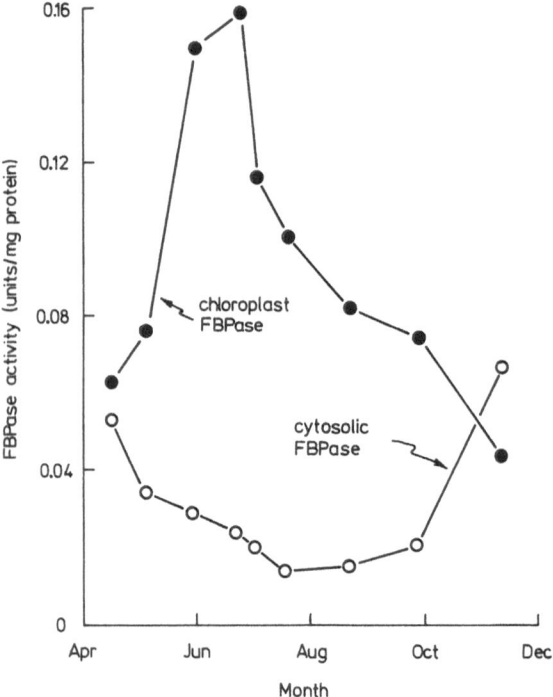

Fig. 1. Contrasting effects of growth temperature on the contents of chloroplast and cytosolic fructosebisphosphatases in field-grown spinach leaves. The preparation of crude extracts and the assay of enzyme activities were essentially as described in (4) ; the cytosolic enzyme is the activity measured at pH 7.5, and the chloroplast enzyme is the difference between the activities measured at pH 8.8 and pH 7.5. The spinach was grown in Weihenstephan, Germany. Parallel experiments with glasshouse-grown spinach gave constant enzyme activities of 0.10 (chloroplast) and 0.04 (cytosolic) units/mg protein, with no significant variation between months. From Zimmermann (15).

THE FRUCTOSEBISPHOSPHATASES IN OTHER PHOTOSYNTHETIC TISSUES

The difference in pH response of the two spinach-leaf fructosebisphosphatases (above) is also apparent in other plants, including pea, bean, and sugar-beet leaves, since in crude extracts of these leaves the activity measured at pH 8.8 is 2–3 times that measured at pH 7.5. However, this situation does not hold for other leaves, and in some cases including moss (Fig. 2) and barley, the activity measured at pH 7.5 actually exceeds that measured at pH 8.8. The reason for this difference between plants is under investigation.

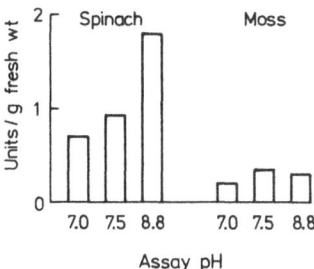

Fig. 2. The activity of fructosebisphosphatase in crude extracts from the leaves of spinach and moss, measured at three pH values. The preparation of crude extracts and the assay of enzyme activities were essentially as described in (4).

PHOSPHORIBULOKINASE

$$\text{D-ribulose-5-P} + \text{ATP} \longrightarrow \text{D-ribulose-1,5-P}_2 + \text{ADP}$$

Phosphoribulokinase is an enzyme found only in the chloroplast (in eukaryotic cells) where it is a component enzyme of the Calvin cycle. Unlike fructosebisphosphatase, it is present with an activity well in excess of that required to support normal rates of CO_2 fixation, hence it was surprising to discover that the reaction it catalyses is usu-ally far from equilibrium *in vivo* (16). An explanation for this anomaly was obtained when it was observed (17) that the enzyme is totally inactive in darkness at pH 6.8 (which approximates the pH of the stroma in darkened chloroplasts). Upon illumi-nation of the chloroplasts, light-mediated activation of the enzyme was clearly de-tected. This activation differs in several respects from the light-mediated activation of fructosebisphosphatase : it occurs very rapidly (17,18), it occurs even with quite low light intensities, and it cannot always be mimicked by addition of dithiothreitol (17). The rapidity and sensitivity of the response of this enzyme to light make it a strong candidate for playing a major role in the "light-on/dark-off" switch of CO_2 fixation.

A second characteristic of phosphoribulokinase worthy of note is its proclivity to bind to thylakoid membranes. Up to half of the enzyme was found bound upon breakage of the chloroplasts, and although inactive in this form, it could be recovered in an active solubilised form simply by washing the membranes with buffer (17). A similar situation has been observed with at least one other chloroplast enzyme that undergoes light-mediated activation, i.e. NADP-linked malate dehydrogenase (19).

PHOSPHOFRUCTOKINASE

$$\text{D-fructose-6-P} + \text{ATP} \longrightarrow \text{D-fructose-1,6-P}_2 + \text{ADP}$$

It is well known that the frontier of influence of phosphofructokinase extends well beyond the reaction it catalyses. As a regulatory enzyme of glycolysis, the plant enzyme responds to changes in the levels of several metabolic intermediates, the most notable being P-enolpyruvate, glycerate-3-P, ATP, P_i, and citrate. In this way it helps to tune the flux through glycolysis to the metabolic needs of the cell.

About 7 years ago, photosynthetic cells were found to contain two phospho-

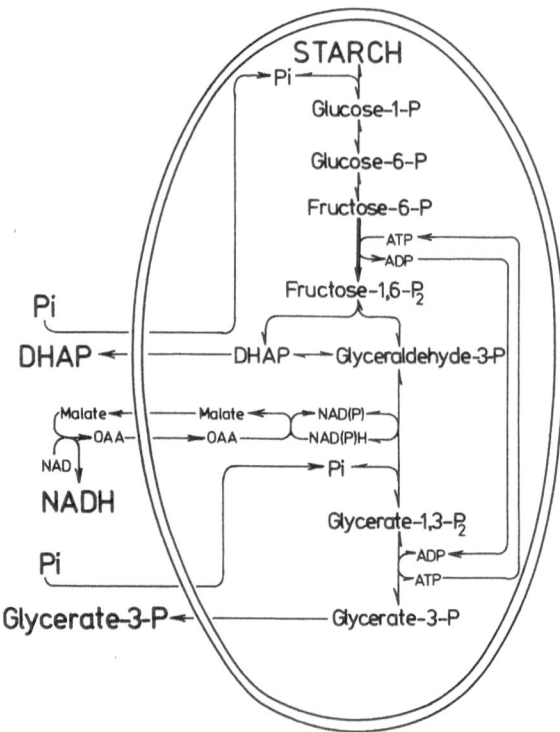

Fig. 3. Mobilisation of chloroplast starch. The sum of the reactions is : $(\text{glucose})_n$ + $2\ P_i$ + $NAD^+ \longrightarrow (\text{glucose})_{n-1}$ + dihydroxyacetone-P + glycerate-3-P + NADH + H^+. The phosphofructokinase reaction is shown by a thick arrow. Abbreviations : DHAP, dihydroxyacetone-P ; OAA, oxaloacetate.

fructokinases, one in the chloroplast and the other in the cytosol (20). The chloroplast enzyme participates in the conversion of starch to products which can be exported from the chloroplast (Fig. 3). This process involves a portion of the glycolytic pathway, and since (a) the chloroplast's content of phosphofructokinase activity is about the same order of magnitude as the rate of starch mobilisation (21) and (b) the enzyme responds to changes in the levels of glycerate-3-P and P_i (22), it seems most probable that this enzyme plays a major role in the regulation of chloroplast starch breakdown. This conclusion is strengthened when the regulatory properties of chloroplast phosphofructokinase and ADP-glucose pyrophosphorylase (the principal regulatory enzyme of starch biosynthesis) are compared (22) (Table I). Effectors found

TABLE I
Influence of Some Effector Metabolites on the Activities of ADP-Glucose Pyrophosphorylase and Phosphofructokinase

Effector	ADP-glucose pyrophosphorylase	Chloroplast phosphofructokinase
Glycerate-3-P	Stimulated	Inhibited
P-enolpyruvate	Stimulated	Inhibited
Glycolate-P	Stimulated	Inhibited
P_i	Inhibited	Stimulated[a]

[a] P_i relieves the inhibitions caused by the other three effectors.
Based on Ghosh and Preiss (23) and Kelly and Latzko (22).

to inhibit one enzyme stimulate the other, suggesting that the two enzymes work in concert in ensuring that starch synthesis and degradation do not occur simultaneously.

Neither the spinach-leaf cytosolic phosphofructokinase, nor the phosphofructokinases from other plant leaves have been investigated to any extent. Exploratory tests have been performed with some partially purified preparations of leaf phosphofructokinase (presumably containing a mixture of the chloroplast and cytosolic enzymes), and during these tests it was found that moss phosphofructokinase did not respond to $400 \mu M$ P-enolpyruvate, whereas other plant phosphofructokinases respond to this concentration of P-enolpyruvate with a complete cessation of activity (24). This case provides a good example of the potential versatility of enzymes with respect to their acquisition of regulatory properties.

PYRUVATE P_i DIKINASE

$$\text{pyruvate} + P_i + ATP \rightleftharpoons \text{P-enolpyruvate} + PP_i + AMP$$

The occurrence of pyruvate P_i dikinase in photosynthetic cells is limited to C_4 plants where it regenerates the CO_2 acceptor (P-enolpyruvate) for the C_4 pathway of photosynthesis. The C_4-plant enzyme has a remarkable ability to mediate regulation of this C_4 pathway according to the external environment, since it ceases activity when leaves are either darkened or cooled (25,26). Current evidence suggests that the enzyme interacts with the increased concentration of ADP that appears upon darkening, and that this interaction induces lower activity (25). On the other hand, cooled enzyme simply dissociates into subunits with consequent loss of activity (26).

Pyruvate P_i dikinase is not absent from C_3 plants. It was recently found in immature wheat grains (27) and in the immature grains of seven other cereal grasses (28). The wheat grain enzyme is not in the chloroplast-containing pericarp tissue, but rather in the aleurone layer. Although having kinetic properties similar to the C_4 plant enzyme, the wheat-grain enzyme must have a different function since the aleurone is not directly involved in photosynthesis. One possible role suggested earlier (27) is that it takes part in a sequence designed to capture CO_2 respired by the seed, thus enabling the synthesis of 4C-amino acids from 3C-amino acids (Fig. 4). Serving this function in a temperate plant, the enzyme would not be expected to have necessarily evolved the ability to respond to light/dark transitions or cool temperatures as

WHEAT GRAIN ALEURONE CELL

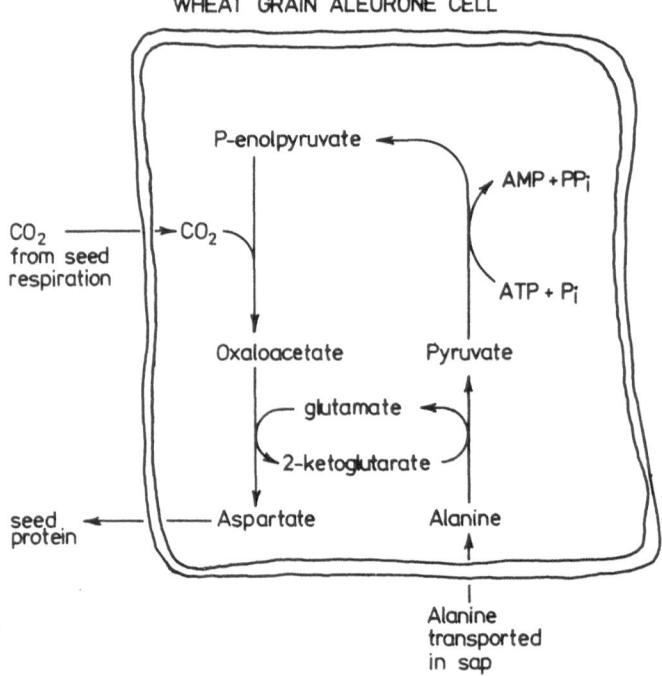

Fig. 4. Proposed role of pyruvate P_i dikinase in the aleurone layer of cereal grains. The sum of the reactions shown is: alanine + CO_2 + ATP + P_i ⟶ aspartate + AMP + PP_i. Cereal grasses have been noted to transport about 25% of their sap nitrogen in the form of alanine (29).

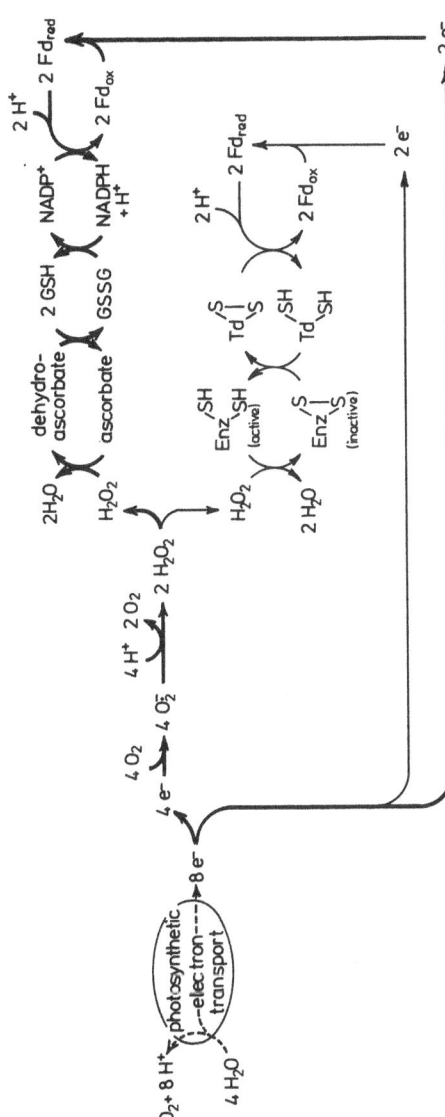

Fig. 5. Relationships between pseudocyclic electron transport, H_2O_2 formation, ascorbate peroxidase, and the reductive-activation/oxidative-deactivation of enzymes in photosynthetic cells. In the theoretical situation shown, one molecule of H_2O_2 is reduced by the ascorbate peroxidase system while another reacts (non-enzymically?) with an enzyme which is thereby oxidatively deactivated. It seems probable that most H_2O_2 would normally be removed by the ascorbate peroxidase system (thicker arrows). Abbreviations : Enz, enzyme ; GSH and GSSG, glutathione (reduced and oxidised) ; Fd, ferredoxin ; Td, thioredoxin. Based on Foyer and Halliwell (33), Nakano and Asada (34), and Buchanan (35).

the C_4 enzyme does. Indeed, no such regulatory function of the wheat-grain enzyme could be found (28), and at present it is not known whether or not the C_3 plant pyruvate P_i dikinase is a regulatory enyzme.

ASCORBATE PEROXIDASE

$$\text{ascorbate} + H_2 O_2 \longrightarrow \text{dehydroascorbate} + 2 H_2 O$$

The discovery of a soluble enzyme capable of catalysing the reduction of $H_2 O_2$ to water using ascorbate as the electron donor has been reported recently for higher plants (30) and for *Euglena* (31). Subsequent investigations have confirmed that the enzyme, named ascorbate peroxidase, is present in all metabolically active tissues of higher plants at levels as high as 50 μmol ascorbate oxidised/min·g fresh weight (32).

The physiological significance of ascorbate peroxidase was immediately apparent because it fitted into an earlier proposed scheme (33) for destroying $H_2 O_2$ by connecting reduction of this $H_2 O_2$ to NADPH oxidation (Fig. 5). Evidence that such a system operates in spinach chloroplasts has been reported recently by Nakano and Asada (34). There are two aspects of this system that involve direct interaction with photosynthetic carbon metabolism. Firstly, the overall process (thicker lines, Fig. 5) can dissipate photosynthetically generated reducing power without either producing or consuming $H_2 O$, O_2, H^+ or CO_2, and it may therefore be looked upon as an alternative, at least in part, to photorespiration, since the current concensus of opinion is that energy dissipation is a principal function of photorespiration (It should be noted, however, that the scheme in Fig.5 cannot dissipate excess energy generated in the form of ATP). Secondly, as shown in Fig. 5, the process in effect competes with alternative reductants for $H_2 O_2$, one of which is the group of enzymes that undergo light-mediated activation by reduction of disulphide linkages to disulphydryls, *e.g.*, chloroplast fructosebisphosphatase discussed earlier in this chapter. The cyclic oxidation and reduction of such enzymes forms a separate system by which photosynthetically generated reducing power may be dissipated (thinner lines, Fig. 5), and although there is no information as to which of the two systems in Fig. 5 predominates, the high content of ascorbate peroxidase and the necessity for enzymes such as fructosebisphosphatase to be active for CO_2 fixation would infer that the system containing ascorbate peroxidase is primarily involved with the disposal of as much $H_2 O_2$ as possible so that the oxidative deactivation of sulphydryl-containing enzymes is kept to a minimum.

While the frontier of direct influence of ascorbate peroxidase is localised to influencing the concentrations of $H_2 O_2$, ascorbate and dehydroascorbate (there is no evidence that it is a regulatory enzyme), the peroxidase is nevertheless a good illustration of the fact that most enzymes, whether regulatory or not, are indispensable to the survival of the cell.

REFERENCES

1. Latzko, E., Zimmermann, G. and Feller, U. : 1974. *Hoppe-Seyler's Z. Physiol. Chem.* **355**, 321–326.
2. Smillie, R. : 1960. *Nature* **187**, 1024–1025.
3. Zimmermann, G., Kelly, G.J. and Latzko, E. : 1976. *Eur. J. Biochem.* **70**, 361–367.
4. Zimmermann, G., Kelly, G.J. and Latzko, E. : 1978. *J. Biol. Chem.* **253**, 5952–5956.
5. Harbron, S., Foyer, C. and Walker, D. : 1981. *Arch. Biochem. Biophys.* **212**, 237–246.
6. Walker, D.A. and Herold, A. : 1977. In Photosynthetic Organelles, Special Issue of Plant and Cell Physiology, No. 3. (Miyachi, S., Katoh, S., Fujita, Y. and Shibata, K., eds.), pp. 295–310, Japan. Soc. Plant Physiol., Japan.
7. Garnier, R.V. and Latzko, E. : 1972. *In* Proc. 2nd Int. Congr. Photosyn. Res., Vol. 3 (Forti, G., Avron, M. and Melandri, A., eds.), pp. 1839–1845, W. Junk, The Hague.
8. Charles, S.A. and Halliwell, B. : 1980. *Biochem. J.* **185**, 689–693.
9. Buchanan, B.B., Schürmann, P. and Kalberer, P.P. : 1971. *J. Biol. Chem.* **246**, 5952–5959.
10. Kelly, G.J., Zimmermann, G. and Latzko, E. : 1976. *Biochem. Biophys. Res. Commun.* **70**, 193–199.
11. Kaiser, W.M. : 1979. *Planta* **145**, 377–382.
12. Leegood, R.C. and Walker, D.A. : 1980. *Arch. Biochem. Biophys.* **200**, 575–582.
13. Charles, S.A. and Halliwell, B. : 1981. *Biochem. J.* **200**, 357–363.
14. Latzko, E. and Kelly, G.J. : 1976. *Prog. Bot.* **38**, 81–99.
15. Zimmermann, G. : 1976. Ph.D. Thesis, Technische Universität München.
16. Bassham, J.A. and Krause, G.H. : 1969. *Biochim. Biophys. Acta* **189**, 207–221.
17. Fischer, K.H. and Latzko, E. : 1979. *Biochem. Biophys. Res. Commun.* **89**, 300–306.
18. Laing, W.A., Stitt, M. and Heldt, H.W.: 1981. *Biochim. Biophys. Acta* **637**, 348–359.
19. Scheibe, R. and Beck, E. : 1979. *Plant Physiol.* **64**, 744–748.
20. Kelly, G.J. and Latzko, E. : 1975. *Nature* **256**, 429–430.
21. Kelly, G.J. and Latzko, E. : 1977. *Plant Physiol.* **60**, 290–294.
22. Kelly, G.J. and Latzko, E. : 1977. *Plant Physiol.* **60**, 295–299.
23. Ghosh, H.P. and Preiss, J. : 1966. *J. Biol. Chem.* **241**, 4491–4504.
24. Kelly, G.J. and Latzko, E. : 1981. *Physiol. Plant.* **52**, 339–342.
25. Sugiyama, T. and Hatch, M.D. : 1981. *Plant Cell Physiol.* **22**, 115–126.
26. Hatch, M.D. : 1979. *Aust. J. Plant Physiol.* **6**, 607–619.
27. Meyer, A.O., Kelly, G.J. and Latzko, E. : 1978. *Plant Sci. Lett.* **12**, 35–40.
28. Meyer, A.O., Kelly, G.J. and Latzko, E. : 1982. *Plant Physiol.* **69**, 7–10.
29. Wallace, W. and Pate, J.S. : 1967. *Ann. Bot.* **31**, 213–228.
30. Kelly, G.J. and Latzko, E. : 1979. *Naturwissenschaften* **66**, 617–618.
31. Shigeoka, S., Nakano, Y. and Kitaoka, S. : 1980. *Biochem. J.* **186**, 377–380.
32. Kelly, G.J. : 1981. Abstr. XIII Int. Bot. Congr., Sydney, 25.
33. Foyer, C.H. and Halliwell, B. : 1976. *Planta* **133**, 21–25.
34. Nakano, Y. and Asada, K. : 1981. *Plant Cell Physiol.* **22**, 867–880.
35. Buchanan, B.B. : 1980. *Annu. Rev. Plant Physiol.* **31**, 341–374.

Plant Phosphorylases: Structure and Function

Toshio Fukui

INTRODUCTION

Starch phosphorylase (α-1,4-glucan:orthophosphate α-D-glucosyltransferase ; EC 2.4.1.1) was discovered in 1940 by Hanes (*1*) in extracts of pea and potato tuber, and was shown to catalyze the reversible phosphorolysis of α-1,4-glucan. Since then, the enzyme has been found in a wide variety of plant tissues including maize, barley, broad bean, and green leaves. Prior to 1960, it was the only enzyme known to be responsible for the synthesis of starch in plants. Leloir *et al.* (*2*), however, found starch synthase (UDPglucose:glycogen 4-α-glucosyltransferase ; EC 2.4.1.11) which was bound to starch granules and used UDPglucose as substrate. Subsequently, it was shown that ADPglucose was a better substrate for this enzyme (*3*), and that the same type of enzyme in a soluble form (ADPglucose:1,4-α-D-glucan 4-α-glucosyltransferase ; EC 2.4.1.21) utilized ADPglucose as a sole substrate (*4–7*). There are therefore at least three possible routes for the biosynthesis of starch in plants. On the basis of available information, however, it appears that the ADPglucose route is predominant for starch synthesis over the other pathways (*8*), although this is not entirely conclusive. On the other hand, there is good evidence for the differentiation in the physiological roles of glycogen phosphorylase and glycogen synthase in animals ; phosphorylase functions solely in the degradation of the polysaccharide, while synthase serves in the synthesis. Moreover, animal phosphorylase has allosteric and covalent control mechanisms and acts as the key enzyme in glycogen metabolism, whereas no plant phosphorylase shows regulatory properties.

Phosphorylase is widely distributed in both plant and animal tissues and has been isolated and characterized from different sources. The enzymes from different sources differ not only in regulatory properties but also in their specificities for substrate glucans. This paper compares the molecular and catalytic properties of phosphorylases isolated from different plant sources using the rabbit muscle enzyme as reference, and discusses the relationship between structure and function.

ISOLATION AND MULTIPLE FORMS

Potato phosphorylase was isolated by Lee (9) in an active form and its activity was unaffected by AMP. It contains no phosphate group other than pyridoxal phosphate and was not phosphorylated by phosphorylase kinase from rabbit muscle (10). Later, the potato enzyme was isolated on a larger scale and its molecular properties were studied in this laboratory (11). It was noted that, in spite of the difference in regulatory properties, the potato and rabbit muscle enzymes were similar in their molecular size, pyridoxal phosphate content, and amino acid composition. Although several other forms of potato phosphorylase have been reported (12-18), the enzyme form we isolated is definitely the major component of phosphorylase in this tissue. One of the forms isolated from this plant tissue was reported to be specific for glycogen and was named glycogen phosphorylase (14,15) ; some of the other forms are actually the proteolytic product of the major component (16). Proteolysis of potato and sweet potato phosphorylases occur in the middle part of their polypeptide chains without loss of enzyme activity, followed by cleavage near the terminus accompanied by a significant decrease in activity (19-21).

Maize kernel contains at least four forms of phosphorylase which are separable on DEAE-cellulose. Their activities vary dependent on the ages of the kernel (22, 23). Burr and Nelson (24) isolated the major component from developing maize seed which had no unusual properties emphasizing a synthetic function. Lee and Braun (25) purified phosphorylase from commercial sweet corn which has no regulatory properties. Phosphorylase activity was shown to be distributed both in the chloroplastic and non-chloroplastic fractions of spinach leaves (26,27). Two types of phosphorylase were separated from the leaves and their molecular and kinetic properties compared (28-32). The physiological significance of the enzyme distributed outside the chloroplasts remains to be studied, as the photosynthesized starch has heretofore been believed to be accumulated and digested inside the chloroplasts in leaf cells.

GENERAL STRUCTURE OF PLANT PHOSPHORYLASES

Rabbit muscle phosphorylase exists as a dimer or tetramer of identical subunits having a molecular weight of 97,400, as calculated from the amino acid composition (33). All the phosphorylases so far isolated from plant tissues exist as a dimer of subunits of similar size. When compared in more detail, however, the molecular weights of plant phosphorylases differ depending on the source of the enzymes. Values of 108,000 and 112,000 were obtained for the subunit molecular weight of potato phosphorylase by two different methods (19). Gel filtration of the potato enzyme under various conditions gave a molecular weight of 220,000 ± 10,000 in its native state (19). A form of potato phosphorylase having high affinity for glycogen has molecular weights of 180,000 and 90,000 in the native and denatured states, respectively (15). Maize phosphorylase appears to be dimeric and has a molecular weight of 223,000 (24), while a larger molecular weight was reported for the sweet corn enzyme (25). Subunit and native molecular weights of spinach leaf chloroplastic phosphorylase were shown to be 110,000 and 200,000–220,000, respectively (31). The spinach leaf non-chloroplastic enzyme exists as a dimer of subunits having a molecular weight of 92,000 (32).

Information on the primary structure of plant phosphorylases is at present limited to that of the potato enzyme. The sequence of a 57-residue segment encompassing the pyridoxal phosphate-binding site was determined (34) and compared with the sequences of corresponding segments from phosphorylases from other sources (33,35,36). This region of the proteins is highly conserved, suggesting that the cofactor has an important enzymatic function. To determine whether such high homologies are restricted to the cofactor-binding site, ten distinct cysteinyl peptides were isolated from potato phosphorylase and sequenced (37). The comparison showed all of the peptides to be homologous with widely distributed specific regions in the rabbit muscle enzyme. In total, 100 of the 169 amino acid residues compared were identical in the two enzymes. This indicates that potato and rabbit muscle phosphorylases are homologous and may have, on the whole, a similar folding of their polypeptide chains.

Nakano et al. (38) further determined the sequence of the amino-terminal 104 residues of potato phosphorylase. A comparison of this sequence with the complete sequence of rabbit muscle phosphorylase (33) indicates that the amino-terminus of the potato enzyme is extended only two residues over the corresponding locus of the rabbit muscle enzyme, and that the amino-terminal 33 residue-regions are completely different from each other in sequences. The remarkable dissimilarity in the structure of this region provides a basis for the difference in the regulatory properties between the two enzymes, since this region is important for both allosteric and covalent controls in the rabbit muscle enzyme (39). It is conceivable that phosphorylases existed originally as a large catalytically active molecule, and that the regulatory mechanism was then evolutionarily formed within the molecule.

The CD spectra in the far-ultraviolet wavelength range were compared in potato and rabbit muscle phosphorylases (40). Although similar in characteristic bands, their molecular ellipticities are different. The calculation of the secondary structure of each enzyme reveals that the potato enzyme contains more β-structure and less α-helix than the muscle enzyme. Since the α-helical segments largely form the surface structure of rabbit muscle phosphorylase (41), the surface structure of potato phosphorylase might be significantly different from that of the muscle enzyme. This idea is not contradictory to the observation that all cysteinyl peptides from the potato enzyme are homologous with some regions in the muscle enzyme (37). These cysteinyl residues of the potato enzyme are all buried within the protein molecule (42) whereas, in rabbit muscle phosphorylase, some of them are exposed to the surface and are highly reactive toward sulfhydryl reagents.

Feldmann et al. (43) reported that matrix-bound monomeric phosphorylase from rabbit muscle had little if any intrinsic activity, but activity appeared upon non-covalent interaction with another subunit. They suggested that the rabbit muscle enzyme in the monomeric state was catalytically inactive. There has been no such evidence for the potato enzyme. However, Tagaya et al. (44) discovered a monomeric intermediate in the reconstitution of potato phosphorylase with pyridoxal phosphate at a low temperature; this intermediate is catalytically inactive and shows no affinity for substrate amylopectin. Upon incubation at a high temperature, catalytic activity and affinity for amylopectin are restored with concomitant dimerization of the protein. The findings support the view that the dimeric form is essential for the catalytic activity of potato phosphorylase.

STRUCTURAL FEATURES OF THE PYRIDOXAL PHOSPHATE-BINDING SITE

Baranowski et al. (45) found pyridoxal phosphate bound to rabbit muscle phosphorylase. Presence of this cofactor was also shown in phosphorylases from plant tissues : potato tuber (10), sweet potato root (21), maize kernel (24), and spinach leaf (29,31). The resolution of the cofactor from potato phosphorylase under mild conditions was achieved in this laboratory (40). Apophosphorylase is catalytically inactive but is reactivated by incubation with pyridoxal phosphate, the degree of reactivation agreeing well with the amount of the phosphate incorporated. Therefore, the cofactor is essential for catalytic activity in potato phosphorylase (40), as it has been shown to be in the animal enzyme. The differential CD spectra (the reconstituted enzyme versus the apoenzyme) of potato and rabbit muscle phosphorylases show similar characteristics (40) : the bands are characteristic of a Schiff base-bonding structure of pyridoxal phosphate surrounded by a strong hydrophobic environment (46).

The pyridoxal phosphate in rabbit muscle phosphorylase is fixed to the enzyme protein upon reduction with sodium borohydride, and the resulting enzyme retains catalytic activity. A similar result has recently been obtained with potato phosphorylase (44). It is thus evident that neither potato nor rabbit muscle phosphorylase requires the 4'-aldehyde group of pyridoxal phosphate to function. The results of reconstitution experiments of rabbit muscle apophosphorylase with various analogues of pyridoxal phosphate show the importance of the 5'-phosphate group in the function of this enzyme.

Shimomura et al. (40,47) synthesized several analogues of pyridoxal phosphate modified at the 5'-position, and examined the reconstitution of potato and rabbit muscle apophosphorylases with these analogues. The CD spectra of phosphorylases reconstituted with pyridoxal and its derivatives having even such a bulky substituent as a benzyl group at the 5'-position were quite similar. The rates of reconstitution vary dependent on the number of anions at the 5'-position. The size of the 5'-substituent of pyridoxal seems not to impose any restriction on the rate of reconstitution. Therefore, the phosphate binding locus of the cofactor site in both phosphorylases is a pocket large enough to adopt a bulky group, and it possesses cationic groups to interact with the phosphate group.

P^1,P^2-bis(5'-pyridoxal)diphosphate, a bifunctional cofactor analogue, was used for affinity labeling of the cofactor site in rabbit muscle phosphorylase (48). The reagent crosslinked the two lysyl residues, namely, Lys-679, the original cofactor-binding residue, and Lys-573, indicating that the location of the ε-amino group of Lys-573 is near the pyridoxal phosphate-binding site. This compound was bound also to potato phosphorylase in the same binding mode as to the rabbit muscle enzyme, and crosslinked between the original cofactor-binding lysyl residue and another lysyl residue, respectively corresponding to Lys-679 and Lys-573 in the muscle enzyme (49). The sequence Lys-573 to Leu-577 in the rabbit muscle enzyme is conserved in the potato enzyme. These results indicate structural similarities in the pyridoxal phosphate-binding site between potato and rabbit muscle phosphorylases. X-ray crystallographic analysis of the rabbit muscle enzyme confirmed the presence of the ε-amino group of Lys-573 near the pyridoxal phosphate, and showed that it may interact with the phosphate group of a substrate glucose 1-phosphate or orthophosphate which binds near the pyridoxal phosphate (50,51).

GLUCAN BINDING MODELS OF POTATO AND RABBIT MUSCLE PHOSPHORYLASES

Based on the finding that glucose 1-phosphate and orthophosphate bound to a common region close to the phosphate group of pyridoxal phosphate, this region was proposed to be the active site in phosphorylase (50). However, maltooligosaccharide which was used for glucan substrate was shown to be bound to another binding site located more than 25 Å distant from the pyridoxal phosphate (50,51). Termed the "glycogen storage site", this provides a point of attachment of the enzyme to the substrate. Kinetic analysis showed that the proposed active site had a much lower affinity for oligosaccharide than the glycogen storage site (52). It is supposed that two chain termini from the same glycogen molecule bind simultaneously to the enzyme, and therefore such a highly branched glucan as glycogen is good substrate for rabbit muscle phosphorylase because of the great multiplicity of two end-group pairs. No oligosaccharide binding has been crystallographically observed at the proposed active site. Existence of the "polysaccharide storage site" in potato phosphorylase could not be expected from its kinetic behavior for this enzyme does not require the branched structure of the substrate. The proper binding site for glucan substrate in the potato enzyme should therefore be the active site itself. If this is really near the cofactor site, especially to the 5'-phosphate group of pyridoxal phosphate, the affinity of the phosphorylase for glucan should be significantly altered by modification at the 5'-position of the cofactor, while that of rabbit muscle enzyme should not be affected by this modification.

Shimomura and Fukui (53) investigated this possibility by comparing the affinities for vaious glucans of potato and rabbit muscle phosphorylases and their derivatives prepared by the reconstitution of apophosphorylase with pyridoxal phosphate analogues modified at the 5'-position. For the muscle enzyme, the interaction with glycogen was not affected at all by modification of the enzyme molecule, as was expected. The finding is reconciled with the crystallographic observation in which the proper binding site for oligosaccharide is away from the cofactor site (50, 51). By contrast, the polysaccharide-binding site in potato phosphorylase was closely related with the cofactor ; introduction of a bulky group on the 5'-position significantly weakened the affinity. The results can be explained by assuming that the non-reducing end of an oligosaccharide chain can bind close to the 5'-phosphate group of the enzyme-bound pyridoxal phosphate.

A single molecule of potato phosphorylase has two cofactors and is composed of two subunits (10,19), suggesting two active sites per molecule. It therefore could be expected from the model of Hu and Gold (54) that the increase in multiplicity of the outer chain of a substrate molecule would strengthen the affinity of the enzyme composed of two active sites. However, our experimental results were contrary to this expectation. This may reflect an inaccessibility of the second chain terminus from the same substrate to the remaining active site after one of the two active sites has been occupied by the first chain terminus. For rabbit muscle phosphorylase, two glycogen storage sites and two active sites are located on the same face of its molecule (55), indicating that the plural termini from the same substrate are able to bind simultaneously. We have therefore proposed a model to explain the inaccessibility of potato phosphorylase to the second chain terminus in which the two active sites of potato phosphorylase are located on reverse sides (Fig.1).

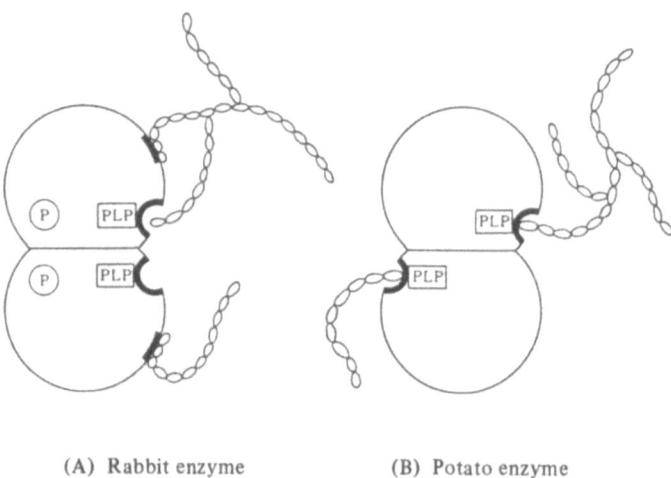

(A) Rabbit enzyme   (B) Potato enzyme

Fig. 1. Schematic model for the dimeric structure and binding of potato and rabbit muscle phosphorylases to linear and branched glucans. PLP, pyridoxal 5'-phosphate; P, Ser-14-phosphate.

AFFINITY FOR GLUCAN OF SPINACH LEAF PHOSPHORYLASES

The two types of spinach leaf phosphorylase differ in their affinity for various glucans. The chloroplastic enzyme gave a dissociation constant of 2.8 mg/ml for amylopectin at $4°C$ and apparently had no affinity for glycogen, as measured by affinity electrophoresis (32). This specificity for glucan is similar to that of potato phosphorylase, which has a dissociation constant of 1.3 mg/ml for amylopectin at $4°C$ and no affinity for glycogen (53), whereas rabbit muscle phosphorylase has a dissociation constant of 1.1 mg/ml at $27°C$ for glycogen (53). The non-chloroplastic enzyme showed much higher affinity for glycogen and amylopectin than the spinach leaf chloroplastic and rabbit muscle enzymes : 18 μg/ml at $19°C$ and 2.5 μg/ml at $4°C$ for glycogen ; 1.5 μg/ml at $19°C$ for amylopectin.

The K_m values for different glucans of phosphorylases vary significantly depending on the source of the enzymes (Table I). The spinach leaf chloroplastic

TABLE I
Comparison of K_m Values for Glucan of Phosphorylases in the Direction of Glucan Phosphorolysis (32)

	K_m (μM of nonreducing end group)				
Glucan	Spinach enzyme		Potato enzyme	Rabbit enzyme	Sweet corn enzyme[a]
	Non-chloroplastic	Chloroplastic			
Amylopectin	0.6	16	46	2.0	24
Glycogen	2.3	1,500	24,000	7.9	5,000
Amylose DP 30	1.6	6.4	7.0	120	
Amylose DP 19	19	13	16		
Maltopentaose	680	78	110	14,000	180

[a] From the data of Lee and Braun (25).

TABLE II
Inhibition Constants for Cyclodextrins of Various Phosphorylases (32) (determined for the phosphorolysis of maltopentaose unless otherwise stated)

Enzyme	K_i (mM)		
	Cyclohexaamylose	Cycloheptaamylose	Cyclooctaamylose
Spinach			
Non-chloroplastic	6.5	0.44	2.2
Chloroplastic	0.091	0.18	1.3
Potato	0.15	0.42	4.9
	0.056[a]	0.19[a]	

[a] Determined for the phosphorolysis of amylopectin.

enzyme hás an extremely high K_m value for glycogen. Potato and sweet corn phosphorylases have similar characteristics, while the non-chloroplastic enzyme can utilize any type of α-1,4-glucan. On the other hand, similar relationships in the V_{max} values for different glucans were found for phosphorylases from spinach leaf, potato and rabbit muscle, although the values themselves differ depending on the enzyme source (32).

Cyclodextrin is a good inhibitor and competitive with respect to glucan, of potato phosphorylase (56,57), but not of rabbit muscle phosphorylase (58). The spinach leaf non-chloroplastic and chloroplastic enzymes showed a normal type of competitive inhibition against maltopentaose (32), a normal one-site competitive type inhibition by a single inhibitor (59). The degree of inhibition is in the order of cyclohexa->cyclohepta->cyclooctaamylose for the chloroplastic enzyme, which agrees with that for the potato enzyme (Table II). On the other hand, the order for the non-chloroplastic enzyme is cyclohepta->cycloocta->cyclohexaamylose. When amylopectin, glycogen or amylose was used in the inhibition experiments, the non-chloroplastic enzyme showed a different mode of inhibition, a multisite competitive type inhibition by multiple inhibitors (59). Potato phosphorylase showed a normal competitive type inhibition when amylopectin was used as a substrate (32). These results suggest different modes in the glucan binding of phosphorylases from different sources.

CLASSIFICATION OF PHOSPHORYLASES IN TERMS OF MOLECULAR SIZE AND GLUCAN SPECIFICITY

Phosphorylases from different sources are classified into three types according to their molecular sizes and glucan specificities (32) :

Type 1 phosphorylase shows the same level of affinities for amylopectin, amylose and maltodextrin, but little affinity for glycogen. Cyclodextrin is a competitive inhibitor with respect to glucan substrate. The two active sites which bind glucan directly are three-dimensionally arranged in a manner that excludes the possibility of the binding of amylopectin by riding on the two active sites. An example of such a binding model is shown in Fig. 1B, as previously proposed for potato phosphorylase (53). This type of phosphorylase exists as a dimer of subunits having a molecular weight of about 110,000, and may be located inside the plastides in plant tissues. It includes phosphorylases from spinach leaf chloroplast, potato tuber and sweet corn.

Type 2 phosphorylase shows much higher affinities for both branched and long linear chain glucans than Type 1. Cyclodextrin is also a good inhibitor for this type of phosphorylase, but their dependences of the degree of polymerization differ. It is unclear at present whether or not this type of phosphorylase has the glucan binding site separated from the active site as in rabbit muscle phosphorylase (Fig. 1A). However, a single substrate molecule containing either multiple outer chains or a single long chain can be bound by riding on the two sites. This type includes the enzymes from the spinach leaf non-chloroplastic fraction and potato tuber (a minor component having a high affinity for glycogen), and has a subunit molecular weight of about 90,000.

Type 3 phosphorylase shows higher affinities for branched glucans than for linear glucans. Cyclodextrin has no effect on the enzyme activity. The active site and the glucan binding site are located separately on a single subunit. A branched glucan molecule can bind by riding on the two sites, although the active site itself shows only very low affinity for glucan. This type includes phosphorylase from rabbit muscle, its subunit molecular weight being intermediate between those of Types 1 and 2.

ROLE OF PYRIDOXAL PHOSPHATE IN THE CATALYTIC MECHANISM OF PHOSPHORYLASE

Our recent studies using the synthetic cofactor analogue pyridoxal (5') diphospho(1)-α-D-glucose, provide solid chemical evidence for the role of pyridoxal 5'-phosphate in the catalytic mechanism of rabbit muscle phosphorylase (*60,61*). The phosphate group of pyridoxal 5'-phosphate acts as a catalyst by direct interaction with the phosphate of substrate glucose 1-phosphate, forming a pyrophosphate-like transition intermediate (Fig. 2) ; positive charges at the active site may constrain the coenzyme phosphate with a configuration which renders phosphorus electrophilic. This facilitates nucleophilic attack by the phosphate of glucose 1-phosphate and cleavage of the glucosidic linkage.

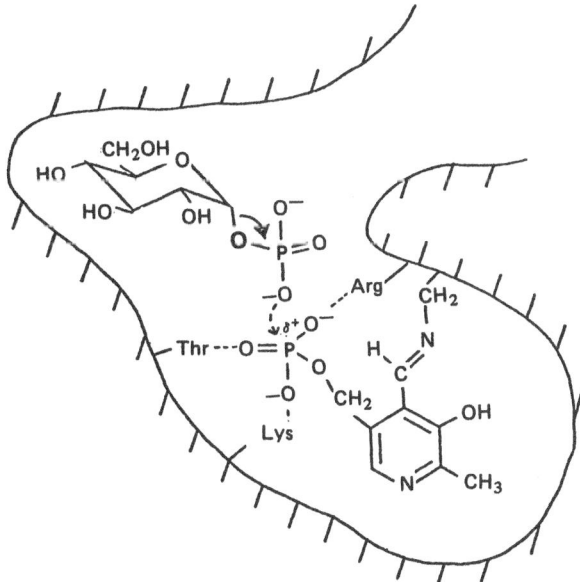

Fig. 2. Scheme illustrating the role of the coenzyme phosphate as an electrophile in the proposed catalytic mechanism (*60*).

We are now studying whether or not the same type of catalytic mechanism also functions in plant phosphorylases. As previously described, however, the structural features of potato and rabbit muscle phosphorylases are quite similar, even though the glucan specificities and the binding models differ. Furthermore, the two enzymes follow a common kinetic mechanism, the rapid equilibrium random Bi Bi kinetics (62). The same mechanism proposed for rabbit muscle phosphorylase may be applicable to the catalysis of potato phosphorylase.

In addition, it was pointed out that the transition-state analogue we discovered in the studies on rabbit muscle phosphorylase is strikingly similar to the chemical structures of ADPglucose and UDPglucose (60). Synthase which contains no pyridoxal phosphate uses these nucleoside diphosphate sugar compounds as substrate, while phosphorylase binds pyridoxal phosphate and uses sugar monophosphate as substrate. Starch synthase and phosphorylase may have a common mechanism of catalytic reaction.

Note Added in Proof

A. Kumar and G.G. Sanwal (1982. *Arch. Biochem. Biophys.* **217**, 341–350; 1982. *Biochemistry* **21**, 4152–4159) have recently claimed the absence of pyridoxal 5'-phosphate in phosphorylases from tapioca and young banana leaves. I consider, however, their results to be due to incomplete separation of phosphorylase from other proteins, especially from ribulose bisphosphate carboxylase/oxidase. This is consistent with the low specific activities compared with other plant phosphorylases and the values of molecular weight they reported for phosphorylases. Further studies will be required in order to prove definitely the absence of the cofactor in those plant phosphorylases.

REFERENCES

1. Hanes, C. S. : 1940. *Proc. Roy. Soc. London* **B129**, 174–208.
2. de Fekete, M. A. R., Leloir, L.F. and Cardini, C.E. : 1960. *Nature* **187**, 918–919.
3. Recondo, E.F. and Leloir, L.F. : 1961. *Biochem. Biophys. Res. Commun.* **6**, 85–88.
4. Doi, A., Doi, K. and Nikuni, Z. : 1964. *Biochim. Biophys. Acta* **92**, 628–630.
5. Frydman, R.B. and Cardini, C.E. : 1964. *Biochem. Biophys. Res. Commun.* **17**, 407–411.
6. Murata, T. and Akazawa, T. : 1964. *Biochem. Biophys. Res. Commun.* **16**, 6–11.
7. Ghosh, H.P. and Preiss, J. : 1965. *J. Biol. Chem.* **240**, 960–962.
8. Preiss, J. and Levi, C. : 1979. Encyclopedia of Plant Physiology, New Series Vol. 6 (Gibbs, M. and Latzko, E., eds.), pp. 282–312, Springer-Verlag, Berlin, Heidelberg and New York.
9. Lee, Y.P. : 1960. *Biochim. Biophys. Acta* **43**, 18–24.
10. Lee. Y.P. : 1960. *Biochim. Biophys. Acta* **43**, 25–30.
11. Kamogawa, A., Fukui, T. and Nikuni, Z. : 1968. *J. Biochem.* **63**, 361–369.
12. Slabanik, F. and Frydman, R.B. : 1970. *Biochem. Biophys. Res. Commun.* **38**, 709–714.
13. Gerbrandy, S.J. and Verleur, J.D. : 1971. *Phytochemistry* **10**, 261–266.
14. Gerbrandy, S.J. and Doorgest, A. : 1972. *Phytochemistry* **11**, 2403–2407.
15. Gerbrandy, S.J. : 1974. *Biochim. Biophys. Acta* **370**, 410–418.
16. Gerbrandy, S.J., Shankar, V., Shivaram, K.N. and Stegemann, H. : 1975. *Phytochemistry* **14**, 2331–2333.
17. Shivaram, K.N. : 1976. *Z. Naturforsch.* **31C**, 424–432.

18. Tandecarz, J.S., Sivak, M.N. and Cardini, C.E. : 1978. *Biochem. Biophys. Res. Commun.* **82**, 157–164.
19. Iwata, S. and Fukui, T. : 1973. *FEBS Lett.* **36**, 222–226.
20. Iwata, S. and Fukui, T. : 1975. *Arch. Biochem. Biophys.* **169**, 58–65.
21. Ariki, M. and Fukui, T. : 1975. *Biochim. Biophys. Acta* **386**, 301–306.
22. Tsai, C.Y. and Nelson, O.E. : 1968. *Plant Physiol.* **43**, 103–112.
23. Tsai, C.Y. and Nelson, O.E. : 1969. *Plant Physiol.* **44**, 159–167.
24. Burr, B. and Nelson, O.E. : *Eur. J. Biochem.* **56**, 539–546.
25. Lee, E.Y.C. and Braun, J.J. : 1973. *Arch. Biochem. Biophys.* **156**, 276–286.
26. Okita, T.W., Greenberg, E., Kuhn, D.N. and Preiss, J. : 1979. *Plant Physiol.* **64**, 187–192.
27. Steup, M. and Latzko, E. : 1979. *Planta* **145**, 69–75.
28. Steup, M., Schächtele, C. and Latzko, E. : 1980. *Z. Pflanzenphysiol.* **96**, 365–374.
29. Steup, M., Schächtele, C. and Latzko, E. : 1980. *Planta* **148**, 168–173.
30. Preiss, J., Okita, T.W. and Greenberg, E. : 1980. *Plant Physiol.* **66**, 864–869.
31. Steup, M. : 1981. *Biochim. Biophys. Acta* **659**, 123–131.
32. Shimomura, S., Nagai, M. and Fukui, T. : 1982. *J. Biochem.* **91**, 703–717.
33. Titani, K., Koide, A., Hermann, J., Ericsson, L.H., Kumar, S., Wade, R.D., Walsh, K.A., Neurath, H. and Fischer, E.H. : 1977. *Proc. Natl. Acad. Sci. U.S.A.* **74**, 4762–4766.
34. Nakano, K., Wakabayashi. S., Hase, T., Matsubara, H. and Fukui, T. : 1978. *J. Biochem.* **83**, 1085–1094.
35. Lerch, K. and Fischer, E.H. : 1975. *Biochemistry* **14**, 2009–2104.
36. Schächtele, K.H., Schiltz, E. and Plam, D. : 1978. *Eur. J. Biochem.* **92**, 427–435.
37. Nakano, K., Fukui, T. and Matsubara, H. : 1980. *J. Biochem.* **87**, 919–927.
38. Nakano, K., Fukui, T. and Matsubara, H. : 1980. *J. Biol. Chem.* **255**, 9255–9261.
39. Fletterick, R.J. and Madsen, N.B. : 1980. *Annu. Rev. Biochem.* **49**, 31–61.
40. Shimomura, S., Emman, K. and Fukui, T. : 1980. *J. Biochem.* **87**, 1043–1052.
41. Sprang, S. and Fletterick, R.J. : 1979. *J. Mol. Biol.* **131**, 523–551.
42. Kamogawa, A., Fukui, T. and Nikuni, Z. : 1971. *Agric. Biol. Chem.* **35**, 248–254.
43. Feldmann, K., Zeigel, H. and Helmreich, E. : 1972. *Proc. Natl. Acad. Sci. U.S.A.* **69**, 2278–2282.
44. Tagaya, M., Shimomura, S., Nakano, K. and Fukui, T. : 1982. *J. Biochem.* **91**, 589–597.
45. Baranowski, T., Illingworth, B., Brown, D.H. and Cori, C.F. : 1957. *Biochim. Biophys. Acta* **25**, 16–21.
46. Shimomura, S. and Fukui, T. : 1977. *J. Biochem.* **81**, 1781–1790.
47. Shimomura, S. and Fukui, T. : 1978. *Biochemistry* **17**, 5359–5367.
48. Shimomura, S., Nakano, K. and Fukui, T. : 1978. *Biochem. Biophys. Res. Commun.* **82**, 462–468.
49. Tagaya, M., Nakano, K., Shimomura, S. and Fukui, T. : 1982. *J. Biochem.* **91**, 599–606.
50. Sygusch, J., Madsen, N.B., Kasvinsky, P.J. and Fletterick, R.J. : 1977. *Proc. Natl. Acad. Sci. U.S.A.* **74**, 4757–4761.
51. Weber, I.T., Johnson, L.N., Wilson, K.S., Yeates, D.G., Wild, D.L. and Jenkins, J.A. : 1978. *Nature* **274**, 433–437.
52. Kasvinsky, P.J., Madsen, N.B., Fletterick, R.J. and Sygusch, J. : 1978. *J. Biol. Chem.* **253**, 1290–1296.
53. Shimomura, S. and Fukui, T. : 1980. *Biochemistry* **19**, 2287–2294.
54. Hu, H.Y. and Gold, A.M. : 1975. *Biochemistry* **14**, 2224–2230.
55. Fletterick, R.J., Sprang, S. and Madsen, N.B. : 1979. *Can. J. Biochem.* **57**, 789–797.
56. Staerk, J. and Schlenk, H. : 1967. *Biochim. Biophys. Acta* **146**, 120–128.
57. Pfannemüller, B. : 1968. *Stärke* **11**, 351–362.
58. Smith, E.E. : 1971. *Arch. Biochem. Biophys.* **146**, 380–390.
59. Segel, I.H. : 1975. Enzyme Kinetics, pp. 465–473, John Wiley and Sons, New York, London, Sydney and Toronto.

60. Withers, S.G., Madsen, N.B., Sykes, B.D., Takagi, M., Shimomura, S. and Fukui, T. : 1981. *J. Biol. Chem.* **256**, 10759–10762.
61. Takagi, M., Fukui, T. and Shimomura, S. : 1982. *Proc. Natl. Acad. Sci. U.S.A.* **79**, 3716–3719.
62. Gold, A.M., Johnson, R.M. and Sanchez, G.R. : 1971. *J. Biol. Chem.* **246**, 3444–3450.

Studies on Cellulose Metabolism (1970–1990)

Gordon Maclachlan

INTRODUCTION

In 1970 the first major review appeared (*1*) in which the word 'metabolism' was used in the title to apply to plant cell wall materials in general, and cellulose in particular. This was a departure from tradition, for such products were widely regarded at the time as simply excreted and deposited in extracellular spaces which contained no endogenous enzymic activities to metabolise them further. Everything outside of protoplasts was commonly described by standard texts as 'non-living' and unreactive. Admittedly, cell walls were useful to plants in structural and protective roles, but they were an irritation to plant biochemists who had to cut or grind their way through these barriers in order to reach the important metabolic events of the cell.

The literature of the time, however, contained many studies of plant cell walls at the ultrastructural level, particularly with shadowing techniques, which showed that most contained an orderly array of microfibrils embedded in an apparently amorphous matrix (*2,3*). The wall begins relatively simply with cell plates separating daughter cells and during development the cellulose framework becomes more extensive with the microfibrils more oriented. The mechanism of re-orientation was generally discussed as a passive process responding to purely physical forces, *e.g.*, turgor pressure from the protoplast. Nevertheless, the first evidence had appeared by 1970 for a more dynamic biochemical turnover (loosening) of wall constituents during growth (*1,4,5*). There was also evidence available to show that wall composition changed markedly during development, although this was considered to be due to differential synthetic activities within the cell rather than to metabolism at the cell surface or within the wall.

Also by 1970, there were many instances recorded from anatomical studies where cell wall materials, including cellulose, were dissolved by healthy plant cells, often in a highly localized and precise manner (reviewed in 6). Thus, for example, the process of abscission could be seen to result from dissolution and weakening of cell walls in a narrow band of petiolar tissue ; many fruits were well known to 'soften' during senescence because of visible disintegration of cell wall structures ; vascular tissue during development forms vessels and sieve tubes by partial breakdown and re-structuring of end walls of preformed elements. This implied not only that cell-wall degrading enzymes must exist naturally in plants, but also that their activity and secretion must be carefully regulated during morphogenesis. Indeed, there were a few reports of endogenous plant cellulase (6–9), pectinase (9), and hemicellulase (9,10) activities, particularly concentrated in growing regions, abscission zones, senescing fruits, etc. All of this justified the notion that plant cell wall materials were indeed subject to metabolism, albeit in a highly controlled manner.

Today, the cell wall is not at all regarded as outside the proper realm for studies of plant metabolism, but rather as a compartment which is part of the natural meta-bolic continuum of cell activities. This paper is a brief survey of the results obtained to date on enzymic activities, primarily in growing regions of the pea epicotyl, which are capable of synthesizing and hydrolysing polysaccharides containing 1,4-β-gluco-side linkages. Past frustrations are described and future directions which research might take are contemplated, in order to achieve a satisfactory understanding of the mechanisms whereby plants conduct and control cellulose metabolism. Several other recent reviews deal with particular aspects of cell wall turnover (11), bio-synthesis (12–14) and hydrolysis (15) of β-glucans in plants, and ultrastructural studies of membranes in relation to cellulose synthesis (16).

ULTRASTRUCTURAL STUDIES

There are few reports of authentic cellulose in any plant cell compartment except the cell wall. One well-documented exception is the green alga *Pleurochrysis,* where in-tricate "scales" of cellulose microfibrils can be seen developing within saccules of the Golgi' apparatus (16,17). These scales are secreted to form the plates which enclose the cells of this unique organism. No such mechanism appears to operate in higher plants. To be sure, microsomal preparations from plants have been reported to con-tain small amounts of polysaccharide, including 1,4-linked glucan (18), but these may derive from part of the wall framework which is carried with fragments of plasma membrane during homogenization, and not from intracellular organelles.

Positive evidence from ultrastructural work which supports the view that cellu-lose is only synthesized at plant cell surfaces by plasma membrane comes from several studies.

a) When naked protoplasts are cultured in a medium containing a suitable sugar source, they often form microfibrils that are partly embedded at first in the outer plasma membrane (19,20). These begin to appear in patches, but quickly coalesce to cover the whole wall. Protoplast microfibrils have been proved to contain 1,4-β-linkages, although the glucan chains are relatively short (DP 300) compared to those in the parent tissue (21,22) and they have a relatively low degree of crystallization (23).

b) Pea stem tissue supplied with radioactive glucose forms insoluble products which can be visualized by radioautography on the inner surface of cell walls that are adpressed against the plasma membrane of living cells (24). When stem slices are supplied with radioactive UDP-glucose, living cells do not incorporate label but cut surfaces do (25,26). The parts of the wall that become labelled are those that retain some membrane flattened against the inner surface. This tissue-slice system is considered to be an example of synthesis in vitro, therefore, with substrate provided to the inner (cytoplasmic) surface of the exposed plasma membrane and primer (acceptor) molecules available in the wall adjacent to the other surface.

c) Highly suggestive associations have been visualized (16,17) between complex particles from the plasma membrane and the ends of cellulose microfibrils. During freeze-fracture electron microscopy, it sometimes happens that microfibrils are lifted through the outer plasma membrane leaflet. These may be seen to terminate in particles which, in turn, resemble many other such particles scattered about the outer surface of the undisturbed membrane (16,26,27). The inner plasma membrane bears rosettes with a frequency similar to particles, and hence the two together have been postulated as constituting a cellulose-synthesizing terminal complex. Several models for how this complex may operate have appeared (28,29). There is no biochemical evidence to prove that the complex actually does synthesizes cellulose as proposed.

d) Cytoplasmic microtubule arrays close to the plasma membrane are often seen in cells which are synthesizing cellulose. The tubules show an orientation that closely parallels that of cellulose microfibrils in the adjacent inner wall surface (30, 31). Treatments which are known to alter microtubules (e.g., colchicine) also change microfibril orientation, and treatments that alter the direction of growth (e.g., ethylene) are often found to change the orientation of both microfibrils and microtubules concurrently. The latter are generally thought to regulate the location of synthesizing complexes in the membrane rather than their activity (32), but no direct physical or biochemical connection has yet been established.

With respect to ultrastructural studies on the location of cellulases in plants, two techniques have been employed. In one, tissue sections are incubated with a soluble substrate (carboxymethyl cellulose) and the production of reducing chain ends by endogenous cellulolytic activity is visualized by reduction of copper using the appropriate reagent (33). In the other technique, antibodies to purified cellulase are linked to ferritin which can then be incubated with fixed tissue slices to locate antigenic cellulase with the EM (34). These techniques were both employed to localize two pea cellulases. One (buffer-soluble, MW 20,000) proved to be confined to the lumen of endoplasmic reticulum vesicles, the other (buffer-insoluble, MW 70,000) was found extracellularly at the inner surface of the wall (34).

BIOCHEMICAL STUDIES OF SYNTHETASE ACTIVITY

The 1970s began with optimism that the synthesis of authentic cellulose would soon be achieved *in vitro* in a defined system at rates comparable to those observed *in vivo*. This had not yet happened to the satisfaction of those working in the field.

Part of the problem is that most β-glucan synthesizing preparations that have been studied are extremely labile whenever attempts are made to abstract them from their natural association with the wall in turgid cells. In the case of pea cell membrane preparations, for example, incorporation from UDP- or GDP-glucose procedes for only a few minutes (*35*), even under conditions of pH, substrate and cofactor levels, which appear optimal. Maximum rates of incorporation are only a small fraction of those observed *in vivo* (*36*). Although early studies had claimed that these activities could be solubilized with detergent, such treatments with the pea enzyme led to even further losses of activity (*36,37*).

One reason for this lability may be the presence in pea extracts of an active protease which, like trypsin or pronase, rapidly inactivates the capacity of membranes to generate 1,4-β-linkages (*3,8,39*). The synthetase machinery appears to be highly accessible to such attack. Another may be the physical dissociation of essential components of the synthetase complex, for this would be expected if the complex extends through the plasma membrane as a multienzyme assembly. The complex may require co-operative transglycosylation from glucose nucleotide provided to the inner (cytoplasmic) surface *via* lipid intermediates to acceptors on the outer surface (*25,40*), in which event fragility during isolation would not be surprising.

Many attempts have been made to overcome lability and fragility. Examples are isolating membranes in the presence of alternate substrates for protease (*37,38*), adding putative lipid acceptors (*40*) or carbohydrate primers (*25*) back to mem-

TABLE I
Initial Rate of Biosynthesis of Alkali-Insoluble β-Glucan from UDP-glucose by Cut Surfaces of Pea Stem Slices

UDP-glucose conc. (μM)	Incorporation	
	1,4-Linked	1,3-Linked
	(pmol glc/2.5 min/5 mm tissue)	
5	5	1
50	24	36
500	190	285
5,000	670	1,350

Elongating regions of week-old pea epicotyls were cut into thin (0.5 mm) slices and incubated in 10 mM $MgCl_2$, Tris buffer pH 7.3 containing UDP [^{14}C]-glucose at concentration indicated. After 2.5 min at 30°C, slices were extracted with methanol, hot water and hot 1N NaOH and ^{14}C was estimated in the residue. Proportions of 1,3- and 1,4-linkages were measured by methylation analysis. Data taken from refs. *38* and *39*.

branes, and conducting reactions in high concentrations of polyols to reduce the water content of mixtures (*37,42*). The most promising procedures have been efforts to increase yields by preserving or reconstituting pea wall and membrane fractions (*36–39*) or by adding a naturally occurring protein acceptor which appears to be glycosylated by membrane (*14,43*). The highest rates of 1,4-β-glucan synthesis yet achieved *in vitro* are found at the cut surfaces of pea stem cells (*36,38,39*) (Table I), where membrane remains in close contact with the wall (*26*).

Another part of the problem has been the fact that all plant preparations that form 1,4-β-linked glucan from sugar nucleotide also form 1,3-β-linked glucan (callose). The 1,3-linkage predominates at high substrate levels in peas (Table I) and in all other preparations that have been tested (see refs. in *39*). Part of this 1,3-linked product may be covalently attached to 1,4-linkages in a mixed-linkage glucan. Much of it is insoluble in alkali (*38,39,42*), and therefore it is not sufficient to identify a product as cellulosic solely on the basis of insolubility.

If plant membranes contain two β-glucan synthetases, one with low K_m that synthesizes β-1,4-linkages and the other with high K_m that forms β-1,3-linkages, it should be possible to physically separate the two systems. Earlier claims to success in separating synthetases by fractionating solubilized enzymes have not been duplicated in recent years, despite many attempts. There is evidence from sucrose gradient analyses of pea membranes that glucan synthetase activities are separable into Golgi- and plasma membrane(PM)-enriched fractions (*44,45*). The latter (I) appears to generate mainly 1,3-β-linkages and the former (II) a product of uncertain properties (xyloglucan?). Other such studies identified β-glucan synthetase activity in the endoplasmic reticulum as well, but this appeared to be a precursor for Golgi and/or PM synthetases (*39,42*).

Thus, it is necessary to consider the possibility that cellulose synthetase can be so "perturbed" during isolation that it converts to a form that generates 1,3-β-linkages (*13*). At this time, there is no strong evidence to argue against such an interpretation. Indeed, callose is well known to form in tissues particularly after 'wounding', and synthetase perturbation may be the reason. Certainly, when cut pea slices are provided with UDP-glucose at levels where the product is mainly 1,3-linked (Table I), this product forms in the periplasmic space between wall and PM at loci which are indistinguishable under the EM from those where 1,4-linked products are deposited (*26*).

Finally, there is a serious problem in identifying products which contain 1,4-β-linkages as authentic cellulose. Apart from the possibility that insoluble mixed-linkage 1,3/1,4-glucan may be formed, there is evidence that the product formed from GDP-glucose is at least in part a linear 1,4-linked glucomannan (*46,47*). It has been suggested (*48,49*) that the product formed by UDP-glucose could be xyloglucan or a precursor to it. This seems possible when UDP-xylose is also present in the reaction medium, but not otherwise. Recent results (*50,51*) indicate that xyloglucan synthetase proceeds by concurrent transfer of the two glycones to a nascent xyloglucan acceptor, and does not operate by xylosyl transfer onto a preformed cellulosic backbone. Thus cellulose synthetase and xyloglucan synthetase appear to be separate enzymes.

1,4-β-GLUCANASE ACTIVITY

Using carboxymethyl cellulose, soluble cellodextrins or mixed-linkage 1,3/1,4-β-glucans as substrates (15,52), many plant tissue extracts have now been shown to contain celluloytic activity, which may reach relatively high concentrations in localized tissues or at particular times in development. In pea stem apices, the activity increases markedly after treatment with the auxin type of growth regulator under conditions where cell expansion occurs concurrently (7). The response is specific to active auxins and can not be duplicated with other hormones. Ethylene also evokes cell enlargement in peas but inhibits cellulase development (36). Inhibitors of RNA and protein but not DNA synthesis prevent both cellulase activity increases and growth responses (8,9), and therefore the enzyme activity appears to be present in expanding parenchyma cells rather than dividing cells.

Upon purification to homogeneity by several criteria, pea cellulase activity was shown (53) to be due to two enzymes with different physical properties but very similar kinetic behaviour towards a variety of substrates (54,55). Antibodies to the two enzymes were useful for demonstrating their intra- and extracellular locations (34). Since the smaller intracellular cellulase activity increased before the larger extracellular cellulase following hormone treatment (55), it was tempting to propose that the former was precursor to the latter. However, antibodies to the two enzymes showed no cross-reactivity in Ouchterlony tests (53). The antibodies were also useful for selectively precipitating polysomal RNA bearing nascent cellulase from hormone-treated tissues. By this means, it was possible to isolate and titrate mRNA for the smaller pea cellulase, and demonstrate that it was generated before cellulase activity appeared in extracts (56). This was the first plant mRNA for a known enzyme to be shown to be truly induced by a hormone.

With respect to the question of the function(s) of induced cellulase activity in pea stems, it was first suggested (6–8) that a direct "loosening" of the cellulosic framework might result after the enzyme was secreted into the wall because of limited hydrolysis of the microfibrilar network. The fibrils closest to the plasma membrane in pea parenchyma are horizontally oriented (57) and the action of cellulase at these loci (33,58) would be expected to lead to lateral cell expansion, as observed (7,8). A marked weakening of the tensile strength of plant cell walls can be demonstrated following treatment of tissues with cellulase from the outside (59).

Another suggestion derives from the observation (60) that β-glucan synthetase activity at the surfaces of cut pea stems can be stimulated several fold by brief pre-incubation of the slices in purified active pea cellulases. This was interpreted (57,58) as resulting from limited hydrolysis of exposed microfibrils, with the consequence that non-reducing chain ends were made available for glycosyl transfer from adjacent pea membranes. In this event, pea cellulase activity could stimulate pea cellulose synthesis provided the hydrolase and synthetase acted co-operatively at the same locus.

A third possibility is that pea 1,4-β-glucanase does not function *in vivo* with cellulose as substrate but with xyloglucan. Peas and many other dicots contain substantial amounts (*e.g.*, 10%) of xyloglucan in their primary walls, where it is thought to be firmly bound to the surfaces of cellulose microfibrils (*61*). It has been proposed that cross-linkage of bound xyloglucan to other wall matrix material contributes rigidity to the wall (*62*). There is evidence to show that auxin evokes a turnover (*63, 64*) and solubilization (*65*) of pea wall xyloglucan. Clearly auxin could do so by inducing cellulase, for fungal cellulase is known to hydrolyse the xyloglucan chain at intervals where there is no substituted xylose unit. The result could be a "weakening" of the primary wall and cell expansion *via* induced xyloglucanase activity.

FUTURE PROSPECTS

There is a pressing need to exploit all known means for increasing the stability and yield of 1,4-β-glucan synthetase activity *in vitro*, and to separate it from activities which lead to the formation of other linkages or heteropolysaccharides. To do so, however, first requires an assay procedure which unequivocally distinguishes between 1,4-β-glucan chains that are at least nascent or, preferably, fully cellulosic. This means that the products should be insoluble and shown to contain 1,4-β-linked glucose only, with a DP of at least 500, *i.e.*, the smallest cellulosic component known in nature (*14,66*). Although DP measurements of plant cellulose have successfully employed a procedure for fractionating cellulose nitrate derivatives from acetone solution, the nitration method is dangerous, delicate, degradative, and very time-consuming. Several new cellulose solvents have been developed in recent years which are simple to use and anhydrous, and therefore do not result in cellulose hydrolysis. These could be employed as solvents in new fractionation procedures, and their use in characterizing products of synthetase activity can be confidently expected.

There is also a need to introduce biochemical parameters into experiments in cell biology, with which putative cellulose synthetase complexes can be tested for their true function. If authentic 1,4-β-glucan synthetase activity were purified *in vitro* and fractionated into its essential constituents, antibodies could be made to these components and used as probes for ultrastructural localization. Alternatively, it may be possible to isolate visually identifiable synthetase complexes in sufficient quantity to conduct direct biochemical tests with them, as has been accomplished, for example, with chitosomes and chitin biosynthesis (*67*).

Ultimately, the aim should be to isolate a completely effective enzyme system for synthesis of authentic cellulose, define the loci in membranes from which it derives, fractionate the system into essential transglycosylases, intermediates, acceptors and activators, and re-assemble the components into a fully active synthetase system. When this is achieved, it will then be possible to ask precise questions about the manner in which cells regulate cellulose metabolism at different stages of their development. It is optimistic to expect that all of this will be accomplished by 1990, as the title of this article suggests, but it is useful to contemplate the magnitude of the task ahead of those in the field. The goals will require interaction of many disciplines from chemistry to cell biology, and progress may well depend on the application of findings from divergent fields. It may turn out, for example, that action of an associated cellulase or xyloglucanase is prerequisite to uncovering of sites for the initiation of cellulose chain-lengthening by the synthetase proper. The role of microtubules, or microtubule-associated protein, in regulating cellulose synthesis needs to be studied on a biochemical level. It appears almost certain that success in these fields will require thinking beyond conventional enzyme solution kinetics about a system with 3-dimensions at the membrane/wall interface.

REFERENCES

1. Lamport, D.T.A : 1970. *Annu. Rev. Plant Physiol.* **21**, 235–270.
2. Muhlethaler, D. : 1967. *Annu. Rev. Plant Physiol.* **18**, 1–24.
3. Preston, R.D. : 1974. The Physical Biology of Plant Cell Walls, Chapman and Hall, London, 491 pp.
4. Maclachlan, G.A. and Duda, C.T. : 1965. *Biochim. Biophys. Acta* **97**, 288–299.
5. Ray, P.M. : 1963. *Biochem. J.* **89**, 144–150.
6. Maclachlan, G.A., Davies, E. and Fan, D.F. : 1968. *In* Biochemistry and Physiology of Plant Growth Substances (Wightman, F. and Setterfield, G.A., eds.), pp. 443–453, Runge Press, Ltd., Ottawa.
7. Fan, D.F. and Maclachlan, G.A. : 1966. *Can. J. Bot.* **44**, 1025–1034.
8. Fan, D.F. and Maclachlan, G.A. : 1967. *Plant Physiol.* **42**, 1114–1122.
9. Datko, A.H. and Maclachlan, G.A. : 1968. *Plant Physiol.* **43**, 735–742.
10. Tanimoto, E. and Masuda, Y. : 1968. *Physiol. Plant* **21**, 820–826.
11. Labavitch, J.M. : 1981. *Annu. Rev. Plant Physiol.* **32**, 385–406.
12. Maclachlan, G. and Fevre, M. : 1982. *In* The Cytoskeleton in Plant Growth and Development (Lloyd, C., ed.), pp.127–146, Academic Press, London.
13. Delmer, D.P. : 1977. *Rec. Adv. Phytochem.* **11**, 45–77.
14. Franz, G. and Heiniger, U. : 1981. Encyclopedia of Plant Physiology New Series **13B**, (Loewus, F.A. and Tanner, W., eds.), pp.47–67, Springer-Verlag, Berlin, Heidelberg and New York.
15. Verma, D.P.S., Kumar, V. and Maclachlan, G. : 1982. *In* Cellulose and Other Natural Polymer Systems (Brown, R.M., ed.), pp. 159–188, Plenum Press, New York.
16. Brown, R.M. and Willison, J.H.M. : 1977. *In* International Cell Biology (Brinkley, B.R. and Porter, K.R., eds.), pp. 267–283, Rockefeller Univ. Press, New York.
17. Brown, R.M. and Romanovicz, D.K. : 1976. *Appl. Polym. Symp.* **28**, 537–585.
18. Satoh, S., Matsuda, K. and Tamari, K. : 1976. *Plant Cell Physiol.* **17**, 1243–1254.
19. Fowke, L.C. and Gamborg, O.L. : 1980. *Int. Rev. Cytol.* **68**, 9–51.
20. Grout, B.W.W. : 1975. *Planta* **123**, 275–282.
21. Blaschek, W., Koehler, H., Semler, V. and Franz, G. : 1982. *Planta* **154**, 550–555.
22. Asamizu, T., Tanaka, K., Takebe, J. and Nishi, A. : 1977. *Physiol. Plant.* **40**, 215–218.
23. Herth, W. and Meyer, Y. : 1977. *Biol. Cell.* **30**, 33–40.
24. Ray, P.M. : 1967. *J. Cell Biol.* **35**, 659–674.
25. Maclachlan, G. : 1982. *In* Cellulose and Other Natural Polymer Systems (Brown, R.M., ed.), pp. 327–339, Plenum Press, New York.
26. Mueller, S.C., Maclachlan, G. and Brown, R.M. : 1982. Advances in Chem. Series, **184**, in press (Timell, T., ed.), Amer. Chem. Soc. Press, Washington.
27. Mueller, S., Brown, R.M. and Scott, T.K. : 1976. *Science* **194**, 949–951.
28. Mueller, S. and Brown, R.M. : 1980. *J. Cell Biol.* **84**, 315–326.

29. Giddings, T.H., Brewer, D.L. and Staehelin, L.A. : 1980. *J. Cell Biol.* **84**, 327–339.
30. Lloyd, C.W., : 1982. The Cytoskeleton in Plant Growth and Development, Academic Press, London.
31. Lloyd, C.W., Slabas, A.R., Powell, A.J. and Lowe, S.B. : 1980. *Planta* **147**, 500–506.
32. Heath, I.B. : 1973. *J. Theor. Biol.* **48**, 445–449.
33. Bal, A.K. : 1973. *Electron Microsc. Enzymes* **3**, 68–76.
34. Bal, A.K., Verma, D.P.S., Byrne, H. and Maclachlan, G. : 1976. *J. Cell Biol.* **69**, 97–105.
35. Spencer, F.S., Ziola, B. and Maclachlan, G. : 1971. *Can. J. Biochem.* **49**, 1326–1332.
36. Maclachlan, G. : 1977. *In* Plant Growth Regulation (Pilet, P.E., ed.), pp.13–20, Springer-Verlag, Berlin.
37. Maclachlan, G., Dürr, M. and Raymond, Y. : 1979. *Methodol. Surv. B. Biochem.* **9**, 147–158.
38. Chao, H.Y. and Maclachlan, G. : 1978. *Plant Physiol.* **61**, 943–948.
39. Raymond, Y., Fincher, G.B. and Maclachlan, G. : 1978. *Plant Physiol.* **611**, 938–942.
40. Dürr, M., Bailey, D.S. and Maclachlan, G. : 1979. *Eur. J. Biochem.* **97**, 445–453.
41. Thomas, D. des S., Smith, J.E. and Stanley, R.G. : 1969. *Can. J. Bot.* **47**, 489–496.
42. Shore, G. and Maclachlan, G. : 1975. *J. Cell Biol.* **64**, 557–571.
43. Franz, G. : 1976. *J. Polym. Sci. C* **28**, 611–621.
44. Ray, P.M., Shininger, T.L. and Ray, M.M. : 1969. *Proc. Natl. Acad. Sci. U.S.A.* **64**, 605–612.
45. Ray, P.M. : 1979. *Methodol. Surv. B. Biochem.* **9**, 135–146.
46. Villemez, C.L. : 1971. *Plant Physiol.* **121**, 151–157.
47. Brett, C. : 1981. *J. Exp. Bot.* **32**, 1067–1077.
48. Ray, P.M., Eisinger, W.R. and Robinson, D.G. : 1976. *Ber. Dtsch. Bot. Ges. Bd.* **89**, 121–146.
49. Ray, P.M. : 1980. *Biochim. Biophys. Acta* **629**, 431–444.
50. Hayashi, T. and Matsuda, K. : 1981. *J. Biol. Chem.* **256**, 11117–11122.
51. Hayashi, T. and Matsuda, K. : 1981. *Plant Cell Physiol.* **22**, 1571–1584.
52. Woodward, J.R. and Fincher, G.B. : 1982. *Eur. J. Biochem.* **121**, 663–669.
53. Byrne, H., Christou, N.V., Verma, D.P.S. and Maclachlan, G.A. : 1975. *J. Biol. Chem.* **250**, 1012–1018.
54. Wong, Y.S., Fincher, G.B. and Maclachlan, G.A. : 1977. *J. Biol. Chem.* **252**, 1402–1407.
55. Maclachlan, G.A. and Wong, Y.S. : 1979. *Adv. Chem. Ser.* **181**, 347–360.
56. Verma, D.P.S., Maclachlan, G.A., Byrne, H. and Ewings, D. : 1975. *J. Biol. Chem.* **250**, 1019–1026.
57. Maclachlan, G.A. : 1976. *Appl. Polym. Symp.* **28**, 645–658.
58. Maclachlan, G.A. : 1977. *Trends Biochem. Sci.* **1**, 226–228.
59. Olson, A.C., Bonner, J. and Morré, D.J. : 1965. *Planta* **66**, 126–134.
60. Wong, Y.S., Fincher, G.B. and Maclachlan, G.A. : 1977. *Science* **195**, 678–680.
61. Valent, B.S. and Albersheim, P. : 1974. *Plant Physiol.* **54**, 105–108.
62. Keegstra, K., Talmadge, K.W., Bauer, W.D. and Albersheim, P. : 1973. *Plant Physiol.* **51**, 188–196.
63. Labavitch, J.M. and Ray, P.M. : 1974. *Plant Physiol.* **53**, 669–673.
64. Gilkes, N.R. and Hall, M.A. : 1977. *New Phytol.* **78**, 1–15.
65. Labavitch, J.M. and Ray, P.M. : 1974. *Plant Physiol.* **54**, 499–502.
66. Spencer, F.S. and Maclachlan, G.A. : 1972. *Plant Physiol.* **49**, 58–63.
67. Bartnicki-Garcia, S., Ruiz-Herrera, J. and Bracker, C.E. : 1979. *In* Fungal Walls and Hyphal Growth (Burnett, J.H. and Trinci, A.P.J., eds.), pp.149–168, Cambridge Univ. Press, London.

Glutamine Metabolism: The Key to the Flow of Nitrogen in Plants

Benjamin J.Miflin

GLUTAMINE METABOLISM — THE DISCOVERY OF THE GLUTAMATE
SYNTHASE CYCLE

No story ever starts in just one place but perhaps the most significant event for me in
this one occurred one day around Christmas 1969 when C. Brown related the details
of some work he had been doing on nitrogen metabolism in bacteria. Brown worked
in the Microbiological Department at Newcastle University, across the road from my
laboratory in the Department of Plant Science. He told me that he and his coworkers,
Tempest and Meers, had discovered that when bacteria were growing on limiting
concentrations of ammonia they assimilated it by a novel pathway. At this stage he
felt that this was probably not of widespread significance and only operated under
conditions of severe nitrogen limitation. However, once this pioneering work was
published (1,2), and other groups became aware of the pathway, it soon became apparent that it was of widespread importance in bacteria (see 3 for review).

The primary features of this pathway, which has become known as the glutamate synthase cycle (4), are shown in Fig. 1 ; it involves the assimilation of ammonia
into the amide position of glutamine, catalyzed by glutamine synthetase, followed by
the reductive transfer of this group to the C-2 of 2-oxoglutarate to give the amino
acid glutamate. This second step is catalyzed by the enzyme glutamate synthase
(or L-glutamate:$NADP^+$ [or ferredoxin] oxidoreductase, EC 1.4.1.13 [or EC1.4.7.
1.]).

Tempest et al.'s discovery of the enzyme and their description of the operation
of the glutamate synthase pathway proved to be a key to understanding the role of

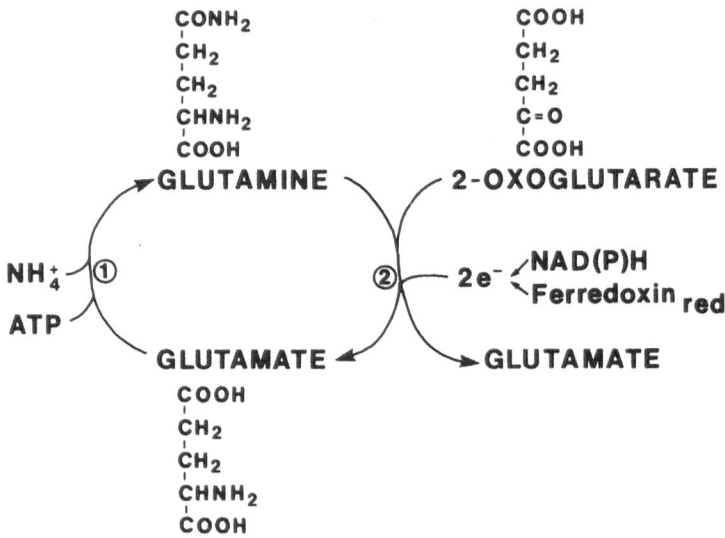

Fig. 1. The glutamate synthase cycle.

glutamine synthetase, and its series of complex regulatory properties, in bacterial nitrogen metabolism (see 5). The finding caused a minor scientific revolution (as defined by Kuhn) (6) and led to the paradigm that "all ammonia assimilation occurred via glutamate dehydrogenase (GDH)" being replaced by a new one which proposed the presence of two alternative pathways operating under different conditions of nitrogen and energy availability. Early attempts to find glutamate synthetase in eukaryotes (yeast, algae, green plants, and animals) were not successful and so for some years the glutamine synthetase/glutamate synthase cycle was considered to be present only in bacteria.

My next involvement in the subject came when I spent a sabbatical period in Harry Beever's laboratory in Santa Cruz. Many threads led to my visit to Santa Cruz (not least its position on the Pacific coast) in that I had worked in the same laboratory as Harry's brother Len, I lectured in Newcastle University from which the Beevers graduated (before my time) and Harry Beevers was the first eminent plant physiologist that I heard lecture when I went to my first American Society of Plant Physiologists Meeting at Purdue as a young graduate student. My aim at Santa Cruz was to learn and develop techniques for isolating plastids that were uncontaminated with microbodies, so as to verify the subcellular localization of nitrite reductase over which there was some dispute (see 7–9). We were successful in developing such techniques (10) and as a result of using these were able to demonstrate unequivocally the presence of nitrite reductase in the chloroplast and its absence from the microbodies (11).

During the period that I was in Santa Cruz I made two significant visits. The first was a return to R. Hageman's laboratory at Urbana where I had done my Masters degree some twelve years previously. I gave a seminar there on work that had been completed before I left England concerning the reduction of nitrite by isolated chloroplasts and immediately afterwards Magalhaes, who was working with Hageman, wanted to know if the work had been published as he had just completed some identical studies. In the event both of us had shown clearly that isolated intact chloroplasts could reduce nitrite to amino acids in a light-dependent reaction ; both papers were published more or less simultaneously in 1974 (*12,13*). The second visit was to the joint meeting of the American and Canadian Societies of Plant Physiologists in Calgary in 1973. I attended this meeting and had a long talk with K. Joy. Our careers had intertwined in that he followed me into Hageman's laboratory in Urbana whilst I moved to Imperial College where he had been working. During our discussions, which I recalled and took place on the way to a barbecue, he told me of his studies that showed that glutamine synthetase was associated with chloroplasts prepared by differential centrifugation (*14*). When I returned to Santa Cruz I quickly confirmed his results using the density gradient methods (*11*). Attempts to show the presence of significant amounts of GDH in the plastids were not successful. An assessment of the position at the end of my stay in Santa Cruz is given in Table I. This showed that chloroplasts had the enzymes to reduce nitrite to ammonia and to incorporate ammonia into glutamine but not for the synthesis of amino-N *via* GDH ; in contrast the studies with intact chloroplasts had shown that they could reduce nitrite to amino acids.

At this point I moved to Rothamsted and met P. Lea. Although he had been working for some years on amino acid analogues and aminoacyl-tRNA formation he

TABLE I
Activities in Chloroplasts of Certain Enzymes Involved in Amino Acid Biosynthesis

Activity	Rate (nmol mg chl^{-1} min^{-1})	Reference
Nitrite reductase (methyl viologen assay)	270	Miflin (*11*)
	155–300	Magalhaes *et al.* (*13*)
Nitrite reduction (light-dependent by chloroplasts)	150	Miflin (*12*)
	200	Magalhaes *et al.* (*13*)
α-Amino-N production (light-dependent by chloroplasts)	145	Miflin (*12*)
	180	Magalhaes *et al.* (*13*)
GDH	40	Miflin (*12*)
	80–100	Leech and Kirk (*73*)
	20–75	Lea and Thurman (*15*)
α-Oxoglutarate consumption (light-dependent by chloroplasts)	10	Givan *et al.* (*74*)
Glutamine synthetase	1,500	Miflin (*11*)

had done his Ph.D. research on GDH and the characteristics of the enzyme associated with chloroplasts (15). As a result of this work he was also disillusioned with the enzyme as a route for ammonia assimilation in chloroplasts. After many hours of discussion we decided to try a new approach to see if the glutamine synthetase/glutamate synthase pathway was operating in chloroplasts. This was based on trying two types of experiment ; one, attempting to find out which nitrogen compounds, other than nitrite, could lead to glutamate formation by intact chloroplasts and two, attempting to find glutamate synthase activity in chloroplast extracts. Since previous attempts of mine and other workers using NAD(P)H had failed to demonstrate the enzyme we decided to use methyl viologen and ferredoxin as electron donors since we knew these were used by nitrite reductase. Over a series of weeks we ran these two types of experiment in parallel, Lea working with the intact plastids and I with the extracts. Once a few teething problems with the glutamate assay were sorted out we quickly found that both approaches pointed firmly to the presence of glutamate synthase in the chloroplasts in that glutamine and oxoglutarate were converted to two molecules of glutamate (Table II) (16). The essential difference between the chloroplast enzyme and that originally described in bacteria was that it was not active with pyridine nucleotides but only with ferredoxin or methyl viologen.

By curious coincidence we went to the International Plant Tissue Conference at Leicester in 1974 whilst our paper was under review in *Nature* and during one lunch break I phoned our office to receive the news that the paper had been accepted. When I returned to the conference hall I found Lea talking to M. Fowler who said that he and D. Dougall had separately produced evidence for a NAD(P)H glutamate synthase in non-green tissue. Dougall was also at the conference and his paper appeared (17) at that time. Discussions showed that there were certain differences between the NAD(P)H (17,18) and the ferredoxin enzymes — most of these have sub-

TABLE II
Glutamate Synthesis by Intact Chloroplasts and Extracts Made from Them

Electron donor[a]	N-donor	N-acceptor	Glutamate formed (μmol mg chl^{-1} hr^{-1})[b]	
			Chloroplast extract	Isolated chloroplasts
Fd or light	None	2-Oxoglutarate	0	0.4
	Glutamine	None	0	0.7
	Glutamine	2-Oxoglutarate	23.2	6.9
NADPH	Glutamine	2-Oxoglutarate	0	–
None or dark	Glutamine	2-Oxoglutarate	0	0.3[c]

[a] The Fd was reduced by dithionite and used for the chloroplast extract experiments except in one case where NADPH was substituted ; chloroplasts were incubated in the light to produce the reductant *in situ*.
[b] The results are from Lea and Miflin (16) except that marked [c] which is from Wallsgrove *et al.* (19).

sequently been confirmed (*19–21*) except for the claim that the NAD(P)H enzyme would work with asparagine (see *22*). The subsequent work has shown that the NADPH enzyme is of different molecular size and is clearly separable from the ferredoxin-dependent enzyme (see *23* for a review). Initial work suggested that the former enzyme was present in non-green tissues whilst the latter was present chiefly in chlorophyllous tissues. It is now clear, however, that both are present in etiolated and green tissue (*24–26*) but that the ratio of the two activities differ.

After we had found the ferredoxin-dependent enzyme in green leaves it became obvious to us that this enzyme was probably present in other chlorophyllous tissue. One particular problem was in the N_2-fixing cyanobacteria. Circumstantial evidence strongly suggested the operation of the glutamate synthase cycle as the organisms contained little GDH and much glutamine synthetase (*27*). Although some studies suggested the presence of minute amounts of NAD(P)H-glutamate synthase in *Anabaena* (*28*) this enzyme was generally considered to be absent from these bacteria (*27,29*). We predicted that the ferredoxin enzyme should be present and set out to show this. We were fortunate that P. Keay and I. Catchpole were growing bulk quantities of *Nostoc* and *Anabaena* at the nearby Luton College of Further Education and were willing to donate material. Using both cell-free and whole-cell assays we demonstrated the presence of significant amounts of ferredoxin-dependent glutamate synthase (*30*). Similar approaches showed that the enzyme was also present in green algae (*31*).

By about the middle of 1975 we had two problems to consider ; one, could our findings of a ferredoxin-dependent glutamate synthase be repeated in other laboratories and, two, what was the significance of our findings. The first problem was fairly rapidly overcome, again through the network of contacts that had been developed. To identify this particular part of the network it is necessary perhaps to go back to the beginning of the 1960s when A. Sims and B. Folkes were working on the assimilation of labelled ammonia ($^{15}NH_3$) in *Candida utilis*. Their work clearly demonstrated a key role for GDH in ammonia assimilation in yeast (*32,33*). The elegance of this work probably lead to other people broadening this conclusion to include the whole plant kingdom and thus leading to a general concept that GDH was the enzyme involved in ammonia assimilation. The other consequence of their work was that they founded a school of graduate students, concerned with nitrogen metabolism, with whom I was pleased to make contact at the first conference on that topic at the Long Ashton Research Station, part of the University of Bristol (*34*). Amongst those students were K. Joy (see above) and also G. Stewart with whom we have developed a close association through our common interest in plant nitrogen metabolism. It was thus natural that he was rapidly informed of our results and he and his graduate student D. Rhodes quickly demonstrated the presence of ferredoxin-dependent glutamate synthase in *Lemna* (*35*).

To try and assess the significance of our findings we considered the two pathways as alternatives. The accepted paradigm for ammonia assimilation prior to 1970 was that virtually all of the primary incorporation into amino-nitrogen occurred into glutamate and was catalyzed by GDH. In the light of Tempest *et al.*'s (*1,2*) findings this was modified by the addition "except under conditions of low concentrations of ammonia and a sufficiency of energy when it occurs *via* the glutamine synthetase/ glutamate synthase cycle" (*e.g.*, see *3*). However if the two pathways are considered as alternatives it is then possible to make a number of predictions. In our original analysis we considered four such predictions (*36*). These were :

1) That, where GDH is the sole route to α-amino-N and the amide-N of glutamine is transferred to only a limited range of compounds, $^{14}NO_3$ and ^{15}N-glutamate should donate their label in the same manner to all amino acids and ^{15}N-amide glutamine should donate its label only to tryptophan, arginine, and histidine. With the operation of the alternative pathway all three compounds should donate their label to amino acids in about the same proportions.

2) If the glutamate dehydrogenase pathway is operating, then ^{15}N from $^{15}NH_3$ should be incorporated primarily into both the amino group of glutamate and the amide group of glutamine ; on transfer back to $^{14}NH_3$ then the amino group of glutamate and the amide group of glutamine should both begin to lose label immediately. For the alternative pathway ^{15}N from labelled ammonia should appear first in the amide group of glutamine and only subsequently in the amino group of glutamate ; on transfer to $^{14}NH_3$ there should be an immediate loss of label from the amide-N of glutamine followed, after a lag, by loss from the amino group of glutamate and glutamine.

3) Methionine sulphoximine (MSO), which is a potent inhibitor of glutamine synthetase (but not of GDH) should inhibit nitrogen assimilation by the alternative pathway but have no effect on that *via* GDH.

4) Similarly to 3) azaserine or diazo-oxonorvaline (DON), which inhibit amide transferases, including glutamate synthase, should not affect the GDH pathway but should block the alternative pathway.

Having made these predictions we searched the literature to see what relevant work had been done. To our amazement we found evidence in favour of the alternative pathway for each of the above predictions. The nature of this evidence is reviewed in detail in Miflin and Lea (*36*) but some of it had been in the literature since 1959 (*37*). This latter paper was particularly interesting in that the data clearly fitted predictions 3) and 4) in favour of the alternative pathway and were obviously not consistent with the involvement of GDH. However, such was the hold of the then current paradigm that the authors, in their discussion, sought to make their observations conform with it. As a contrast similar results were produced by Stewart and Rhodes (*38*) with the multicellular plant *Lemna* after the glutamine synthetase/ glutamate synthase pathway had been proposed and were interpreted in an entirely different fashion.

In the subsequent years many further experiments have been done in a number of countries including Japan and the results of these point overwhelmingly to the operation of the alternative pathway (*39–43*). Perhaps the most significant of these later results, particularly in relation to this story, are those of Rhodes *et al.* (*44*). By this time Sims and Folkes had both independently moved to the University of East Anglia and were joined by Rhodes who had obtained his Ph.D. under

G. Stewart. They then set out to use a sophisticated version of their original approach with yeast to determine quantitatively the path of NH_4 assimilation in *Lemna*. Their results were entirely consistent with predictions 2), 3) and 4) for the alternative pathway, both under normal growth conditions and under conditions of elevated ammonia concentrations and low light intensity. They therefore concluded that the glutamate synthase pathway was the only one that was concerned with ammonia assimilation in *Lemna minor*.

As a result of these various experiments Lea and I concluded, in a review of the literature up to 1980 (*45*), as follows "since the proposal was first made that eukaryotic green plants used the glutamine synthetase/glutamate synthase pathway for ammonia assimilation (*16,36*) a large amount of evidence has occurred to support that contention. Studies on the distribution of the enzymes, the activity of partial systems (*e.g.*, chloroplasts), and the incorporation of labelled N, both in the absence and presence of inhibitors, have produced results entirely compatible with the operation of the glutamate synthase cycle and inconsistent with any significant role for GDH in ammonia assimilation. This conclusion has proved true for both the autotrophic and heterotrophic parts of plants. The only remaining exceptions to this general picture are probably certain species of algae that contain a high-affinity GDH."

THE LOCALIZATION OF THE GLUTAMATE SYNTHASE CYCLE ENZYMES

The above quote refers to enzyme distribution and I would like to briefly deal with this topic. R. Wallsgrove in the Rothamsted laboratory became interested in a question left over from the earlier localization studies (*11*) that of determining if *all* of the cellular activity of certain enzymes was in the chloroplast. This question had been almost impossible to resolve with mechanical homogenates of leaves because of the low recovery of intact chloroplasts in even the best preparations. We now changed to the techniques of G. Edwards and his colleagues (*46*), in which protoplasts are first made from leaf tissue and then gently lysed. We were fortunate that Edwards was in the U.K. at the time and that we could persuade him to spend a few days in our laboratory. Using this approach we were able to recover over 80% of the chloroplasts intact and to demonstrate that they contained all of the cell's nitrite reductase and glutamate synthase but only part (about 60%) of the glutamine synthetase. We therefore had a system in which we believed ammonia assimilation could occur in the cytoplasm and the chloroplast but that amino-N synthesis occurred only in the chloroplast (*47*). Before these results were published we went to hear Rhodes, Sims, and Folkes present their ^{15}N analyses to a meeting of the Society of Experimental Biology. To our great excitement they proposed, on the basis of their computer analysis of the labelling kinetics, a two-compartment model in which in one compartment both glutamine and glutamate synthesis could occur and in the other only glutamine synthesis. Both groups were more than pleased to find independent and concurrent corroboration of their proposals using widely different approaches.

AMMONIA RECYCLING

The role of the cytoplasmic glutamine synthetase in leaves then became a matter
for conjecture. Certainly its presence was consistent with ammonia feeding experiments in light and dark by Canvin and Atkins (48) and by Ito et al. (49). However,
ammonia is not usually thought to be a source of inorganic nitrogen nutrition for
leaves. But we were aware of another source of ammonia in leaf tissue. This part of
the story also involves weaving in another connecting thread reaching back to my Ph.
D. studies at Imperial College where I was concerned with demonstrating that a part
of the flux of carbon from photosynthetic $^{14}CO_2$ fixation was metabolized via the
glycollate pathway through glycine to serine and eventually sucrose (50). At that
time, in the neighbouring laboratory, A. Keys and his colleagues I. Bird and M.
Cornelius were studying different problems but all three were shortly to move to
the Botany Department at Rothamsted, with C.Whittingham, to work on the photorespiratory (glycollate) pathway. By 1976 Keys and his colleagues were very concerned with measuring the rate of conversion of glycine to serine during photorespiration (51,52). This reaction results in not only the evolution of CO_2, which had
been the main focus of research interest, but also of stoichiometric amounts of
ammonia ; a fact often omitted in text books. Since the estimated rates of the conversion of glycine to serine and of CO_2 (and thus ammonia) released exceeded the
known rates of primary ammonia assimilation, Lea and I were concerned as to the
fate of this ammonia. The fact that Keys, Lea, and I often made up slightly greater
than 25% of the Rothamsted cricket team gave plenty of opportunities for discussion of this uncomfortable anomaly and led to the initiation of a number of
^{15}N-labelling experiments being set up. Before these were fully analyzed another
important meeting took place – the Fourth International Photosynthesis Congress at Reading. At this meeting J. Berry briefly discussed results he, Woo and
Osmond had produced (later published in detail Woo et al.) (53) which clearly
showed the reassimilation of the ammonia released from glycine. Although they were
clearly thinking along the same lines as us regarding the role of the glutamate synthase
cycle in reassimilation, they had apparently no clear evidence for its involvement. In
any event this presentation caused some dismay in the Rothamsted camp as our ^{15}N-
glycine experiments, as they then stood, were not going to be worth publishing. After
the Congress we met to decide whether or not to continue the project and if so to
design some definitive experiments. We decided to go ahead and to test directly the
idea that cytoplasmic glutamine synthetase was capable of reassimilating ammonia
released from glycine. Experiments with a reconstituted system using partially purified
enzyme and isolated mitochondria demonstrated that it was (43) ; conversely, isolated mitochondria alone were unable to reassimilate the ammonia even when the electron transport chain was blocked (43,54). Further experiments with MSO suggested
that the system also worked in vivo and consequently we proposed that a cycle – the
photorespiratory nitrogen cycle – normally operated to prevent the net loss of
ammonia from organic-N during the operation of the photorespiratory pathway

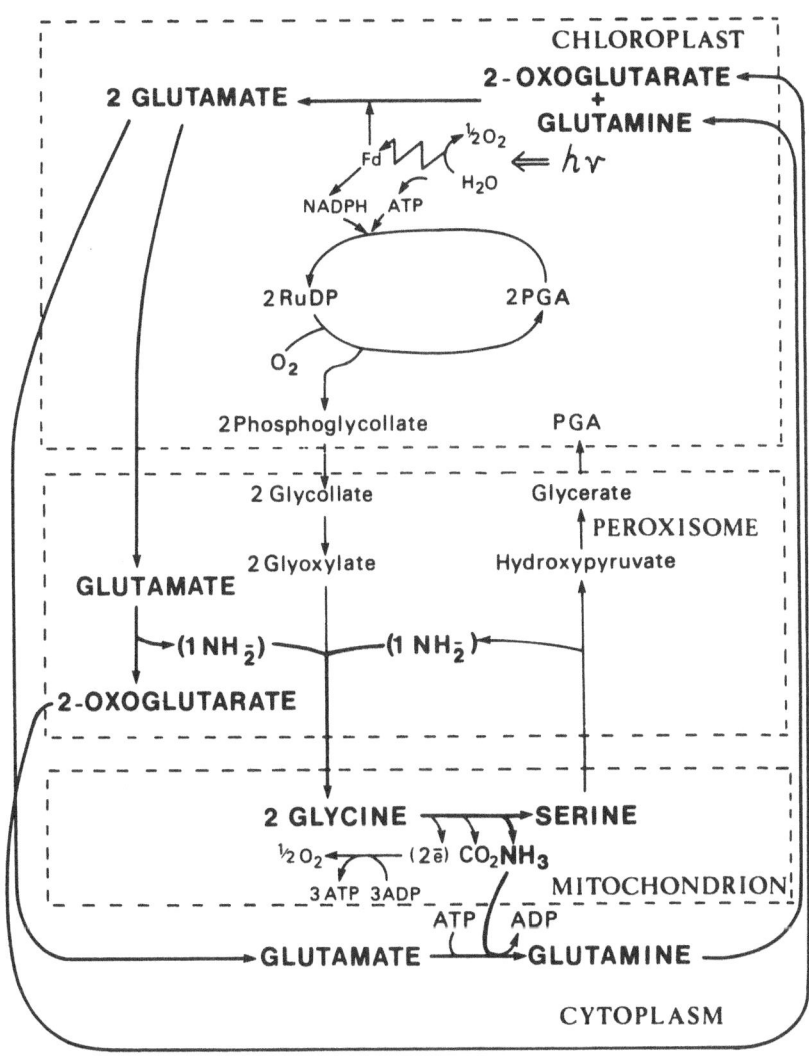

Fig. 2. The photorespiratory nitrogen cycle.

(Fig. 2). Conclusive proof for this cycle was provided by Somerville and Ogren (55), working at the University of Illinois, who isolated a mutant lacking ferredoxin-dependent glutamate synthase (but not the NAD(P)H enzyme) and which accumulated ammonia under photorespiratory conditions.

A major inference of the pathways in Fig. 2 is that the plant can routinely cycle N into and out of organic combination. Interestingly we were reaching the same conclusion based on other work in the department. This thread reached back to an earlier interest of P. Lea in asparagine metabolism (e.g., 56). Both the synthesis and metabolism of asparagine were (and to some extent still are) poorly understood. Consequently the suggestion that asparagine might act as a donor for amide-transfer reactions was an attractive concept ; we were disappointed when our results (22) failed to confirm the idea. Subsequent work showed the presence of asparaginases both unaffected by (57), and strictly dependent on, K^+ ions (58) in developing legume seeds. Concurrent work by Joy and colleagues showed that a different enzyme, asparagine transaminase, was probably responsible for asparagine breakdown in leaves (59,60). Consequently it appears that the amide-N of asparagine, which in many species is a major nitrogen transport compound, is only utilized to form other N-compounds after it has been released as ammonia and then reassimilated. Analysis of the probable routes of metabolism of other N-transport compounds suggests that this release and reassimilation of ammonia is probably widespread (61,62).

CONCLUSION

In the above review I have tried to show how, through my own work and that of a number of close associates, I came to the view that is embodied in the title of this essay. By introducing a personal thread to this discussion I have sought to show how that final view was the product of the interaction of a number of people and not of any one person. Indeed in the development of many of the ideas it is not clear exactly who generated any particular concept. For example, many of the discussions we have had at Rothamsted seem to have progressed by means of sentences started by one person and finished by another ; the ideas, experiments and interpretation appeared to grow almost of themselves. Also in this account I have tried to trace some of the network of contacts stretching back and forth between laboratories in different countries, which have been fostered by meetings at conferences and traffic of personnel between laboratories and that have contributed immeasurably to the progress of the work. This continual process of meeting and exchange of ideas is far from being an excuse for tourism, as some administrators appear to think ; rather it is the life blood of scientific advance.

In reviewing the development of the concepts recorded here I have been struck by how much that development accords with Kuhn's theory of scientific revolution (6). The central paradigm − that ammonia assimilation occurred *via* GDH − was entrenched in the text books for many years. Despite the production of contradictory results the strength of the paradigm and the absence of an alternative led to these being assimilated within the central theory. The onset of the "crisis" or "revolution" was triggered by the beautifully simple and intelligent experiments of Tempest *et al.* (1,2). These were given the helping hand of fate because they chose the organism which in hindsight shows the reported phenomena in their most dramatic form. The presence of this alternative hypothesis in bacteria enhanced the sense of crisis arising from experiments with higher plants and led to the drive to find glutamate synthase in higher plants. Once its presence was established it became apparent that the evidence for the alternative pathway had been in the literature since at least 1959, thus providing a good example of Kuhn's contention that an existing paradigm or theory is only put to the test after a sense of crisis has evoked a competing or alternative theory. The existence of the alternative pathway not only allows many, previously incompatible, pieces of the jigsaw to fit instantly but also stimulates the activities of normal science to fill out the theory by experimental verification of the various predictions that can now be made. Interestingly the degree to which the glutamate synthase cycle rather than GDH is the major assimilatory route varies according to the group of plants under consideration. In the yeast and fungi the evidence is still strongly in favour of the operation of the latter under almost all (but not every) condition ; in bacteria the operation of the two pathways is balanced and in most species can swing from one extreme to the other in response to the external environment ; in higher eukaryotic plants assimilation is almost completely by the glutamate synthase cycle with virtually no assimilatory role envisaged for GDH.

Finally, what of the future? The establishment of a new paradigm is usually followed by a period in which various loose ends are tied up. This has and will continue to take place. However the identification of glutamine as the central compound and glutamine synthetase as the central enzyme of nitrogen metabolism allows some fundamental questions to be asked. To me the most interesting are :

1) How is the flux controlled – particularly in relation to the energy and carbohydrate status of the cell? Various ways in which the activity of chloroplastic glutamine synthetase may be related to the energy status (as determined by illumination) have been discussed (23,63) and some mechanisms of control of glutamine synthetase have been probed at the enzymic level (44,64,65). Further down the line the questions revolve around how nitrogen is shunted into the different transport compounds (e.g., asparagine, ureides etc.)?

2) How does the plant respond to the generation of a localized source of ammonia as is produced in root nodules? We know that as effective nodulation proceeds and nitrogenase is formed the amount of glutamine synthetase in the nodule increases markedly (66,67). This enzyme is present in the cytoplasmic part of the plant portion of the nodule (68) and recent work has shown that, at least for Phaseolus vulgaris, this increased activity is due to the synthesis of new forms of the enzyme not previously present in the uninoculated plant root (69,70). Consequently there appears to be a coordinated pattern of gene expression occurring in both the Rhizobium and in the plant – how this is organized is a question that currently fascinates us.

One further line of research that follows out of our work discussed above is the relationship between carbon and nitrogen fluxes in photorespiration. There have been many studies suggesting a relationship between nitrate and ammonia metabolism and the rate of photorespiration (see 71 for a discussion). Recently Emes and Erismann (72) have produced evidence suggesting a direct effect of the type of nitrogen nutrition on the activity and molecular organization of glycollate oxidase.

These future questions are all concerned with uncovering how the plant regulates its nitrogen metabolism and integrates it within the rest of the metabolic fluxes. This goal of understanding the regulation of plant metabolism is a fundamental aim of plant biochemistry.

Acknowledgements
I hope that it is obvious to all that the work described above has been totally dependent on the unselfish collaboration of a number of people but particularly Dr. Peter Lea, Dr. Roger Wallsgrove, Dr. Alfred Keys, Dr. Ken Joy, and Dr. George Stewart, to whom I express my gratitude. I would also like to thank my wife Hilary for her help throughout this work but especially for many discussions on Kuhn's philosophy of scientific discovery.

REFERENCES

1. Tempest, D.W., Meers, J.L. and Brown, C.M. : 1970. *Biochem. J.* 117, 405–407.
2. Tempest, D.W., Meers, J.L. and Brown, C.M. : 1970. *J. Gen. Microbiol.* 64, 187–194.
3. Brown, C.M., MacDonald, Brown, D.S. and Meers, J.L. : 1974. *Adv. Microbiol. Physiol.* 11, 1–45.
4. Rhodes, D., Sims, A.P. and Folkes, B.F. : 1980. *Phytochemistry* 19, 357–365.
5. Prusinier, S. and Stadtman, E.R. (eds.) : 1973. The Enzymes of Glutamine Metabolism, Academic Press, New York.
6. Kuhn, T.S. : 1962. The Structure of Scientific Revolution, Univ. Chicago Press, Chicago.
7. Lips, S.H. and Avissar, Y. : 1972. *Eur. J. Biochem.* 29, 20–24.
8. Miflin, B.J. : 1970. *Planta* 93, 160–170.
9. Dalling, M.J., Tolbert, N.E. and Hageman, R.H. : 1972. *Biochim. Biophys. Acta* 283, 513–519.
10. Miflin, B.J. and Beevers, H. : 1974. *Plant Physiol.* 53, 870–874.
11. Miflin, B.J. : 1974. *Plant Physiol.* 54, 550–555.
12. Miflin, B.J. : 1974. *Planta* 116, 187–196.
13. Magalhaes, A.C., Neyra, C.A. and Hageman, R.H. : 1974. *Plant Physiol.* 53, 411–415.
14. O'Neal, D. and Joy, K.W. : 1973. *Nature* 246, 61–62.
15. Lea, P.J. and Thurman, D.A. : 1972. *J. Exp. Bot.* 23, 440–449.
16. Lea, P.J. and Miflin, B.J.: 1974. *Nature* 251, 614–616.
17. Dougall, D.K. : 1974. *Biochem. Biophys. Res. Commun.* 58, 639–646.
18. Fowler, M.W., Jessup, W. and Sarkissian, G.S. : 1974. *FEBS Lett.* 46, 340–342.
19. Wallsgrove, R.M., Harel, E., Lea, P.J. and Miflin, B.J. : 1977. *J. Exp. Bot.* 28, 588–596.
20. Boland, M.J. and Benny, A.G. : 1977. *Eur. J. Biochem.* 79, 355–362.
21. Chiu, J.Y. and Shargool, P.D. : 1979. *Plant Physiol.* 63, 409–415.
22. Miflin, B.J. and Lea, P.J. : 1975. *Biochem. J.* 149, 403–409.
23. Stewart, G.R., Mann, A.F. and Fentem, P.A. : 1980. *In* Amino Acids and Their Derivatives (Miflin, B.J., ed.), Vol. 5 of 'The Biochemistry of Plants', pp.272–327. Academic Press, New York.
24. Matoh, T., Ida, S. and Takahashi, E. : 1980. *Plant Cell Physiol.* 21, 1461–1477.
25. Cullimore, J.V. and Sims, A.P. : 1981. *Phytochemistry* 20, 597–600.
26. Wallsgrove, R.M., Lea, P.J. and Miflin, B.J. : 1982. *Planta* 154, 473–474.
27. Haystead, A., Dharmawardene, N.W.N. and Stewart, W.D.P. : 1973. *Plant Sci. Lett.* 1, 439–445.
28. Dharmawardene, M.W.N., Haystead, A. and Stewart, W.D.P. : 1972. *Arch. Mikrobiol.* 90, 281–295.
29. Nielson, A.H. and Doudoroff, M. : 1973. *Arch Mikrobiol.* 89, 15–22.
30. Lea, P.J. and Miflin, B.J. : 1975. *Biochem. Soc. Trans.* 3, 381–384.
31. Lea, P.J. and Miflin, B.J. : 1975. *Biochem. Biophys. Res. Commun.* 64, 856–862.
32. Folkes, B.F. : 1959. *In* Utilization of Nitrogen and Its Compounds by Plants, pp.126–147.

Symposia of the Society for Expt. Biol. XIII, Cambridge Univ. Press, London.
33. Sims, A.P. and Folkes, B.F. : 1964. *Proc. Roy. Soc. Lond. Ser. B.* **159**, 479–502.
34. Hewitt, E.J. and Cutting, C.V. : 1968. Recent Aspects of Nitrogen Metabolism in Plant, Academic Press, London.
35. Rhodes, D.A., Rendon, G.A. and Stewart, G.R. : 1976. *Planta* **129**, 203–210.
36. Miflin, B.J. and Lea, P.J. : 1976. *Phytochemistry* **15**, 873–885.
37. Meulen, P.Y.F. van der and Bassham, J.A. : 1959. *J. Am. Chem. Soc.* **81**, 2233–2336.
38. Stewart, G.R. and Rhodes, D. : 1976. *FEBS Lett.* **64**, 296–300.
39. Yoneyama, T. and Kumazawa, K. : 1975. *Plant Cell Physiol.* **16**, 21–26.
40. Thomas, J., Meeks, J.C., Wolk, C.P., Shaffer, P.W., Austin, S.M. and Chien, W.S. : 1977. *J. Bacteriol.* **129**, 1554–1565.
41. Lewis, O.A.M. and Probyn, T.A. : 1978. *New Phytol.* **81**, 519–526.
42. Meeks, J.A., Wolk, C.P., Schilling, N., Shaffer, P.W., Avissar, Y. and Chien, W.S. : 1978. *Plant Physiol.* **61**, 980–983.
43. Keys, A.J., Bird, I.F., Cornelius, M.J., Lea, P.J., Wallsgrove, R.M. and Miflin, B.J. : 1978. *Nature* **275**, 741–743.
44. Rhodes, D.A., Sims, A.P. and Stewart, G.R. : 1979. *In* Nitrogen Assimilation in Plants (Hewitt, E.D. and Cutting, C.V., eds.), pp.501–520, Academic Press, New York.
45. Miflin, B.J. and Lea, P.J. : 1980. *In* Amino Acids and Their Derivatives (Miflin, B.J., ed.), Vol. 5 of 'The Biochemistry of Plants', pp.169–202, Academic Press, New York.
46. Rathnam, C.K.M. and Edwards, G.E. : 1976. *Plant Cell Physiol.* **17**, 177–186.
47. Wallsgrove, R.M., Lea, P.J. and Miflin, B.J. : 1979. *Plant Physiol.* **63**, 232–236.
48. Canvin, D.T. and Atkins, C.A. : 1974. *Planta* **116**, 207–224.
49. Ito, O., Yoneyama, T. and Kumazawa, K. : 1978. *Plant Cell Physiol.* **19**, 1109–1119.
50. Miflin, B.J., Marker, A.F. and Whittingham, C.P. : 1966. *Biochim. Biophys. Acta* **120**, 266–273.
51. Kumarasinghe, K.S., Keys, A.J. and Whittingham, C.P. : 1977. *J. Exp. Bot.* **28**, 1163–1166.
52. Kumarasinghe, K.S., Keys, A.J. and Whittingham, C.P. : 1977. *J. Exp. Bot.* **28**, 1247–1257.
53. Woo, K.C., Berry, J.A. and Turner, G.L. : 1978. *Carnegie Inst. Yearbook* **77**, 240–245.
54. Wallsgrove, R.M., Keys, A.J., Bird, I.F., Cornelius, M.J., Lea, P.J. and Miflin, B.J. : 1980. *J. Exp. Bot.* **31**, 1005–1017.
55. Somerville, C.R. and Ogren, W.L. : 1980. *Nature* **286**, 257–259.
56. Lea, P.J. and Fowden, K. : 1975. *Proc. Roy. Soc. Lond. Ser. B.* **192**, 13 26.
57. Lea, P.J., Fowden, L. and Miflin, B.J. : 1978. *Phytochemistry* **17**, 217–222.
58. Sodek, L., Lea, P.J. and Miflin, B.J. : 1980. *Plant Physiol.* **65**, 22–26.
59. Lloyd, N.D.H. and Joy, K.W. : 1978. *Biochem. Biophys. Res. Commun.* **81**, 186–192.
60. Ireland, R.J. and Joy, K.W. : 1981. *Planta* **151**, 289–292.
61. Lea, P.J. and Miflin, B.J. : 1980. *In* Amino Acids and Their Derivatives (Miflin, B.J., ed.), Vol. 5 of 'The Biochemistry of Plants', pp.569–607, Academic Press, New York.
62. Miflin, B.J., Lea, P.J. and Wallsgrove, R.M. : 1980. *In* Glutamine : Metabolism, Enzymology and Regulation (Mora, J. and Palacios, R., eds.), pp.213–234, Academic Press, New York.
63. Miflin, B.J. : 1977. *In* Regulation of Enzyme Synthesis and Activity in Higher Plants (Smith, H., ed.), pp.23–40, Academic Press, London.
64. Tischner, R. and Hutterman, A. : 1980. *Plant Physiol.* **66**, 805–808.
65. Cullimore, J.V. : 1981. *Planta* **152**, 587–591.
66. Robertson, J.G., Warburton, M.P. and Farnden, K.J.F. : 1975. *FEBS Lett.* **55**, 33–37.
67. Werner, D., Morschel, E., Stripf, R. and Winchenbach, B. : 1980. *Planta* **147**, 320–329.
68. Awonaike, K.O., Lea, P.J. and Miflin, B.J. : 1981. *Plant Sci. Lett.* **23**, 189–195.
69. Cullimore, J.V., Lea, P.J. and Miflin, B.J. : 1982. *Israel J. Bot.*, in press.
70. Lea, P.J., Awonaike, K.O., Cullimore, J.V. and Miflin, B.J. : 1982. *Israel J. Bot.*, in press.

71. Yamada, Y. and Ikeda, M. : 1980. *In* Carbon-nitrogen Interaction in Crop Production, pp.41–52, Jap. Soc. for Promotion of Science, Tokyo.
72. Emes, M.J. and Erismann, K.H. : 1982. *Plant Sci. Lett.* 27, 103–109.
73. Leech, R.M. and Kirk, P.R. : 1968. *Biochem. Biophys. Res. Commun.* 32, 685–690.
74. Givan, C.V., Givan, A. and Leech, R.M. : 1970. *Plant Physiol.* 45, 624–630.

Protein Synthesis in Chloroplasts

Carl A. Price, M. Elizabeth Miller
and Ellen M. Reardon

It has been known for many years that the development of normal chloroplasts requires the interaction of both nuclear and plastid genomes (*cf. 1* for review). Information on the location of the genes and the sites of synthesis for specific chloroplast proteins has come more slowly. We shall begin by outlining the various strategies that have been employed to obtain this information. We shall then summarize the information obtained to date and finally report briefly on the regulation of plastid gene expression.

IDENTIFICATION OF PLASTID TRANSLATION PRODUCTS

The earliest strategy for distinguishing between translation products of the cytoplasm and those of chloroplasts (and mitochondria) was the use of selective inhibitors (*cf. 2–5*). It was known that translation on 80S types of ribosomes, located in the cytoplasm, are sensitive to the antibiotic cycloheximide, whereas translation on the 70S types of ribosomes found in plastids (and mitochondria) are sensitive to D-*threo*-chloramphenicol, lincomycin, and certain other antibiotics. This selectivity corresponds to the prokaryotic-like character of the organellar ribosomes. Cross sensitivity under appropriate conditions is less than 5%.

    The strategy has been that separate batches of plant cells or tissues are treated *in vivo* with cycloheximide (CHI) or chloramphenicol (CAP) for an interval, and functional or compositional changes in the plastids (or mitochondria) determined. If the synthesis of a protein was inhibited by CHI and not inhibited by CAP, it is inferred that the protein was synthesized in the cytoplasm.

107

The procedure yields data that is abundant, but often ambiguous: the selectivity of the two classes of inhibitors *in vivo* may be less distinct than that *in vitro*: both classes of inhibitors may affect the same protein or function; or the synthesis of one protein may be stringently linked to that of another. We see a probable example of stringency in an effect of CHI on the synthesis of the large subunit (LS) of ribulose bisphosphate carboxylase (RuBPC) (see Fig. 6 below). Without an independent determination of the effect of a "selective inhibitor", data obtained with inhibitors *in vivo* can only be suggestive.

The most direct means of determining the site of synthesis of a protein is by the incorporation of labelled amino acids by isolated chloroplasts. Although attempted by others, Blair & Ellis (6) were the first to use this procedure successfully. The key to their success was in employing methods for chloroplast isolation in which full photosynthetic activity was retained and in the use of high-resolution SDS-gel electrophoresis for the separation of specific polypeptides. Ellis (*cf.* 1) subsequently detected over one hundred translation products in the stroma alone, but it is not known how many of these are discrete; *i.e.*, some of the translation products might be precursors of others or artifacts of premature termination. The Morgenthalers in this laboratory added the tactic of employing chloroplasts that were rigorously puri-

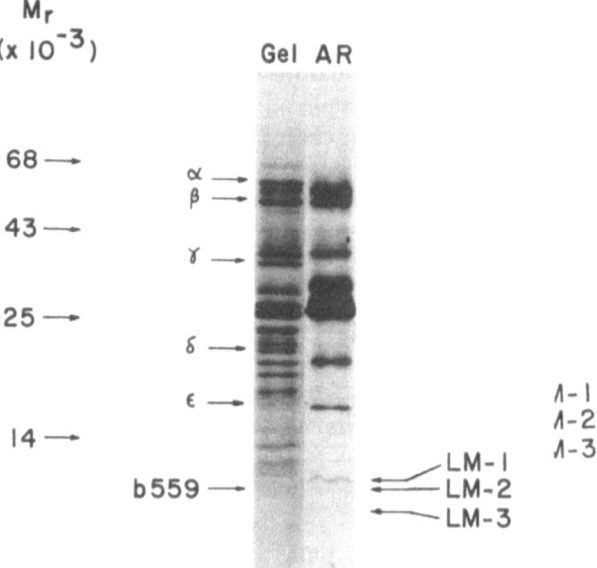

Fig. 1. Synthesis of thylakoid proteins by isolated chloroplasts of spinach (*11*). Chloroplasts were isolated on silica gradients (*7,12*) and incubated in the light with [^{35}S]-Met. Thylakoids were obtained from lysed chloroplasts by sedimentation through step gradients of sucrose and subjected to SDS-polyacrylamide gel electrophoresis. The stained gel ("gel" in the figure was autoradiographed ("AR") to visualize the components synthesized *in organello*. The symbols α, β . . . refer to the subunits of CF_1; LM-1, . . . refer to low-molecular-weight compounds.

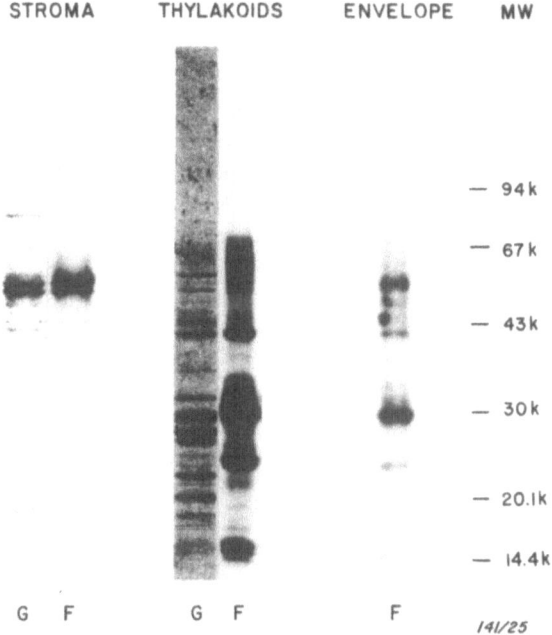

Fig. 2. Synthesis of soluble and membrane proteins by isolated chloroplasts of *Euglena* (*13*). Chloroplasts were isolated on silica gradients (*12,14*) and incubated in the light with [^{35}S]-Met. The chloroplasts were fractionated into stroma, thylakoids, and envelopes by sedimentation on step gradients of sucrose, and subjected to SDS-polyacrylamide gel electrophoresis. Products of chloroplast translation are visualized by fluorography. The stained gels (G) and fluorographs (F) show the distribution of polypeptides and translation products in the three fractions. The envelope fraction is clearly contaminated with thylakoids.

fied by density gradient centrifugation (*7–9*). This strategy minimizes possible contributions to the translation products from cytoplasmic ribosomes that contaminate crude suspensions of chloroplasts (*10*). Examples of translation *in organello* are shown in Figs. 1 and 2. Although a number of translation products do not coincide with stainable polypeptides, because they are synthesized as precursors, are present in too small quantities to be stained, or are artifacts, *etc.*, we estimate that about one-third of the thylakoid polypeptides are synthesized in the plastid.

A third line of evidence to detect plastid translation products is to isolate mRNA's from plastids, translate the mRNA's *in vitro* (*cf.* Fig. 4), and identify the labelled polypeptides, usually by immunoprecipitation. Some means of positive identification is essential, since a certain (but unknown) fraction of the translation products in the plastid are formed as precursors of higher molecular weight. One can not, therefore, rely solely on electrophoretic mobility to identify mature proteins.

It is conceivable that an mRNA might move across the chloroplast envelope. Although there have been claims for such transport (*15*), the evidence is internally inconsistent and difficult to reconcile with others in the field (*cf. 1*).

IDENTIFICATION OF GENES FOR PLASTID PROTEINS

In all cases for which there is hard evidence, a polypeptide which is translated in plastids is also coded on plastid DNA. Similarly, as noted above, no cases are known where mRNA's are produced in the nucleus and transported to plastids or the reverse. Evidence that a polypeptide is synthesized in plastids has become, therefore, strongly suggestive of a plastid location for the corresponding gene.

The earliest line of evidence for specific plastidic genes was a maternal (or uni-parental) pattern of inheritance of a structural mutant of a plastid protein. This was adduced for the LS of RuBPC by Hirai & Wildman (16) and for a plastid ribosomal protein (cf.17).

A second method is to employ cloned fragments of plastid DNA for hybrid-selected or hybrid-arrested translation of plastid RNA's. This method is generally applicable, provided that one has a means of identifying the *in vitro* translation product. It has been used successfully to map genes for the LS of RuBPC (cf. 18), P_{700}-chl *a* protein, and cytochromes *f* and b_6 (19).

The most recent strategy is to probe plastid DNA with DNA fragments containing bacterial genes thought to be functionally similar. This procedure has proven successful with the genes for the β and β' subunits of RNA polymerase, the elongation factor Tu, and a ribosomal protein (20), initiation factors, and the α, β and ϵ subunits for chloroplast coupling factor 1 (CF_1) (cf. 19). The location of genes by heterologous hybridization under controlled stringency is proving to be extremely powerful in allowing the extension of findings from one genome to other, widely separated ones, such as among eubacteria, cyanobacteria, algae from different phyla, and higher plants.

CURRENT UNDERSTANDING OF THE PLASTID GENOME

Table I is a compilation of the protein genes or translation products of plastids which have been identified thus far*. More than 15 polypeptides have been identified as specific translation products. Of these, RuBPC and ATP synthase are *hybrid oligomers* in which some of the subunits are synthesized by the plastid and the remainder by the cytoplasm (cf. 1,30).

Including the genes for ribosomal RNA's and transfer RNA's (cf. 31), about 22% of the plastid DNA of spinach has been mapped. Herrmann *et al.* (19) estimates that about half the proteins of the thylakoids are coded by plastid genes. The remainder are imported as precursors (cf. 32).

---

* This review was completed in September 1982.

TABLE I
Proteins Synthesized in Plastids

Protein of polypeptide	Translation product	Gene mapped
LS of RuBPC	Pea (6)	Chlamydomonas (27)
Coupling factor subunits		
α	Spinach (9)	Spinach (28)
β	Spinach (9)	Spinach (28)
ϵ	Spinach (9)	Spinach (28)
I	Spinach (22)	Spinach (28)
III	Spinach (22)	Spinach (28)
Elongation factors		
G	Spinach (23)	–
Tu	Spinach (23)	Spinach (19)
Cytochrome b_{559}	Spinach (11)	
f	Spinach (24)	Spinach (19)
b_{563}	Spinach (19)	Spinach (19)
tmpA (= 32 kd protein)	Pea (25)	maize (18)
P_{700}-chl a protein	Spinach (11,25,29)	Spinach (11)
Rieske protein	Spinach (19)	Spinach (19)
Ribosomal proteins	Spinach (26)	Chlamydomonas (17)

The species are listed in which the protein was first identified as a plastid translation product or the gene mapped.

REGULATION OF PROTEIN SYNTHESIS IN PLASTIDS

There are three clear-cut conditions where the synthesis of plastid proteins exhibits dramatic regulation: Light-induced development, the differentiation of C_4 plastids into mesophyll and bundle sheath cells; and the differentiation of plastids into amyloplasts, chromoplasts, and other non-green plastids.

We have examined protein synthesis in organello during light-induced plastid development in Euglena (33; Miller et al., in preparation). We find that translation rates of plastids increase sharply upon exposure of dark-grown cells to light (Fig.3). Proplastids incorporate [^{35}S]-methionine at very low rates, but most of the principal translation products of the proplastid are identical with the plastids of fully greened cells. Plastids isolated from cells exposed to one to three hours of light synthesize a set of novel polypeptides of high molecular weight which are not detectable at earlier or later stages of development.

We have also examined the in vitro translation of RNA's from plastids during the same interval of development (Fig.4). It appears that the rabbit reticulocyte lysate translation system responds to some but not all plastid messages. In particular, it fails to synthesize any polypeptides corresponding to the putative photogene product (ca. 30 kd in Euglena). RNA's from proplastids are almost completely inactive in directing protein synthesis in that system. Messages for other polypeptides appear

Time (hrs) 24 12 9 6 3 I 0 24 24 12   M$_r$

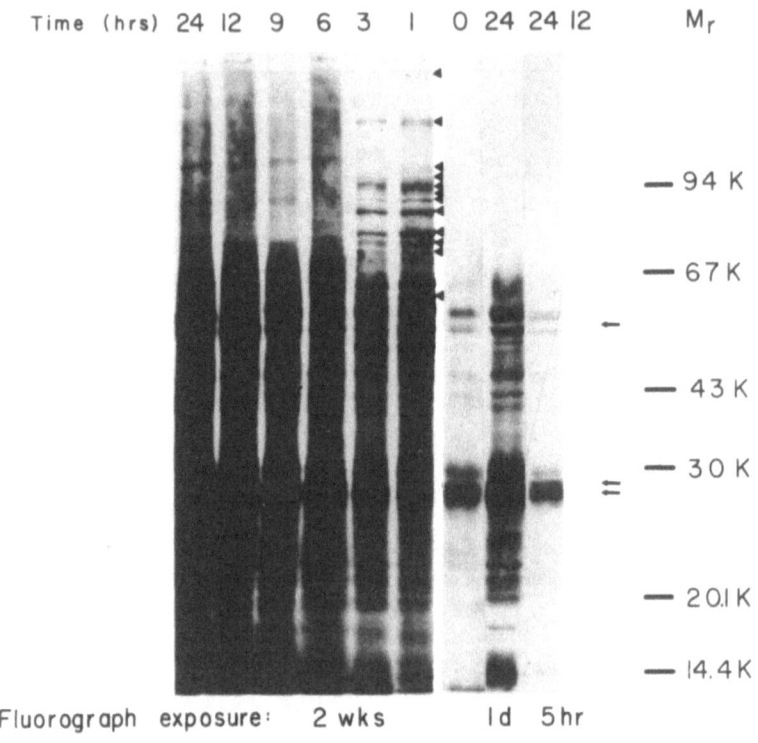

Fluorograph exposure: 2 wks   I d   5 hr

Fig. 3. Translation *in organello* during light-induced development of plastids in *Euglena* (*33*). Plastids were isolated from dark-grown *Euglena gracilis* at the indicated intervals after exposure to light. They were then incubated with [^{35}S]-Met and analyzed as described in Fig. 2. "N" refers to control cells grown continuously in the light. The data show quantitative and qualitative regulation of plastid translation during chloroplast development.

at various stages of light induction. We conclude therefore that a number of plastid genes are regulated transcriptionally.

In C$_4$ plants, chloroplasts of bundle sheath cells contain relatively normal amounts of RuBPC, whereas chloroplasts from mesophyll cells are essentially devoid of the enzyme. Bedbrook *et al.* (*34*) have shown by Southern blots that the mRNA's for the LS of RuBPC are present in bundle sheath cells and absent in mesophyll cells. In this case, also, regulation occurs at the level of transcription.

There is almost nothing known about the molecular aspects of the differentiation of plastids into organelles other than chloroplasts. The sole published information is that amyloplasts of potato contain the same plastid DNA as that in chloroplasts (*35*). No information is available on gene expression in amyloplasts nor in any of the other differentiation states of the plastid.

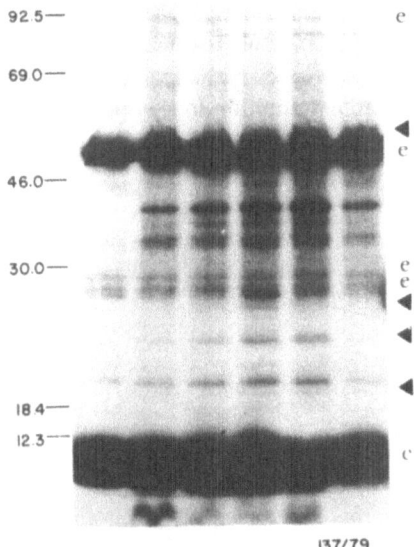

Fig. 4. Translation *in vitro* during light-induced development in *Euglena* (Miller *et al.*, in preparation). Total RNA was isolated from plastids similar to those described in Fig. 3. The RNA was then used to direct protein synthesis *in vitro* by the rabbit reticulocyte system. The increased mRNA activity during light-induced development indicates quantitative and qualitative regulation at the level of transcription.

INTERACTIONS BETWEEN THE NUCLEUS AND PLASTID

There can be no doubt that the plastid and the nucleocytoplasmic systems communicate with one another. The only known informational molecules that traverse the chloroplast envelope are the various precursors of plastid proteins that are synthesized in the cytoplasm. The question of how the nucleus controls the plastid has been asked repeatedly without substantial result.

Among the first products of chloroplast translation to be recognized were the LS of RuBPC and the α, β, and ε subunits of CF1. RuBPC and CF1 were thus seen to be hybrid oligomers, and it was proposed (*cf. 9,36*) that the cytoplasmically derived subunits might somehow regulate transcription or translation of the corresponding oligomers that are synthesized in the chloroplast. There remains no specific evidence for this possibility and, since no further examples of hybrid oligomers have been identified among plastid translation products, this supposed mode of regulation is probably not applicable generally.

In addition to supplying precursors to plastid structures, the nucleocytoplasmic system may also contribute to plastid development by providing one or more enzymes for proteolytic processing of precursor proteins. Treatment of *Euglena* cells with cycloheximide prior to isolation of the plastids, for example, results in the synthesis of some chloroplast proteins that appear to have higher molecular weight than in controls (Fig.5). A phenotypically similar result is caused by a nuclear mutation in maize (*cf.37*).

Cycloheximide also causes a strong inhibition in the subsequent synthesis of LS

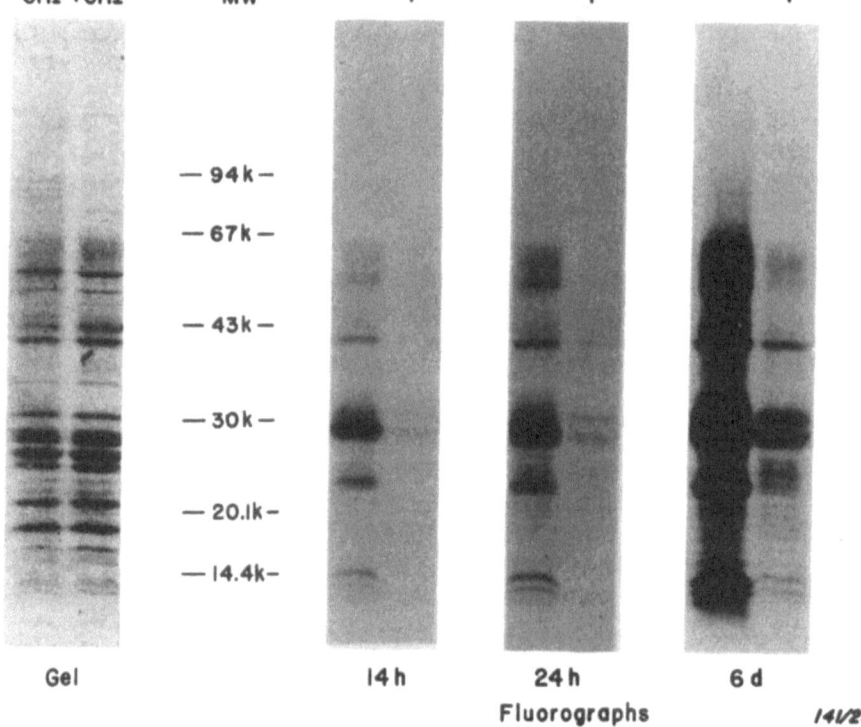

Fig. 5. Synthesis of thylakoid protein *in organello* by chloroplasts from *Euglena* treated with cycloheximide (*13*). Chloroplasts were isolated from *Euglena gracilis* which had been treated for 4 hr prior to harvest with cycloheximide (+CHI) and allowed to incorporate [^{35}S]-Met. Fluorographs (right) of the thylakoid polypeptides (left) were exposed to X-ray film for varying intervals. Prior treatment of cells with cycloheximide strongly inhibits subsequent protein synthesis by the isolated chloroplasts. In addition to the quantitative effects, there are qualitative changes. In several instances, notably in the pair of polypeptides around 30 kd, there is a relative accumulation of polypeptides of higher molecular weight with a concommitant reduction in the more rapidly migrating polypeptide. We interpret this phenomenon as an accumulation of precursors.

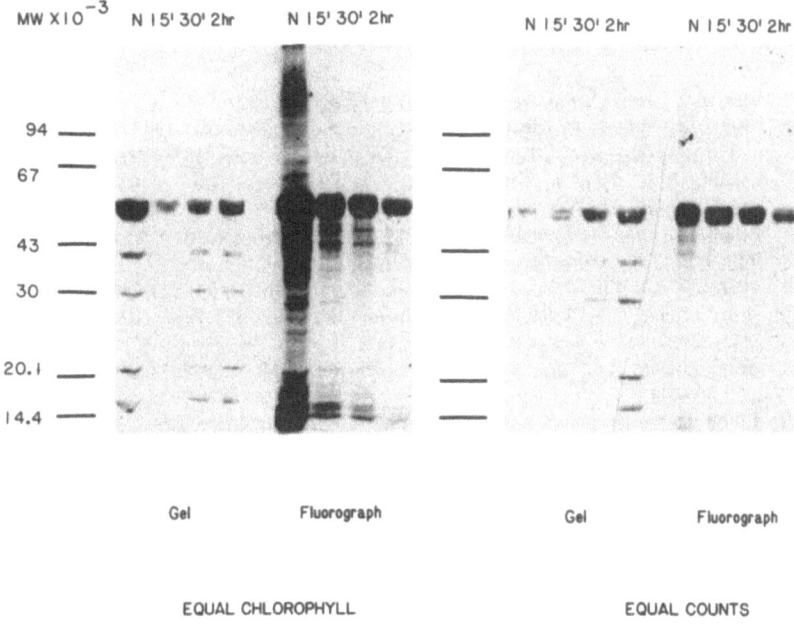

Fig. 6. Synthesis of stromal proteins *in organello* by chloroplasts from *Euglena* treated with cycloheximide *(13)*. The experiment is similar to that described in Fig.5, except that the interval of exposure to CHI was varied and the stromal proteins were analyzed. In this case, there are large and early effect of CHI inhibition on the subsequent synthesis of stromal proteins, especially of the LS of RuBPC. No qualitative effects were seen among stromal translation products.

in chloroplasts isolated from treated cells (Fig.6). We interpret this as indicating a stringent mode of regulation by a yet unknown mechanism.

CONCLUSION

The plastid genome codes for a significant fraction of the polypeptides that constitute the mature chloroplast. About 15 polypeptides have been identified as specific plastid translation products and the genes for most of these have been mapped on the plastid genome. Quantitative and qualitative regulation of gene expression in the plastid have been demonstrated and appear to be vital to the normal development of the organelle.

Acknowledgement
    The research reported here was supported in part by grants from the United States Department of Agriculture, the National Science Foundation, and the Charles and Johanna Busch Memorial Fund.

116   C.A. PRICE *ET AL.*

REFERENCES

1. Ellis, R.J. : 1981. *Annu. Rev. Plant Physiol.* **32**, 111–137.
2. Ciferri, O., Tiboni, O., Munoz-Calvo, M.L. and Camerino, G. : 1977. *In* Acides Nucleiques et Synthese (Bogorad, L. and Weil, J.H., eds.), pp.155–166, NRS, Paris.
3. Criddle, R.S., Dan, B., Kleinkopf, G.E. and Huffaker, R.C. : 1970. *Biochem. Biophys. Res. Commun.* **41**, 621–627.
4. Iwanij, V., Chua, N.-H. and Siekewitz, P. : 1975. *J. Cell Biol.* **64**, 572–585.
5. Ellis, R.J. : 1977. *Biochim. Biophys. Acta* **463**, 185–215.
6. Blair, E.J. and Ellis, R.J. : 1973. *Biochim. Biophys. Acta* **319**, 223–234.
7. Morgenthaler, J.-J., Price, C.A., Robinson, J.M. and Gibbs, M. : 1974. *Plant Physiol.* **54**, 532–534.
8. Morgenthaler, J.-J. and Mendiola-Morgenthaler, L.M. : 1976. *Arch. Biochem. Biophys.* **172**, 51–58.
9. Mendiola-Morgenthaler, L.M., Morgenthaler, J.-J. and Price, C.A. : 1976. *FEBS Lett.* **62**, 96–100.
10. Bartolf, M. and Price, C.A. : 1979. *Biochemistry* **18**, 1677–1680.
11. Zielinski, R.E. and Price, C.A. : 1980. *J. Cell Biol.* **85**, 435–445.
12. Price, C.A. and Reardon, E.M. : 1982. *In* Methods in Chloroplast Molecular Biology (Edelman, M., Hallick, R. and Chua, N.H., eds.), pp.189–209, Elsevier Biomedical.
13. Reardon, E.M. : 1982. Ph.D. Thesis, Rutgers University.
14. Ortiz, W., Reardon, E.M. and Price, C.A. : 1980. *Plant Physiol.* **66**, 291–294.
15. Geetha, V. and Gnanam, A. : 1980. *J. Biol. Chem.* **255**, 492–497.
16. Hirai, A. and Wildman, S.G. : 1977. *Biochim. Biophys. Acta* **479**, 39–52.
17. Bogorad, L. : 1977. *In* International Cell Biology 1976–1977 (Brinkley, B.R. and Porter, K.R., eds.), pp.175–182, Rockefeller Univ. Press.
18. Bogorad, L. : 1981. *J. Cell Biol.* **91**, 2565–2705.
19. Herrmann, R.G., Alt, J., Bisanz, C., Hauska, G., Nelson, N., Westhoff, P. and Winter, P. : 1983. *In* Structure and Function of Plant Genomes (Ciferri, O. and Dure, L., eds.), Pergamon, in press.
20. Watson, J.C. and Surzycki, S.J. : 1982. *Proc. Natl. Acad. Sci. U.S.A.* **79**, 2264–2267.
21. Morgenthaler, J.-J. and Mendiola-Morgenthaler, L.R. : 1976. *Arch. Biochem. Biophys.* **172**, 51–58.
22. Nelson, N., Nelson, H. and Schatz, G. : 1980. *Proc. Natl. Acad. Sci. U.S.A.* **77**, 1361–1364.
23. Ciferri, O., Di Pasquale, G. and Tiboni, O. : 1980. *Eur. J. Biochem.* **102**, 331–335.
24. Doherty, A. and Gray, J.C. : 1979. *Eur. J. Biochem.* **98**, 89–92.
25. Eaglesham, A.R.J. and Ellis, R.J. : 1974. *Biochim. Biophys. Acta* **335**, 396–407.
26. Mache, R., Dorne, A.M., Lescure, A.M. and Eneas-Filho, J. : 1983. *In* Structure and Function of Plant Genomes (Ciferri, O. and Dure, L., eds.), Pergamon, in press.
27. Rochaix, J.-D. and Malnoe, P. : 1978. *In* Chloroplast Development (Akoyunoglou, G. and Argyroudi-Akoyunoglou, J.H., eds.), pp.581–586, Elsevier/North Holland, Amsterdam.
28. Westhoff, P., Nelson, N., Buenemann, H. and Herrmann, R.G. : 1981. *Curr. Genet.* **4**, 109–120.
29. Cederblad, A.V. and Vasconcelos, A.C. : 1978. *Plant Physiol.* **61** (Supple) , 83.
30. Watanabe, A. and Price, C.A. : 1982. *Proc. Natl. Acad. Sci. U.S.A.* **79**, 6304–6308.
31. Weil, J., Mubumbila, M., Kuntz, M., Keller, M., Steinmetz, A., Crouse, E.J., Burkard, G., Guillemaut, P., Selden, R., McIntosh, L., Bogorad, L., Loeffelhardt, W., Mucke, H., Bohnert, H.J., Dietrich, A., Souciet, G., Colas, B., Imbault, P. and Sarantoglou, V. : 1983. *In* Structure and Function of Plant Genomes (Ciferri, O. and Dure, L., eds.), Pergamon, in press.
32. Grossman, A.R., Bartlett, S.G., Schmidt, G.W., Mullet, J.E. and Chua, N.-H. : 1982. *J. Biol. Chem.* **257**, 1558–1563.

33. Miller, M.E. and Price, C.A. : 1982. *FEBS Lett.* **147**, 156–160.
34. Bedbrook, J.R., Link, G., Coen, D.M., Bogorad, L. and Rich, A. : 1978. *Proc. Natl. Acad. Sci. U.S.A.* **75**, 3060–3064.
35. Karimov, M., Nasyrov, Yu. S. and Vinetskii, Yu. P. : 1979. *Dokl. Akad. Nauk. Tadzh. SSR* **22**, 204–207.
36. Ellis, R.J. : 1975. *Phytochemistry* **14**, 89–93.
37. Leto, K.J. and Miles, D. : 1980. *Plant Physiol.* **66**, 18–24.

Biogenesis of Mitochondria in Higher Plant Cells

Tadashi Asahi and Masayoshi Maeshima

INTRODUCTION

Various metabolic changes are induced in higher plant tissues in response to changes in physiological and environmental conditions. In many cases, such changes are closely related to induction of biogenesis, development, transformation, degradation or deterioration of cell organelles. For instance, in greening leaves, chloroplasts develop from either proplastids or etioplasts, which causes the development of photosynthetic activity. Leaf peroxisomes are also formed in greening leaves, and their biogenesis is responsible for the appearance of photorespiration (See 4 for a review). In the storage tissues of fatty seeds, glyoxysomes develop rapidly during germination undergoing degradation in later stages. The development and degradation of glyoxysomes are causal to the induction and loss of the ability to synthesize sugars from stored fat, respectively (see 4 and 38 for reviews). In many fruits, the chloroplasts are transformed into chromoplasts during maturation, which runs parallel to the decrease in photosynthetic activity and the increase in carotene content (see 11 for a review). Higher plant cell organelles have the characteristic ability to induce such dramatic, dynamic changes when the cells undergo physiological changes or receive stress.

Higher plant mitochondria are also susceptible to changes in the physiological and environmental conditions of cells. When dry seeds imbibe, active development of mitochondria is observed in the storage tissues (25). The development includes increases in mitochondrial membrane protein, respiratory capacity, and various enzyme activities (25). In this case, there seems to be no increase in the number of mitochondria in spite of the active formation of the inner membrane. After seeds

119

have fully imbibed, the number of mitochondria in the storage tissues seems to increase, that is, mitochondrial biogenesis occurs (5). Active formation of the mitochondrial inner membrane without any accompanying increase in the number of
mitochondria also occurs in elongating root cells (37). When higher plant tissues are
mechanically injured or infected with pathogens, active formation of the mitochondrial inner membrane, probably together with an increase in the number of mitochondria, is induced in the tissue parts adjacent to the injured or infected ones (1,
42). In addition, mitochondrial deterioration has been observed in higher plant
tissues during senescence or chilling (22,27,45). Such biogenesis, development,
degradation or deterioration of mitochondria seems to contribute to changes in the
respiratory activity of cells. It should be noted that such dynamic changes in mitochondria take place in non-dividing cells. In organisms other than higher plants, such
dynamic changes in mitochondria have rarely been observed except for yeast, in
which the mitochondria develop or are degraded in response to changes in the concentrations of oxygen and glucose.

A question arises as to why higher plant mitochondria have the ability to conduct
such dynamic changes in response to alterations in the physiological and environmental conditions of cells. In order to answer this question, the mechanisms of the
dynamic changes should first be elucidated. Our group has investigated the mechanisms of mitochondrial biogenesis in wounded tissues and of mitochondrial development in imbibing seeds. In recent years, great progress has been made in studies on
the mechanism of mitochondrial biogenesis with microorganisms and animal cells
(see 41 for a review). We have compared the mechanism of the synthesis of mitochondrial membrane enzymes in such higher plant tissues with that in other organisms to find whether there are similar characteristics.

Only a few reports are available of detailed biochemical studies on the biogenesis
of mitochondria in higher plant cells (see 12 for a review), though much information
has accumulated on the biochemical mechanism of chloroplast biogenesis. We expect
that our studies will not only elucidate some characteristics of the synthesis of mitochondrial enzymes in wounded higher plant tissues and imbibing seeds but will also
provide new information about its general mechanism.

MITOCHONDRIAL BIOGENESIS IN WOUNDED SWEET POTATO ROOT TISSUE

When slices of sweet potato root tissue are incubated in air at moderate temperatures
under moist conditions, active biogenesis of mitochondria takes place without accompanying cell division: namely, the number of mitochondria, the content of the
inner membrane, and the activities of mitochondrial respiration and enzymes per cell
increase (1,42). The injury-induced mitochondrial biogenesis is stimulated when
tissue slices are incubated either in ethylene-containing air or after inoculation with
fungi (18,42). Our recent attention has been directed to the mechanisms of increase
in the activities of mitochondrial membrane enzymes such as cytochrome c oxidase
and succinate dehydrogenase.

The Mechanism of Increase in Cytochrome c Oxidase Activity

Cytochrome c oxidase activity increases after a lag period lasting about 12 hr to attain a maximum (about twice the original activity) after 2 days when slices of sweet potato root tissue are incubated at 30°C in air. The increase in the activity is inhibited when tissue slices are incubated in a D-*threo*-chloramphenicol solution for 1 hr before incubation in air (*16*). Interestingly, the increase is not inhibited when tissue slices are treated with a cycloheximide solution in a similar way (Fig. 1). It is well known that cytochrome c oxidase from animals and fungi is composed of seven to eight subunits, of which the largest three are synthesized on the mitochondrial ribosomes under the control of mitochondrial genes and the others are on the cytoplasmic ribosomes under the control of nuclear genes (*41*). Thus, the lack of inhibition with cycloheximide leads us to postulate the following two possibilities.

(i) Sweet potato cytochrome c oxidase consists of only the subunits synthesized on the mitochondrial ribosomes.

(ii) There is a precursor composed of the cytoplasmic origin of cytochrome c oxidase subunits in intact tissue, and the precursor is assembled with the

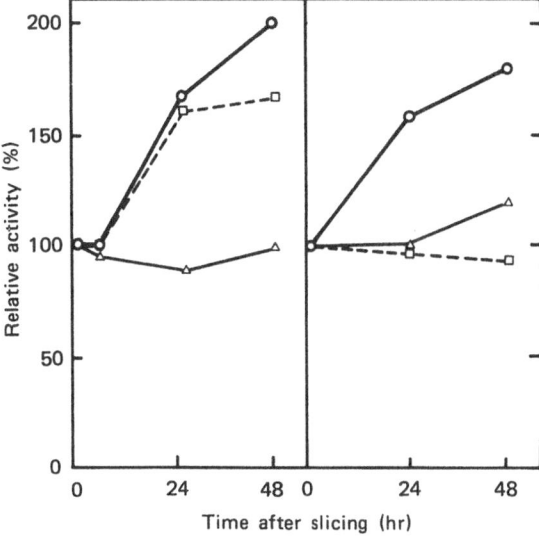

Fig. 1. Effects of D-*threo*-chloramphenicol and cycloheximide on the increase in cytochrome c oxidase and succinate dehydrogenase activities in wounded sweet potato root tissue (a combined figure of Fig. 1 in ref. *16* and Fig. 1 in ref. *7*). Tissue slices (3 mm thick and 19 mm in diameter) were incubated in 1mM phosphate buffer, pH 6.5, containing 6×10^{-3} M D-*threo*-chloramphenicol or 10^{-5} M cycloheximide at room temperature with continuous stirring for 1 hr, then incubated in air at 30°C for the times indicated. Controls were run with buffer containing no inhibitors. Left, cytochrome c oxidase activity; right, succinate dehydrogenase activity. ○, control; △, D-*threo*-chloramphenicol; ⊓, cycloheximide.

subunits that are newly synthesized on the mitochondrial ribosomes in wounded tissue.

The first possibility is improbable because purified sweet potato cytochrome c oxidase is composed of at least five subunits with different molecular weights : 39,000 (subunit I), 33,500 (subunit II), 26,000 (subunit III), 20,000 (subunit IV), and 5,700 (subunit V) (cf. Fig. 2) (15). When tissue slices are supplied with ^{14}C-leucine, the largest three (subunits I, II and III) of five cytochrome c oxidase sub-units, but not the others (subunits IV and V), are labeled (16). This supports the second possibility. An alternate possibility that the enzyme in intact tissue contains excess amounts of subunits IV and V so as to provide the newly formed subunits (I, II and III) with parts of this excess is excluded because of the following findings. Cytochrome c oxidase purified from wounded tissue is identical to that from intact tissue with respect to its sedimentation velocity and subunit stoichiometry (16). Similarly, there is no difference in the subunit stoichiometry between immuno-precipitates with anti-sweet potato cytochrome c oxidase antibody from Triton X-100 extracts[*1] of the submitochondrial particles from intact and wounded tissues.

We found that there was a protein precipitated with anti-cytochrome c oxidase antibody in a deoxycholate extract (cf. footnote* 1) of the submitochondrial particles from intact tissue, though the extract showed no cytochrome c oxidase activity (17). This protein, which we call X-protein, decreases during incubation of tissue slices, suggesting that it is a precursor of cytochrome c oxidase (17). It should be noted that X-protein is associated more weakly with the mitochondrial inner membrane than cytochrome c oxidase, because the former is solubilized by a low concentration of a detergent by which the latter remains associated with the membrane. We have suc-ceeded in purifying X-protein and preparing a specific antibody to this protein (un-published data). Our anti-sweet potato cytochrome c oxidase antibody reacts with subunits II, IV and V of the enzyme in electrophoretic, immunological blotting tests[*2] (Maeshima and Asahi, unpublished data). X-protein competes with purified sweet potato cytochrome c oxidase for precipitating with the anti-cytochrome c oxidase antibody (17) as tested by the double immunodiffusion method (29). Anti-X-protein antibody forms no precipitin line with purified sweet potato cytochrome c oxidase in double immunodiffusion tests but reacts with smaller subunits of the

---

[*1] The extracts were prepared in the following way. Submitochondrial particles were treated with 1 M KCl and 0.4 mg deoxycholate per mg protein, then centrifuged at $80,000 \times g$ for 1 hr. The pellet was treated with 0.1 M KCl and 1.5% Triton X-100, then centrifuged as described above. All the activity of cytochrome c oxidase in the submitochondrial particles was recovered in the supernatant form in the second centrifugation (Triton X-100 extract).

[*2] The method of Towbin et al. (39) was used. First, cytochrome c oxidase preparations under-went electrophoresis on slab polyacrylamide gels containing sodium dodecylsulfate (cf. 15 for details). Then, the polypeptides on the gels were transferred to nitrocellulose paper by electro-phoresis. The nitrocellulose paper was incubated in a solution of anti-cytochrome c oxidase anti-body, then in a fluoresceine-labeled goat anti-rabbit immunoglobulin G antibody. The blots on the paper were detected under ultraviolet light.

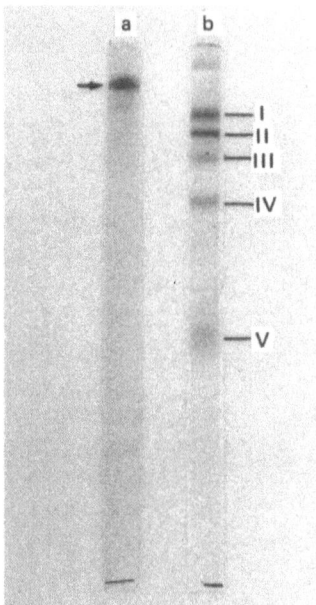

Fig. 2. Sodium dodecylsulfate-urea polyacrylamide gel electrophoresis of X-protein (a) and purified sweet potato cytochrome *c* oxidase (b) (Fig. 7 in ref. *17*). Proteins were dissociated into the subunits by treatment with 2% sodium dodecylsulfate, 2% β-mercaptoethanol, and 2 M urea; they then underwent electrophoresis on 10% polyacrylamide gels containing 0.1% sodium dodecylsulfate and 8 M urea following the method of Swank and Mundres (*36*). Gels were stained with Coomassie brilliant blue R.

enzyme in electrophoretic, immunological blotting tests (Maeshima and Asahi, unpublished data). All the results described above suggest that X-protein is a precursor composed of the smaller subunits (that is, the cytoplasmic origin of subunits) of cytochrome *c* oxidase. We therefore tested to see whether this was true. However, only one polypeptide with a high molecular weight (57,000) was detected after electrophoresis of X-protein on a sodium dodecylsulfate-urea polyacrylamide gel (Fig. 2).

Now we propose the model shown in Fig. 3 for the mechanism of the increase in cytochrome *c* oxidase in wounded sweet potato root tissue. There is a large pool of X-protein in the mitochondria in intact tissue. The protein consists of only one kind of polypeptide that has subunits IV and V in a single polypeptide chain and is synthesized on the cytoplasmic ribosomes under the control of a nuclear gene. It is associated very weakly with the mitochondrial inner membrane. When the tissue receives injury, X-protein is processed into subunits IV and V, which are assembled with subunits I, II and III newly synthesized on the mitochondrial ribosomes to form active cytochrome *c* oxidase protein. Poyton and McKemmie (*32,33*) proposed from their work with yeast that all the cytoplasmic origin of cytochrome *c* oxidase subunits are

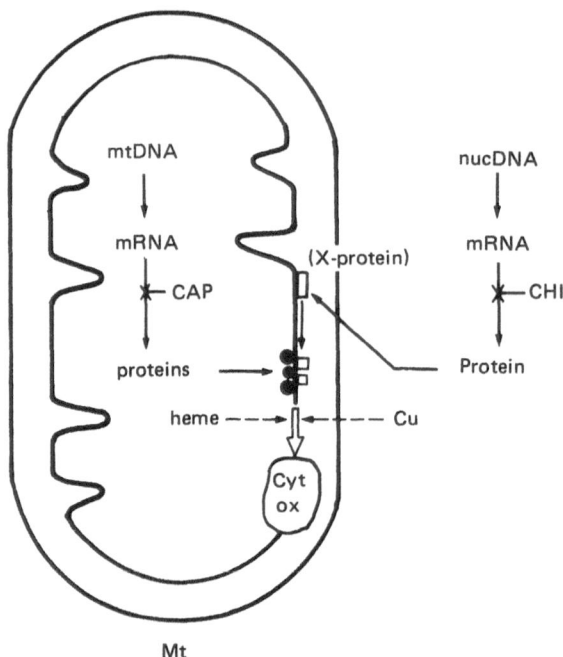

Fig. 3. Model for the mechanism of the increase in cytochrome *c* oxidase in wounded sweet potato root tissue. Abbreviations : CAP, D-*threo*-chloramphenicol; CHI, cyclo-heximide; Cyt ox, cytochrome *c* oxidase; Mt, mitochondrion; mt, mitochondrial; nuc, nuclear. See text for details.

synthesized as a single polypeptide chain, that is, as a polyprotein precursor. This proposal, however, has been proven wrong by many workers. In *in vitro* translation experiments, the cytoplasmic origin of the subunits are synthesized in separate poly-peptide chains with slightly higher molecular weights than the respective mature sub-units, so-called larger precursors, and their mRNAs are too small in size to code all the subunits in a single mRNA molecule (*13,21,28,35*). The antibody against yeast cytochrome *c* oxidase used by Poyton and McKemmie seemed not to be mono-specific to the enzyme. In our case, however, X-protein certainly resembles cyto-chrome *c* oxidase with respect to immunological properties. We propose that at least in higher plant cells, the cytoplasmic origin of cytochrome *c* oxidase subunits are synthesized as a polyprotein precursor, which ensures synthesizing and transporting into the mitochondria the subunits in a correct stoichiometric ratio. Alternatively, X-protein may be an unusual precursor of cytochrome *c* oxidase, that is, it may be of a particular storage form.

The Mechanism of Increase in Succinate Dehydrogenase Activity
　　　The activity of succinate dehydrogenase also increases during incubation of slices of sweet potato root tissue in a manner similar to the increase in cytochrome *c* oxidase

activity. The increase in succinate dehydrogenase activity is inhibited when tissue slices are treated with either D-*threo*-chloramphenicol or cycloheximide (Fig. 1). Succinate dehydrogenase has been shown to be synthesized on the cytoplasmic ribosomes under the control of nuclear genes in yeast (*10,30,43*). Sweet potato succinate dehydrogenase consists of two subunits (molecular weights : 65,000 and 26,000) (*6*) just as the enzyme from other sources. Using antibody against the large subunits of the sweet potato enzyme, the changes in activity have been shown to parallel those in the amount of the large subunit (*7*). Our group interprets the inhibition of the synthesis of the large subunit by both inhibitors to mean that a mitochondrially made protein controls either the synthesis or the transport into the mitochondria of succinate dehydrogenase subunits. The concept that the nucleocytoplasmic and mitochondrial genetic systems regulate each other during synthesis of mitochondrial proteins is now generally accepted, and many studies that show the regulation of protein synthesis on mitochondrial ribosomes by the nucleocytoplasmic genetic system have been made (for instance, references *14* and *40*). However, there is only one report that the synthesis of mitochondrial protein on cytoplasmic ribosomes is regulated by the mitochondrial genetic system; Poucholet and Shore (*31*) found that in tadpoles accumulation of carbamylphosphate synthetase induced with thyroxine was inhibited by D-*threo*-chloramphenicol. They showed that an increase in mRNA for the enzyme was depressed by the inhibitor. The work in our laboratory is the second to show the regulation of mitochondrial protein synthesis on cytoplasmic ribosomes by the mitochondrial genetic system, and further detailed studies on the synthesis of succinate dehydrogenase in wounded sweet potato root tissue are of great interest from the standpoint of the regulation of the synthesis of mitochondrial proteins.

MITOCHONDRIAL DEVELOPMENT IN IMBIBING PEA COTYLEDONS

Rapid development of mitochondria occurs in pea cotyledons in very early stages of germination. Cotyledon mitochondria in dry pea seeds are very fragile and most of them are destroyed during isolation, while those in fully hydrated seeds are rather stable (*25*). The inner membrane of the former contains high amounts of phospholipids relative to protein content and lacks some membrane proteins as compared with the latter (*34*). Mitochondria from dry pea cotyledons show almost no succinate oxidase activity (*25*). As seeds imbibe, cotyledon mitochondria rapidly become active in succinate oxidation, and their succinate oxidase activity attains a maximum after imbibition for about 18 hr (*25*). This increase in activity is accompanied by an increase in proteins and enzyme activities in the mitochondrial inner membrane (*25, 34*).

Figure 4 illustrates a model for the development of mitochondria in pea cotyledons during seed hydration proposed from the results described above together with electron microscopic pictures reported by Bain and Mercer (*2,3*). The mitochondria in dry pea cotyledons possess only a few cristae and slightly electron-dense matrix, and their inner membrane is very rich in phospholipids. As seeds imbibe, the

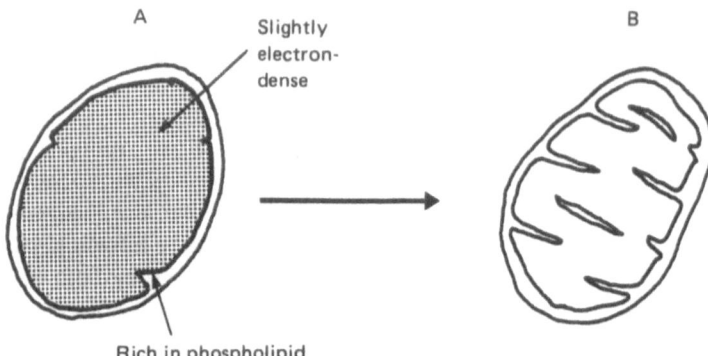

Fig. 4. Schematic illustration of mitochondrial development in imbibing pea cotyledons. A, mitochondrion in dry cotyledons; B, that in fully hydrated ones.

cristae develop, which results in the increase in respiratory activity of the mitochondria.

Interestingly, the mitochondrial development in imbibing pea cotyledons is also observed even when seeds are forced to imbibe either a D-*threo*-chloramphenicol or a cycloheximide solution instead of water (26). There is then a possibility that the development requires no *de novo* synthesis of mitochondrial proteins. In the matrix of cotyledon mitochondria in dry pea seeds, there may be precursors that await seed hydration to be assembled into the phospholipid-rich mitochondrial inner membrane. This proposal has been supported by the following detailed studies on the mechanisms of increase in succinate dehydrogenase and cytochrome *c* oxidase activities in imbibing pea cotyledons.

The Mechanism of Increase in Succinate Dehydrogenase Activity in Imbibing Pea Cotyledons

When pea seeds imbibe for 18 hr, there is about a three-fold increase in the activity of succinate dehydrogenase associated with the mitochondrial inner membrane (23). However, only a 1.5-fold activity increase is observed in crude extract of cotyledons during seed hydration for the same period (23). These increase in activity are inhibited by neither D-*threo*-chloramphenicol nor cycloheximide (23). In the case of dry pea cotyledons, succinate dehydrogenase activity in the crude extract is about twice that in the mitochondrial inner membrane fraction, and there is a soluble form of succinate dehydrogenase which disappears during seed hydration (23). The results suggest that the soluble form of succinate dehydrogenase in dry cotyledons, probably in the mitochondrial matrix, is assembled to the phospholipid-rich mitochondrial inner membrane during seed hydration accompanied by about a two-fold increase in enzyme activity.

Our group has partially purified the soluble form of succinate dehydrogenase in dry pea cotyledons as well as the enzyme associated with the mitochondrial inner membrane in fully hydrated ones (23,24). There are differences in the K_m value for succinate and the relative electrophoretic mobility, but not in the molecular weight, between the soluble form of the enzyme and the enzyme purified partially from the mitochondrial inner membrane of fully hydrated cotyledons in the presence of suc-

TABLE I
Comparison between Soluble and Membrane-bound Succinate Dehydrogenases from Pea Cotyledons

Form of succinate dehydrogenase	Succinate[a]	K_m for succinate (mM)	Molecular weight ($\times 10^3$)	Relative mobility[b]
Soluble[c]	−	4.65	92	0.61
Membrane-bound[d]	−	0.24	113	ND
Membrane-bound[d]	+	0.50	92	0.51

[a]Presence (+) or absence (−) of 20 mM succinate in all media used for partial purification.
[b]Relative mobility to bromophenol blue on 7% polyacrylamide gels.
[c]Soluble form of the enzyme partially purified from dry pea cotyledons.
[d]Enzyme partially purified from an Emulgen 810 extract of the mitochondrial inner membrane in the cotyledons of pea seeds imbibing for 1 day.
ND, not determined.

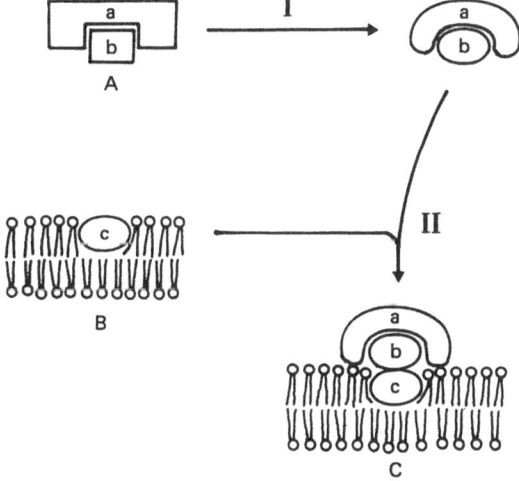

Fig. 5. Schematic illustration of the assembly of a soluble form of succinate dehydrogenase into the mitochondrial inner membrane in pea cotyledons during seed hydration. The form of the enzyme in the membrane is speculated on the basis of the report of Merli et al. (20). Process I, a conformational change; process II, assembly of the soluble form of the enzyme into the membrane. A, the soluble form of the enzyme; B, the phospholipid-rich mitochondrial inner membrane in dry cotyledons; C, the mitochondrial inner membrane in fully hydrated cotyledons. a, the large subunit of the enzyme; b, the small subunit of the enzyme; c, a polypeptide with a molecular weight of 21,000.

cinate (Table I). This suggests that the two forms of succinate dehydrogenase are of charge isomers with different conformations. When the membrane-bound enzyme is solubilized and partially purified in the absence of succinate, its molecular weight is about 113,000 (Table I). Probably the enzyme, which consists of two subunits, the large (65,000) and the small (27,000) ones, exists in a complex with polypeptide with a molecular weight of about 21,000 in the mitochondrial inner membrane. The polypeptide may participate in the association of succinate dehydrogenase with the membrane. Such a situation has been observed with succinate dehydrogenase in *Bacillus subtilis* and *Neurospora crassa* (*9,44*). In the case of the mammalian enzyme, the situation is quite different from that for the pea and microorganism enzyme, as the former exists in a large complex with at least seven polypeptides of different molecular weights (*8*).

Figure 5 schematically illustrates our proposed model for the assembly of the soluble form of succinate dehydrogenase into the mitochondrial inner membrane in pea cotyledons during seed hydration. The soluble form of the enzyme existing in the mitochondrial matrix of dry cotyledons takes a conformation with a low specific activity. When seeds imbibe, it undergoes a conformational change which facilitates its association with the 21,000-dalton polypeptide in the mitochondrial inner membrane and causes a roughly two-fold increase in the specific activity of the enzyme.

The Mechanism of Increase in Cytochrome c Oxidase Activity in Imbibing Pea Cotyledons

Cytochrome *c* oxidase activity in pea cotyledons increases about twice during seed hydration for 12 hr. In this case, even with dry cotyledons, the activity is detected only in the mitochondrial inner membrane fraction and not in the soluble fraction. The increase is also inhibited by neither D-*threo*-chloramphenicol nor cycloheximide, suggesting the presence in dry pea cotyledons of a precursor that waits seed hydration to become the active enzyme protein.

In order to obtain evidence of the presence of a cytochrome *c* oxidase precursor in dry pea cotyledons, attempts were first made to purify pea cytochrome *c* oxidase and to prepare its antibody (*19*). The enzyme purified from the shoots of etiolated pea seedlings consists of five subunits with different molecular weight : 39,000 (subunit I), 33,000 (subunit II), 28,500 (subunit III), 16,500 (subunit IV), and 8,000–6,000 (subunit V). Thus, in general, higher plant cytochrome *c* oxidase is composed of five kinds of subunits, whereas the enzyme in other eukaryotes is seven to eight*. The antibody raised in a rabbit by the purified enzyme forms a single precipitin line with a detergent extract of pea shoot mitochondria in double immunodiffusion tests (*19*). It recognizes only the largest subunit (subunit I) of pea cyto-

---

* In both pea shoots and sweet potato root tissue, when an immunoprecipitate from a detergent extract of the mitochondrial inner membrane with anti-cytochrome *c* oxidase antibody is electrophoresed as described for Fig. 2, two additional polypeptides are detected besides the five subunits and the heavy and light chains of immunoglobulin G. Thus, higher plant cytochrome *c* oxidase in the mitochondrial inner membrane may contain seven subunits, two of which are very dissociable and do not participate in the oxidation of reduced cytochrome *c* with oxygen.

chrome c oxidase in electrophoretic, immunological blotting tests (Matsuoka and Asahi, unpublished data). Anti-sweet potato cytochrome c oxidase antibody also forms a precipitin line with the pea enzyme in double immunodiffusion tests; it reacts with subunits II, IV and V in electrophoretic, immunological blotting tests (Matsuoka and Asahi, unpublished data).

A series of experiments have been conducted with both antibodies to examine whether there is a precursor of cytochrome c oxidase in dry pea cotyledons (Matsuoka and Asahi, unpublished experiments). Proteins immunoreactive to anti-pea cytochrome c oxidase antibody in the inner membrane of cotyledon mitochondria do not change in amount during seed hydration despite the two-fold increase in enzyme activity. This suggests the presence of a cytochrome c oxidase precursor containing at least subunit I in the membrane in dry cotyledons but not in fully hydrated ones. In contrast, proteins immunoreactive to anti-sweet potato cytochrome c oxidase antibody in the membrane increase in parallel to the increase in enzyme activity during seed hydration. This means that another precursor containing subunits II, IV and/or V exists in a soluble form in dry pea cotyledons and is transformed into active cytochrome c oxidase protein tightly associated with the mitochondrial inner membrane during seed hydration.

The membrane-bound form of the precursor has been separated from active cytochrome c oxidase by differential solubilization (the precursor is solubilized from the membrane by a low concentration of Triton X-100 but the active enzyme is not) and precipitated by anti-pea cytochrome c oxidase antibody. When the immunoprecipitate is analyzed by sodium dodecylsulfate-urea polyacrylamide gel electrophoresis, only a single polypeptide band is detected in the position corresponding to subunit I besides the bands for the heavy and light chains of immunoglobulin G. This indicates that the membrane-bound precursor contains subunit I only in a free form and not in a complex form with other subunits. On the other hand, a similar analysis with an immunoprecipitate from the soluble fraction of dry cotyledons with anti-

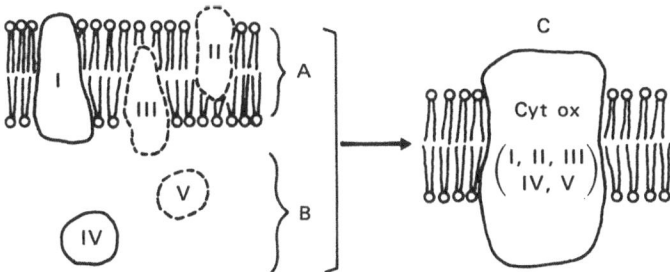

Fig. 6. Schematic illustration of the assembly of free forms of cytochrome c oxidase subunits into the active enzyme protein in pea cotyledons during seed hydration. A, the mitochondrial inner membrane of dry cotyledons; B, the mitochondrial matrix of dry cotyledons; C, the mitochondrial inner membrane of fully hydrated cotyledons. Cyt ox, cytochrome c oxidase. Roman numerals show the number of the subunits.

sweet potato cytochrome c oxidase antibody has shown that the soluble form of precursor contains only a free form of subunit IV.

On the basis of the finding described above, a schematic model shown in Fig. 6 has been proposed for the mechanism of the increase in cytochrome c oxidase activity in pea cotyledons during seed hydration. In dry cotyledons, there are free forms of cytochrome c oxidase subunits; some of them are associated relatively weakly with the phospholipid-rich mitochondrial inner membrane, while others exist in soluble forms in the mitochondrial matrix. It remains to be investigated which of the compartments, the membrane or the matrix, subunits II, III and V in the free forms. In the model, however, subunits II and III are proposed to be present in the membrane because they are considered to be very hydrophobic polypeptides synthesized on the mitochondrial ribosomes just as is subunit I. Subunit V may be present in the matrix because of its low hydrophobicity as subunit IV. The reason no free forms of subunits II and V have been detected with anti-sweet potato cytochrome c oxidase antibody is probably that the antibody can recognize the subunits but makes no precipitates with them. During seed hydration, the free forms of subunits are assembled together to form active cytochrome c oxidase protein and become tightly associated with the mitochondrial inner membrane.

The model poses two important problems. First, how can the subunits take such free forms? These have never been detected in other organisms except for very small pools of free cytochrome c oxidase subunits found in labeling experiments. It is of great interest to compare the molecular forms of the free subunits with those of the subunits in the active enzyme protein. The comparison may elucidate why the subunits can be present in forms free of the association with other subunits although they are tightly bound to each other in the active enzyme protein. Second, how are the free forms assembled with each other to form the active enzyme protein with a definite subunit arrangement, and tightly associated with the mitochondrial inner membrane in a definite topological arrangement? This question is closely related to the basic problem of how membrane protein composed of several subunits is formed and localized in a membrane.

CONCLUSION

Recent work in our laboratory has shown that in higher plant cells some enzymes in the mitochondrial inner membranes are formed by characteristic mechanisms that have never been reported with other organisms. A special cytochrome c oxidase precursor composed of only the cytoplasmic origin of subunits exists in the mitochondrial inner membrane of intact sweet potato root tissue. In dry pea cotyledons, there is a soluble form of succinate dehydrogenase that awaits seed hydration to be associated with the mitochondrial inner membrane. In particular, it is of great interest that free forms of cytochrome c oxidase subunits exist in dry pea cotyledons and are assembled with each other to form the active enzyme protein in response to seed hydration. We propose that the facts are related to the ability of higher plant mitochondria to make drastic and dynamic changes in response to changes in the physiological and environmental conditions of cells. In wounded tissue, cytochrome c oxidase is formed without new synthesis of the subunits on the cytoplasmic ribosomes, which facilitates rapid development in the respiratory enzyme system. In imbibing seeds, active formation of the mitochondrial inner membrane takes place through assembly of the respiration enzyme precursors into the membrane, which provides rapid and easy development of the mitochondria. The forms of precursor proteins are probably more stable than those of active proteins in a dry state.

We expect that further detailed studies on the phenomena described above will provide new information on the general molecular mechanism of mitochondrial biogenesis. For instance, studies on the formation of cytochrome c oxidase from its precursor and the subunits newly synthesized on the mitochondrial ribosomes in wounded sweet potato root tissue would be helpful in understanding the general mechanism of the subunit assembly into an active enzyme protein and of its localization into the mitochondrial inner membrane. Similar expectation is held for studies on the formation of cytochrome c oxidase and the association of succinate dehydrogenase with the mitochondrial inner membrane in imbibing pea cotyledons. In addition, studies on the synthesis of succinate dehydrogenase in wounded sweet potato root tissue would provide us with new knowledge of the regulation of mitochondrial biogenesis.

REFERENCES

1. Asahi, T. : 1978. Biochemistry of Wounded Plant Tissues (Kahl, G., ed.), pp. 391–419, Walter de Gruyter & Co., Berlin.
2. Bain, J.M. and Mercer, F.V. : 1966. *Aust. J. Biol. Sci.* 19, 49–67.
3. Bain, J.M. and Mercer, F.V. : 1966. *Aust. J. Biol. Sci.* 19, 69–84.
4. Beevers, H. : 1979. *Annu. Rev. Plant Physiol.* 30, 159–193.
5. Breidenbach, R.W., Castelfranco P. and Criddle, R.S. : 1967. *Plant Physiol.* 42, 1035–1041.
6. Hattori, T. and Asahi, T. : 1982. *Plant Cell Physiol.* 23, 515–523.
7. Hattori, T. and Asahi, T. : 1982. *Plant Cell Physiol.* 23, 525–532.
8. Hatefi, Y. : 1978. *Methods Enzymol.* 53, 27–35.
9. Hederstedt, L., Holmgren, E. and Rutberg, L. : 1979. *J. Bacteriol.* 138, 370–376.
10. Kim, I.C. and Beattie, D.S. : 1973. *Eur. J. Biochem.* 36, 509–518.
11. Kirk, J.T.O. : 1978. The Plastids (Kirk, J.T.O. and Tilney-Bassett, R.A.E., eds.), pp.788–872, Elsevier/North-Holland Biomedical Press, Amsterdam.
12. Leaver, C.J. and Gray, M.W. : 1982. *Annu. Rev. Plant Physiol.* 33, 373–402.
13. Lewin, A.S., Gregor, I., Mason, T.L., Nelson, N. and Schatz, G. : 1980. *Proc. Natl. Acad. Sci. U.S.A.* 77, 3998–4002.
14. Lin, L.F.H., Kim, I.C. and Beattie, D.S. : 1974. *Arch. Biochem. Biophys.* 160, 458–464.
15. Maeshima, M. and Asahi, T. : 1978. *Arch. Biochem. Biophys.* 187, 423–430.
16. Maeshima, M. and Asahi, T. : 1981. *J. Biochem.* 90, 391–397.
17. Maeshima, M. and Asahi, T. : 1981. *J. Biochem.* 90, 399–406.
18. Makimoto, N. and Asahi, T. : 1981. *Plant Cell Physiol.* 22, 1051–1058.
19. Matsuoka, M., Maeshima, M. and Asahi, T. : 1981. *J. Biochem.* 90, 649–655.
20. Merli, A., Capaldi, R.A., Ackrell, B.A.C. and Kearney, E.B. : 1979. *Biochemistry* 18, 1393–1400.
21. Mihara, K. and Blobel, G. : 1980. *Proc. Natl. Acad. Sci. U.S.A.* 77, 4160–4164.
22. Nakayama, N., Iwatsuki, N. and Asahi, T. : 1978. *Plant Cell Physiol.* 19, 51–60.

23. Nakayama, N., Sugimoto, I. and Asahi, T. : 1980. *Plant Physiol.* **65**, 229–233.
24. Nakayama, N. and Asahi, T. : 1981. *Plant Cell Physiol.* **22**, 79–89.
25. Nawa, Y. and Asahi, T. : 1971. *Plant Physiol.* **48**, 671–674.
26. Nawa, Y. and Asahi, T. 1973. *Plant Physiol.* **51**, 833–838.
27. Nawa, Y., Izawa, Y. and Asahi, T. : 1973. *Plant Cell Physiol.* **15**, 1073–1080.
28. Northemann, W., Schmelzer E. and Heinrich, P.C. : 1981. *Eur. J. Biochem.* **119**, 203–208.
29. Ouchterlony, O. and Nilson, L. Å. : 1973. Handbook of Experimental Immunology (Weir, D.M., ed.), pp.19.1–19.39, Blackwell Scientific Publications, Oxford.
30. Perlman, P.S. and Mahler, H.R. : 1974. *Arch. Biochem. Biophys.* **162**, 248–271.
31. Poucholet, M. and Shore, G.C. : 1981. *Biochim. Biophys. Acta* **654**, 67–76.
32. Poyton, R.O. and McKemmie, E. : 1979. *J. Biol. Chem.* **254**, 6763–6771.
33. Poyton, R.O. and McKemmie, E. : 1979. *J. Biol. Chem.* **254**, 6772–6780.
34. Sato, S. and Asahi, T. : 1975. *Plant Physiol.* **56**, 816–820.
35. Schmelzer, E., Northeman, W., Kadenbach, B. and Heinrich, P.C. : 1982. *Eur. J. Biochem.* **127**, 177–183.
36. Swank, R.T. and Munkres, K.D. : 1971. *Anal. Biochem.* **39**, 462–477.
37. Tanaka, Y. and Asahi, T. : 1973. *Plant Cell Physiol.* **14**, 965–971.
38. Tolbert, N.E. : 1981. *Annu. Rev. Biochem.* **50**, 133–157.
39. Towbin, H., Staehelin, T. and Gordon, J. : 1979. *Proc. Natl. Acad. Sci. U.S.A.* **76**, 4350–4354.
40. Tzagoloff, A. : 1971. *J. Biol. Chem.* **246**, 3050–3056.
41. Tzagoloff, A. and Macino, G. : 1979. *Annu. Rev. Biochem.* **48**, 419–441.
42. Uritani, I. and Asahi, T. : 1980. The Biochemistry of Plants (Stumpf, P.K. and Conn, E.E., eds.), pp.463–485, Academic Press Inc., New York.
43. Vary, M.J., Edwards, C.L. and Stewart, P.R. : 1969. *Arch. Biochem. Biophys.* **130**, 235–243.
44. Weiss, H. and Kolb, H.J. : 1979. *Eur. J. Biochem.* **99**, 139–149.
45. Yamaki, S. and Uritani, I. : 1973. *Plant Physiol.* **51**, 883–888.

Mechanism and Regulation of Ethylene Biosynthesis

Shang Fa Yang

INTRODUCTION

Ethylene is a plant hormone which initiates fruit ripening and regulates many aspects of plant growth, development, and senescence (1). The ability of certain gaseous agents to stimulate fruits to ripen has been known for many years. The Chinese knew for hundreds of years that picked fruits would ripen more quickly if placed in a chamber containing buring incense. This effect is undoubtedly attributable to the action of ethylene. The use of smoke to induce pineapple plants to bloom out of season was discovered in 1893 as a result of an accidental fire (2) ; to the surprise of the grower, the plants burst into flower instead of being damaged. Interest in ethylene as a plant growth regulator has extended over many years. although it was clear even in the mid-nineteenth century that the presence of gaseous materials in the air could modify the growth of plants (3), the identity of ethylene as an active agent was not established until the turn of the century. Neljubow (4) reported that ethylene was the component of illuminating gas which caused reduction of elongation, increased radial expansion, and horizontal growth in pea seedlings.

The first suggestion that biological material may release a gas which affects the growth of nearby plant material was reported in 1910 by Cousins (5), who observed that gases from oranges caused early ripening of bananas in mixed commercial ship-

Abbreviations : MTA, 5'-methylthioadenosine ; SAM, S-adenosylmethionine ; MTR, 5-methylthioribose ; ACC, 1-aminocyclopropane-1-carboxylic acid ; AVG, aminoethoxyvinylglycine or 2-amino-4-aminoethoxy-$trans$-3-butenoic acid ; AOA, aminoxyacetic acid ; AEC, 1-amino-2-ethylcyclopropane-1-carboxylic acid.

ments. It is now certain, however, that the ethylene produced did not come from the oranges but from fungus growing on the oranges, since uninfected oranges produced very little ethylene. This copious production of ethylene by infected oranges led Biale (6) and Miller et al. (7) to discover that the green mold, Penicillium digitatum, produces ethylene at high rates. Elmer (8) showed that the presence of ripe apples and pears caused abnormal growth of potato sprouts. Independently, Kidd and West (9) observed that a gaseous emanation from ripe apples could stimulate the ripening of unripe apples, an effect caused by ethylene also. In 1934 Gane (10) provided chemical proof that ethylene was indeed produced by ripe apples. Following the advent of sensitive gas chromatographic instruments in the early 1960's, postharvest physiologists demonstrated that ethylene functions as a ripening hormone (1,11,12). Demonstration of the natural function of ethylene in fruit ripening has stimulated the search for other functions. As the realization has grown that ethylene is a naturally produced plant hormone, so has the list of processes which it is known to regulate. The various effects of ethylene on plants or plant parts include the breaking of seed and bud dormancy, root initiation, stem strengthening, lateral branching, leaf epinasty, flower initiation, modification of flower sex expression, flower and fruitlet thinning, fruit growth stimulation, initiation of fruit ripening and flower senescence, fruit and leaf degreening, and leaf abscission and senescence. Nearly all plant tissues appear to be capable of producing ethylene, although the production rate is normally low. As part of the normal life of a plant, ethylene production is induced during certain stages of growth, such as seed germination, fruit ripening, leaf and flower senescence, and abscission (1). Ethylene production can also be induced by many external factors, including the application of auxins, physical wounding and cutting ; chilling injury, drought and water flooding (13). It has been recognized that this increased ethylene production in turn brings about many important physiological consequences. In this paper I shall summarize the state of current knowledge of ethylene biosynthesis and review the available information concerning the biochemical basis of the regulation of ethylene production in plant tissues.

PATHWAY OF ETHYLENE BIOSYNTHESIS

Methionine was first suggested as a possible precursor of ethylene by Lieberman and Mapson (14) who observed that the amino acid was readily converted into ethylene in a model chemical system consisting of Cu^{2+} and ascorbic acid. It is now established that methionine serves as a biological precursor of ethylene in all higher plant tissues. The biochemical mechanism by which ethylene is formed from methionine in the chemical system is different, however, from that in the biological system. In the former system, ethylene is formed from methionine via methional as an intermediate, and the CH_3S- group of methionine yields methyl disulfide (15). In apple tissue, methionine, but not methional, serves as the ethylene precursor, and the CH_3S- group is not converted into volatile methyl disulfide but is retained in the tissue (15). In this tissue, a substantial rate of ethylene production is maintained for extended periods, but the methionine level is quite low. This led Baur and Yang (*16*) to suggest that the sulfur of methionine must be recycled to maintain this continuous ethylene production. Subsequently, Adams and Yang (*17*) demonstrated that during ethylene production the CH_3S- group of methionine was released as 5'-methylthioadenosine (MTA) from S-adenosylmethionine (SAM), and the MTA was rapidly hydrolyzed to 5-methylthioribose (MTR). Since the CH_3 group and the S atom of MTR were incorporated into methionine in apple tissue with equal efficiency, it was concluded that CH_3S- group of MTR was converted into methionine as a unit (*17*). Although the source of 2-aminobutyrate moiety of methionine was not investigated, Adams and Yang (*17*) proposed that MTR donates its CH_3S- group to a 4-carbon acceptor, probably homoserine or its related analog, to form methionine, while the ribose moiety of MTR is split off. Recent work with yeast cells (*18*), rat liver homogenates (*19*) and cell-free extracts of *Enterobacter aerogenenes* (*20*) demonstrated that the ribose unit of MTA or MTR is directly incorporated into methionine along with the CH_3S- group. This has prompted us to reexamine the fate

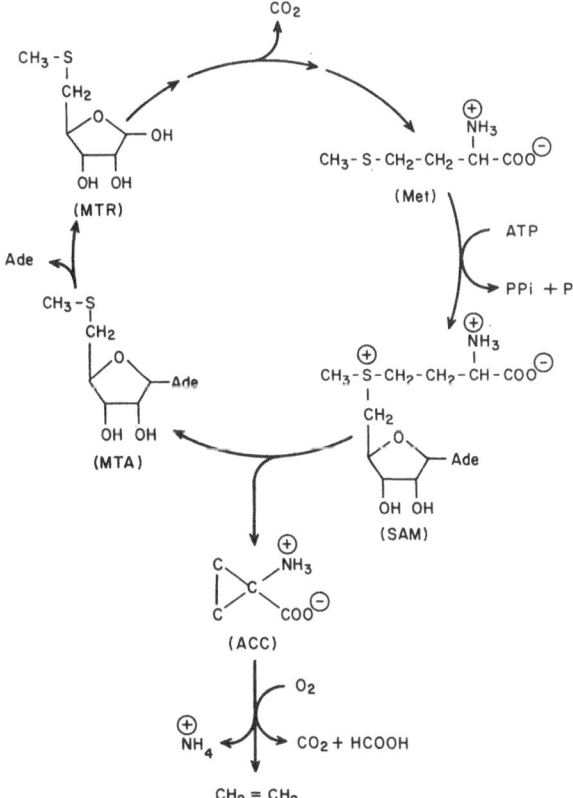

Fig. 1. Mthionine-sulfur cycle in relation to the biosynthesis of ethylene in apple tissue. From Adams and Yang (*17*) and Yung *et al.* (*21*).

of MTR in the recycling process which leads to methionine formation in apple tissue. Yung *et al.* (*21*) fed to apple fruit MTR dually labeled with [^{3}H] on the methyl group and [^{14}C] on the ribose moiety. They showed that the primary fate of MTR was the formation of methionine, and that both the CH_3S- group and the ribose portion of MTR were equally incorporated into methionine. Thus the pathway for methionine synthesis involves modification of the ribose portion of MTR into the 2-aminobutyrate portion of methionine, with the CH_3S- group remaining intact. The recycling of the CH_3S- group for the synthesis of methionine in relation to ethylene synthesis in apple tissue is diagrammed in Fig. 1. The overall result of this cycle is that the CH_3S- group of methionine is conserved for continued synthesis of ethylene, and that the 4-carbon moiety of methionine, from which the ethylene molecule is derived, is ultimately replenished from the ribose moiety of ATP. This cycle is particularly important in those plant tissues which produce ethylene at high rates but contain very low concentrations of endogenous methionine.

It is evident from the chemical structures of SAM and of ethylene that SAM cannot be converted into ethylene in a single step. It has been known for some time that both endogenous ethylene production and the conversion of methionine to ethylene, cease in plant tissue placed in an anaerobic atmosphere, but that a surge of ethylene production occurs upon reexposure of the tissue to air. These observations are interpreted to indicate that an intermediate accumulates during anaerobic incubation and is subsequently converted to ethylene upon exposure to oxygen. Adams and Yang (*22*) compared the metabolism of methionine in air and in a nitrogen atmosphere. In air, methionine was efficiently converted to ethylene ; in nitrogen, however, it was metabolized not to ethylene but to MTR and a compound later identified as 1-aminocyclopropane-1-carboxylic acid (ACC). In the presence of air ACC was rapidly converted to ethylene, indicating that ACC is an intermediate and that the conversion of ACC to ethylene is oxygen-dependent. These data indicate the following sequence for the pathway of ethylene biosynthesis in apple tissue :

Methionine → SAM → ACC → ethylene

Although aminoethoxyvinylglycine (2-amino-4-aminoethoxy-*trans*-3-butenoic acid; AVG) strongly inhibits the conversion of methionine to ethylene, it did not block the conversion of methionine to SAM or the conversion of ACC to ethylene. However, it effectively blocked the conversion of methionine to ACC. Thus, AVG must exert its inhibitory effect by blocking the conversion of SAM to ACC. Soon after the pathway for ethylene biosynthesis became known, Boller *et al.* (*23*) reported that cell-free extracts prepared from tomato fruit were capable of converting SAM to ACC. They found that the enzyme was soluble and was strongly inhibited by AVG, as predicted by Adams and Yang. The K_m for SAM was estimated to be 13 μM, and the enzyme utilized SAM specifically as substrate. Employing labeled SAM, Yu *et al.* (*24*) confirmed that SAM was converted to ACC and MTA by the enzyme preparation from tomato fruit. In addition , they demonstrated that low concentrations of pyridoxal-phosphate could activate ACC synthase and that the enzyme was strongly inhibited by aminoxyacetic acid (AOA), another well-known inhibitor of pyridoxal enzymes. These observations strongly support the view that ACC synthase is a pyridoxal enzyme.

The conversion of SAM to ACC and MTA is a typical γ-elimination (1,3-elimination) reaction. Organic chemists have shown that γ-eliminations proceed readily with a carbanion as depicted below :

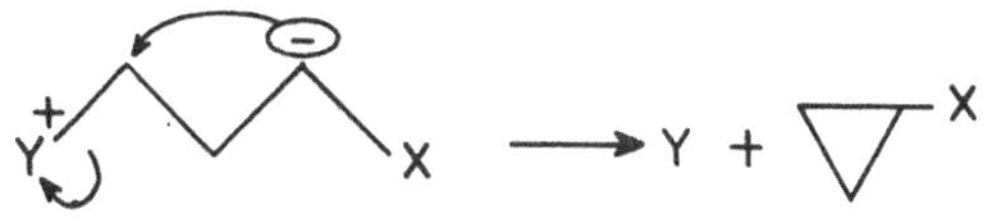

It is known that a pyridoxal enzyme (Py·E) facilitates elimination of the proton from the α-carbon of an amino acid, yielding a carbanion (25). The positive sulfonium group of SAM is an excellent leaving group and once the carbanion is formed an intramolecular nucleophilic displacement reaction can occur, resulting in the elimination of MTA and the formation of ACC.

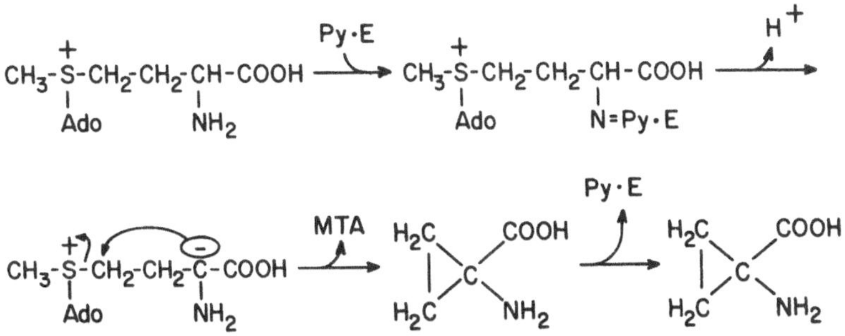

From the above discussion it can be reasoned that there exists a 2-fold biological significance in the functioning of SAM serving as an intermediate in the biosynthesis of ethylene from methionine. It provides (a) a positive sulfonium function which is an excellent leaving group facilitating the γ-elimination reaction in the formation of ACC, and (b) the eliminated product, MTA, which is readily recycled back to methionine (Fig. 1) thus conserving the sulfur atom for continued synthesis of methionine.

Although Konze and Kende (26) have recently reported an enzymatic conversion of ACC to ethylene by a pea seedling extract, it does not appear to be the physiological system active in vivo, since its characteristics do not resemble those of the natural system. Based on in vivo observations, it is shown that the reaction requires oxygen for the conversion (17), and is inhibited by high temperature (27), by uncouplers of oxidative phosphorylation (27), by Co^{2+} (28) and by osmotic shock (29). Because of the enzyme system which converts ACC to ethylene is labile, and subject to disruption by various chemicals and treatments, it is thought that it is membrane-associated.

Although the ACC molecule possesses reflective symmetry, the two methylene groups are not geometrically equivalent and can be distinguished by a stereospecific enzyme. Ethyl substitution of each of the four methylene hydrogens of ACC results in four stereoisomers of 1-amino-2-ethylcyclopropane-1-carboxylic acid (AEC) whose absolute configurations are : $(1R,2R)$, $(1S,2S)$, $(1R,2S)$, and $(1S,2R)$. In order to determine whether ACC conversion to ethylene by plant tissues proceeds in stereospecific fashion, Hoffman et al. (30) have recently administered the four stereoisomers of AEC to postclimacteric apple and to etiolated mungbean hypocotyls. In both tissues, $(1R,2S)$-AEC was the preferred substrate yielding 1-butene. By chemical oxidation using NaOCl, in contrast, all AEC isomers were converted with equal efficiency to butene. ACC and AEC appear to be degraded by the same enzyme since both reactions are inhibited to the same extent by nitrogen atmosphere or by Co^{2+}, and since, when both substrates are present simultaneously, each acts as an inhibitor with respect to the other. The aforementioned observations indicate that ACC is stereospecifically converted to ethylene. Thus, for AEC to be an active precursor of 1-butene, the ethyl substituent should be *trans* to the carboxyl group and the *pro-(S)* methylene group should be unsubstituted. It is suggested that the enzyme interacts with amino, carboxyl, and *pro-(S)* methylene groups, a configuration corresponding to an L-amino acid. This view is consistent with our observation that L-form of alanine or methionine inhibits the conversion of ACC to ethylene more than the corresponding D-amino acid in the mungbean hypocotyl system (30).

Until the enzyme system responsible for the conversion of ACC to ethylene is isolated and characterized, it is premature to discuss the biochemical reaction. Nevertheless it is interesting to note that the chemical oxidation of substituted cyclopropylamine to ethylene *via* a nitrenium ion intermediate has been documented (31). By analogy with the chemical oxidation, it is suggested that ACC is oxidized enzymatically either by hydroxylation followed by dehydration, or by dehydrogenation of the amino group yielding a nitrenium intermediate which then fragments spontaneously into ethylene and cyanoformic acid. Cyanoformic acid would be then decarboxylated to CO_2 (derived from the carboxyl group of ACC) and HCN (derived from C-1 of ACC) ; HCN would then be metabolized into formic acid, CO_2 and other products. The validity of this pathway in plant tissues remains to be established.

When ACC was applied to various plant organs, including root, stem, leaf, inflorescence, and fruit, a remarkable increase in ethylene production was observed (32). This suggests that the formation of ACC is the rate-limiting reaction in ethylene biosynthesis. Independently, Lürssen et al. (33) discovered from screening tests that

ACC is able to enhance ethylene production in a number of plants or plant organs. Although the intermediate role of ACC in ethylene biosynthesis was not demonstrated, they hypothesized that ACC was derived from SAM.

Because of the important role of ACC in ethylene biosynthesis, a sensitive and simple assay for ACC in plant tissues is essential. Lizada and Yang (34) have developed an assay in which ACC is degraded to ethylene with NaOCl and the ethylene evolved is determined by gas chromatography. The method is highly specific to ACC and can detect as little as 5 pmol.

Before the recent recognition of ACC as the immediate precursor of ethylene, ACC was simply one of the non-protein amino acids that have been isolated from plants. ACC was isolated more than 20 years ago from perry pears and cider apples by Burroughs (35) and from ripe fruits of the cowberry by Vahatalo and Virtanen (36). Although Burroughs (37) observed that the amount of ACC in perry pears increased during storage, he could not detect ACC in other varieties of apple or pear he examined. Undoubtedly this was due to the insensitive method employed at that time.

REGULATION OF ETHYLENE PRODUCTION IN RIPENING FRUITS AND SENESCING FLOWERS

Climacteric fruits are characterized by a surge of ethylene production at the onset of ripening, and it is recognized that ethylene plays an essential role in the ripening process (12). To study the regulation of ethylene production during the ripening of fruits, Hoffman and Yang (38) examined the changes in internal ACC content during ripening, as well as the effect of exogenous ACC on ethylene production in preclimacteric fruits. In preclimacteric fruits of avocado, banana, and tomato, the ACC content was very low (less than 0.1 nmol/g), but a massive increase occurred at the time vigorous ethylene production commenced. The relationship between the change in ACC content and the change in the ethylene production rate with ripening of avocado is illustrated in Fig. 2. ACC increased to 45 nmol/g in the later stage of the climacteric rise of respiration, then declined to 5 nmol/g, and later increased again, while ethylene production declined and remained at a low rate. The ability to convert ACC to ethylene was obviously impaired in these overripe fruit tissues, and this is thought to be the cause of the second increase of ACC. These data indicate that the inability of preclimacteric fruit tissue to produce ethylene is due to its lack of ability to form ACC. If ACC synthase is the only enzyme in the pathway which is limiting in the preclimacteric fruits, administration of ACC should greatly increase ethylene production, as was observed in vegetative tissues. Preclimacteric apple and cantaloupe tissues, however, exhibit only a slight (less than 5-fold) increase in ethylene production following the application of ACC, while the increase in ethylene production at the climacteric peak is several hundred-fold. Although preclimacteric fruits are capable of converting methionine to SAM, they have little ability to convert SAM to ACC and to convert ACC to ethylene (17,22). It should be noted, however, that application of ACC to intact preclimacteric tomato fruit significantly en-

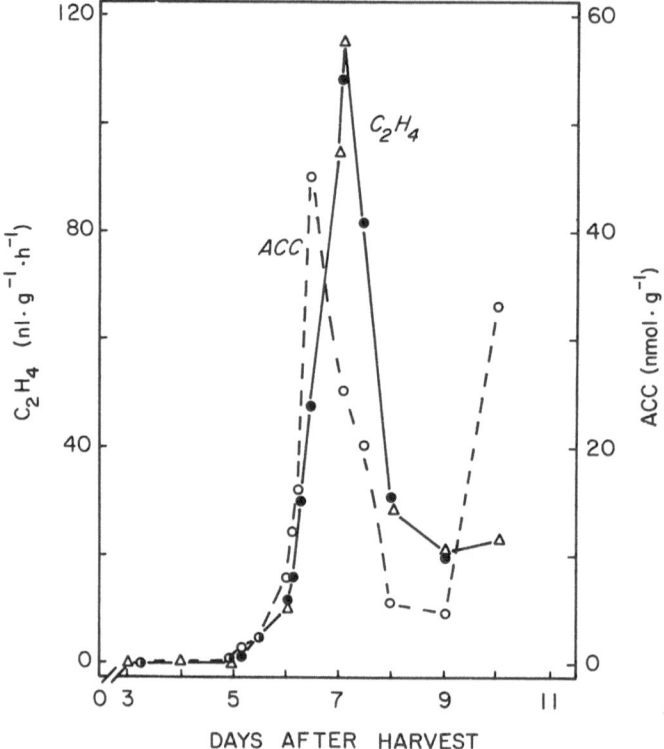

Fig. 2. Change in ACC content of avocado fruit at various stages of ripeness. Each ACC value is from a single fruit which had been monitored for ethylene production and assigned an arbitrary stage of ripeness by comparison with the established climacteric patterns of ethylene production. From Hoffman and Yang (38).

hanced the ripening process (A. C. Cameron and S. F. Yang, unpublished). This suggests that the slight increase in ethylene production caused by the application of ACC may be enough to trigger the ripening process. Autocatalysis of ethylene production is a common feature of ripening fruits and some senescing tissues (1,12,39) in which an increased synthesis of ethylene is triggered by exposure to ethylene at a concentration above a low threshold level. Recently Hoffman and Yang (40) have studied the effect of exogenous ethylene on biosynthesis of ethylene in excised tissue of unripe cantaloupe. Within a short (12 hr) incubation period, ethylene treatment increased the rate of conversion of ACC to ethylene to a level comparable to that of ripe fruit, but failed to cause a marked increase in ACC content. It has been suggested that there exists in unripe fruit an endogenous ripening inhibitor which interferes with the ethylene synthesis and/or action and which is gradually destroyed during the maturation process (41,42). We speculate that this ripening inhibitor system may include a repressor which restricts the synthesis of ACC synthase.

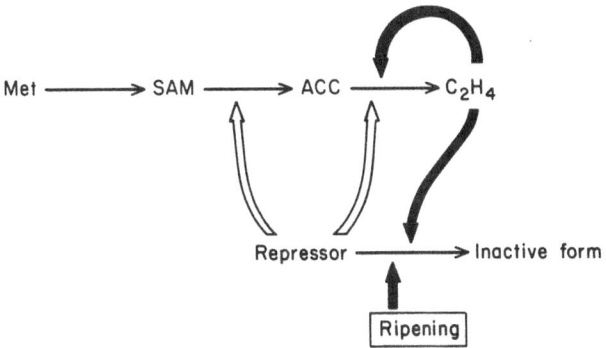

Fig. 3. Regulation of ethylene biosynthesis in ripening fruits. ⟶, catalytic reaction ;
➡, stimulation of the reactions ;⟹, inhibition of the reaction.

The role of ethylene in autocatalytic ethylene production would then be to promote
the synthesis of the enzyme responsible for the conversion of ACC to ethylene and to
accelerate the destruction of the ACC-synthase repressor. Thus, as the fruit approaches the ripening stage, there occurs a slight degradation or inactivation of the
repressor and a corresponding development of ACC synthase and the enzyme converting ACC to ethylene. The direct result is the formation of a small amount of
"System 1" ethylene. This small amount of "System 1" ethylene would then further
induce the synthesis of the enzyme responsible for the conversion of ACC to ethylene
and further accelerate the destruction of the repressor, resulting in massive synthesis
of ACC and of "System 2" ethylene. The regulation of ethylene synthesis by
ethylene during the ripening process is illustrated in Fig. 3.

In vegetative tissue, ethylene production is known to be regulated by another
class of hormone, the auxins. The action of auxin in relation to ethylene biosynthesis
has been studied by several investigators (43 15) and will be discussed in detail in
this volume by Imaseki. These studies indicate that auxin stimulates ethylene production by inducing ACC synthase, and that this step is rate-limiting for ethylene
production. Unlike vegetative tissues, the onset of ethylene production in ripening
fruits is not regulated by auxin. Thus, although fruit tissues and vegetative tissues
share a common pathway for ethylene biosynthesis, the mode of regulation is different.

The senescence of carnation flowers is accompanied by a marked increase in
the synthesis of ethylene, and a concomitant climacteric rise in respiration. The
changes in ACC content of cut carnation flowers in relation to their senescence were
studied by Bufler et al. (46). In freshly harvested carnation flowers, ACC content and
ethylene production rates were very low. With the onset of senescence and the accompanying autocatalylic rise in ethylene production there was a rapid increase in
ACC content as shown in Fig. 4. As senescence progressed ethylene production fell,
but the ACC content of the tissue remained high. This could be the result of a
more rapid fall in the rate of ACC conversion to ethylene than in the rate of ACC

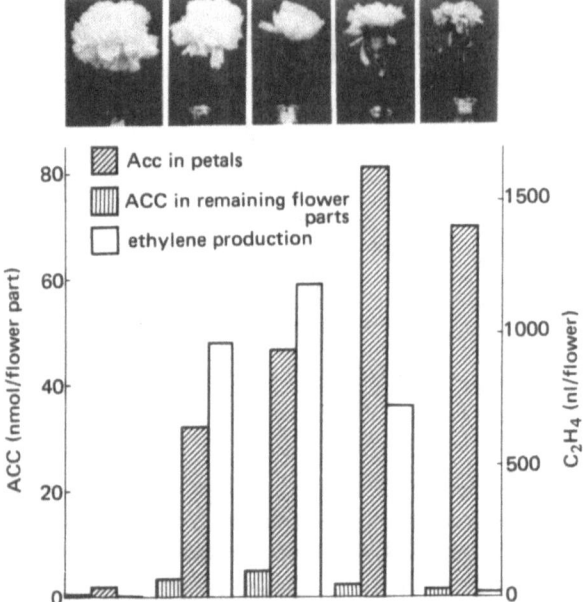

Fig. 4. Ethylene production and ACC content of carnation flowers at different stages of senescence were photographed and their ethylene production was measured before estimation of ACC content of the petals and of the remainder of the flower. From Bufler *et al.* (*46*).

synthesis. This is expected if the enzyme converting ACC to ethylene is associated with membrane which is disintegrated during the later stage of ripening. Pretreatment with Ag^+ (an anti-ethylene agent) or with AOA (an inhibitor of ACC synthase) retarded flower senescence and prevented the ACC accumulation.

REGULATION OF STRESS-INDUCED ETHYLENE PRODUCTION

It is well documented that various kinds of stress, including chemicals, temperature extremes, drought, waterlogging, radiation, insect damage, disease or mechanical wounding, result in increased ethylene production (*1,13*). The biochemistry of ethylene formation in wounded tissues has been studied in a number of plant materials, and in all of them methionine appears to be the precursor (*13*). To determine whether wound ethylene is also synthesized through the ACC pathway, Yu and Yang (*47*) and Hoffman and Yang (*40*) studied stress-ethylene production induced by Cu^{2+} treatment of mungbean hypocotyls, or by slicing citrus albedo tissue, immature cantaloupe or tomato fruits. Both ethylene production and ACC content were very low before or immediately following wounding, but a rapid increase in ACC occurred during subsequent incubation, paralleling the rise in ethylene production. Stress did not cause a change in the quantity of endogenous SAM, however, suggesting that the conversion of SAM to ACC is a key reaction controlling the production of stress ethylene. This possibility was further supported by the observation that application of AVG, a potent inhibitor of ACC synthase, eliminated the increase in ACC formation and the production of stress ethylene. In preclimacteric tomato tissue a marked increase in ACC synthase activity was associated with an increase in ACC formation and stress ethylene production. Infiltration with cycloheximide, an inhibitor of protein synthesis, completely blocked all wound-induced development of ACC synthase activity, ACC formation and ethylene production. These data indicate that wounding induces the synthesis of ACC synthase, which in turn causes accumulation of ACC and the onset of stress ethylene production.

Cycloheximide induces wound ethylene production when applied to intact citrus fruits (48). Because of this property cycloheximide has been used commercially to hasten the abscission of orange fruit for mechanical harvesting (48). Riov and Yang (49) has shown that the induction of ACC synthesis was responsible for the great increase in ethylene production following cycloheximide application. It is intriguing that cycloheximide should exert such conflicting effects on ethylene biosynthesis : it inhibits excision-induced ethylene production in excised discs, but enhances ethylene production in intact fruits. It may be interpreted that cycloheximide has a dual effect on ethylene biosynthesis : it causes inhibition of ethylene production by interfering with the synthesis of ACC as well as the conversion of ACC to ethylene, but it also causes chemical injury of the cells of the flavedo tissue, resulting in the formation of a wound signal. As cycloheximide is rapidly metabolized in intact fruit, its concentration would eventually fall below the level required for the inhibition of ethylene synthesis. When this happens, such a wound signal, which is distributed to a network of adjacent cells, would then induce formation of ACC and ethylene.

Drought is known to cause plants to increase ethylene production, which in turn promotes abscission and thereby reduces water loss (50,51). The pathway and the step at which water deficit-induced ethylene production is regulated appear to be identical to those of other systems of stress ethylene production. The increase in ethylene production in response to wilting is caused by an increase in ACC synthesis. Fig. 5 shows the relationship of water loss, ACC accumulation and ethylene production in excised wheat leaves (52). It is worth mentioning that the stimulatory influence of water-deficit stress on ethylene biosynthesis was not restricted to its effect on ACC synthesis. In addition to increased ACC synthesis, the capacity of the system which converts ACC to ethylene was also increased. When wheat leaves were stressed by wilting, both the ACC content and ethylene synthesis increased markedly, although both declined later. Apelbaum and Yang (52) observed that the decrease in quantity of ACC in wilted leaves was much greater than the amount of ACC converted to ethylene. These observations led them to suggest that ACC is metabolized by some pathway other than ethylene production. Hoffman et al. (53) recently examined the metabolism of ACC by feeding [2-^{14}C] ACC to light-grown wheat leaves and discovered that ACC was primarily converted into a nonvolatile metabolite, which was identified as N-malonyl-ACC. The natural occurrence of N-malonyl-ACC in the wilted wheat leaves was confirmed by gas chromatography-mass spectrometry

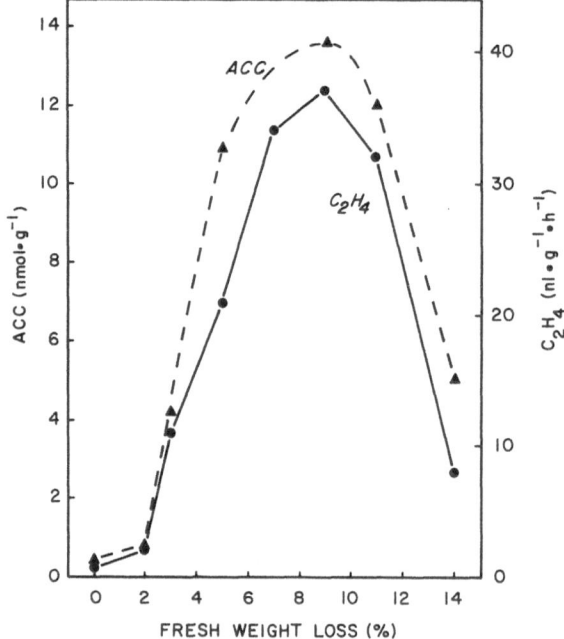

Fig. 5. Effect of the severity of water stress on ethylene production and ACC content in wheat leaves. Ethylene production and ACC content were determined 4 hrs after the end of the water stress treatment by which they reached the indicated water loss. From Apelbaum and Yang (52).

(53). Independently, Amrhein and co-workers (54) identified N-malonyl-ACC from etiolated buckwheat seedlings after feeding exogenous ACC. Other than that this bound form of ACC shows little ability to serve as an ethylene precursor, we know very little about its physiological significance.

It has long been recognized that waterlogging creates anaerobic conditions in the root zone and thus causes elevated synthesis of ethylene by the shoot in many plants. This elevated ethylene production in turn causes adventitious root formation, leaf senescence, and epinastic growth of the petioles (55). Since ethylene is not synthesized in the anaerobic root, and, due to its gaseous nature, cannot be transported to the shoot in significant quantity (56), it has been suggested that a "signal" is synthesized in the anaerobic root and transported to the shoot, where it stimulates ethylene synthesis (55). Since ACC accumulates under anaerobic conditions and is readily converted to ethylene in aerobic tissues, ACC is a logical candidate to serve as the "signal". Bradford and Yang (57) collected the xylem sap from de-topped tomato plants and determined the ACC content of the sap. ACC content was very low in xylem sap of the control plants, but increased markedly in response to waterlogging and root anaerobiosis. This appearance of ACC in the xylem sap from the roots of de-topped flooded plants preceded both the increase in ethylene production

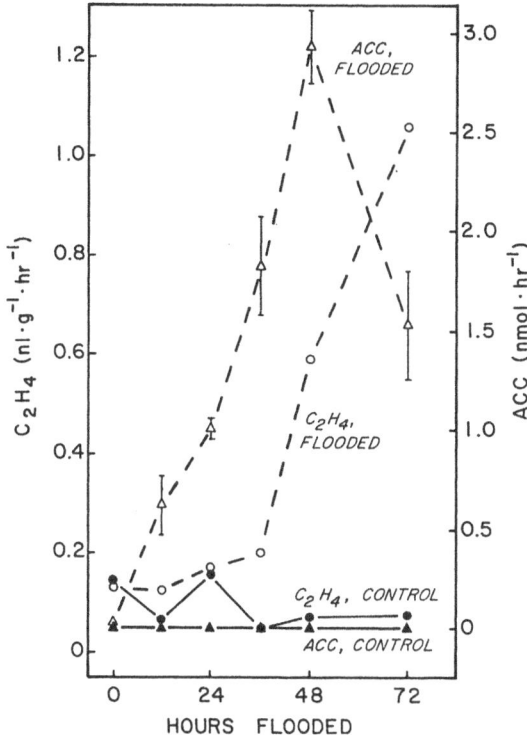

Fig. 6. Time course of changes in the quantity of ACC exuded in the xylem sap and of changes in petiolar ethylene production following flooding of tomato plants. From Bradford and Yang (57).

by excised petioles and epinastic growth in intact flooded plants (Fig. 6). When ACC was supplied through the base of the cut stem of tomato shoots at concentrations comparable to those found in xylem sap, increased ethylene production and epinasty of the shoot were observed. Since the amount of ACC exported from anaerobic roots exceeded the amount of ethylene synthesized by roots maintained continuously in aerobic condition, a mere blockage of ethylene synthesis in the root by anaerobiosis is insufficient to account for the increased quantity of ACC transported. Anaerobic stress must therefore not only block conversion of ACC to ethylene in the root, but also stimulate ACC synthesis.

By classical definition, a hormone is an endogenous compound which is syn-thesized at one site and is transported to another site where it exerts a physiological effect. Due to its gaseous nature, ethylene exerts a physiological effect only at or near a site where it is synthesized. For this reason, the classical definition of a hormone does not apply to ethylene. This work has shown, however, that ACC can be synthe-sized in one part of the plant, be readily transported, and exert its effect through conversion to ethylene in another part of the plant.

REGULATION OF ETHYLENE BIOSYNTHESIS BY ETHYLENE

Autocatalysis of ethylene production is a common feature of ripening fruits and other senescing tissues in which a massive increase in ethylene production is triggered by exposure to ethylene above a threshold level. Autoinhibition of ethylene production has also been recognized in a number of fruits and vegetative tissues. The characteristics of autoinhibition of ethylene production in the flavedo tissue of citrus fruits have been examined by Riov and Yang (58). Slicing of flavedo tissue of citrus fruit resulted in wound ethylene production, but this ethylene production was greatly inhibited by exogenous ethylene, as shown in Fig. 7. The increase in wound ethylene production in the absence of exogenous ethylene was paralleled by an increase in ACC content, whereas in ethylene-treated discs there was little increase in ACC content. Ethylene production in ethylene-pretreated discs was completely restored by application of ACC, indicating that the conversion of ACC to ethylene is not impaired by the presence of ethylene. Autoinhibition of ethylene synthesis was thus exerted by limiting the availability of ACC. Ethylene treatment resulted in a slight decrease in extractable ACC synthase activity, which was too small to account for the marked inhibition of ACC production. The data suggest that the autoinhi-

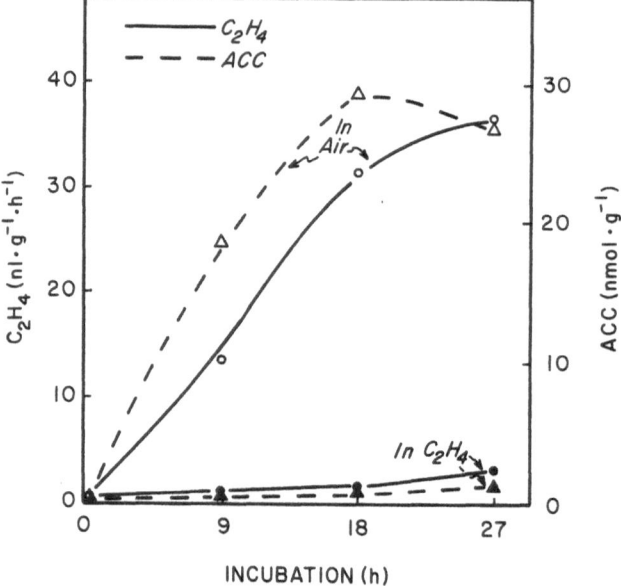

Fig. 7. Effect of exogenous ethylene on inhibition of ethylene production and ACC accumulation in grapefruit flavedo discs. Rates of ethylene production by discs which had been incubated for various periods under air or ethylene (20 μl/liter) as indicated, were determined under ethylene-free air. From Riov and Yang (58).

bition of ethylene production in citrus flavedo discs results from suppression of ACC formation primarily through inhibition of ACC synthase activity.

Autocatalysis of ethylene production in intact citrus leaves and in citrus leaf discs was studied by Riov and Yang (59). Ethylene production was stimulated by exposure to exogenous ethylene for 24 hrs or more, as shown in Fig. 8. This promotive effect of exogenous ethylene resulted from its enhancement of ACC formation, as well as from acceleration of the conversion of ACC to ethylene. It is suggested that autocatalysis involves increased synthesis of ACC synthase and of the enzyme

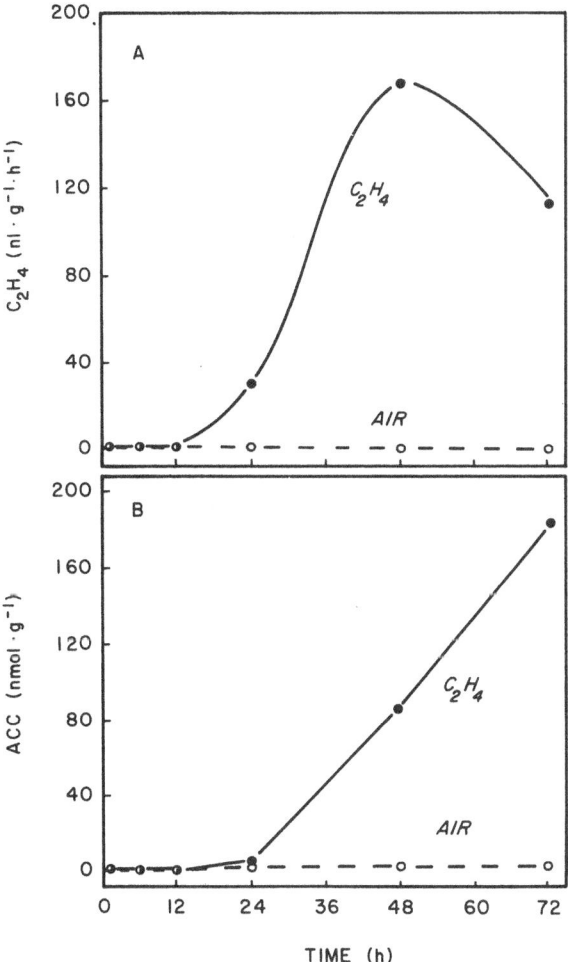

Fig. 8. Effect of exogenous ethylene on promotion of (A) ethylene production and (B) ACC accumulation in citrus leaf discs. Ethylene production rates of discs which had been incubated for various periods under air or ethylene (12 μl/liter) as indicated, were determined under ethylene-free air. From Riov and Yang (59).

responsible for the conversion of ACC to ethylene, whereas autoinhibition exerts its
action primarily by suppression of the activity of ACC synthase.

CONCLUSION

Ethylene regulates many aspects of plant growth and development, whether it is
of endogenous origin or is applied exogenously. Depending upon where and when it
occurs, it may have important beneficial or detrimental effects upon the growth
or development of plants. Ethylene action has long been recognized to be regulated
by the levels of oxygen and CO_2 in the atmosphere in which harvested fruits, vegetables or flowers are stored. Such principles have been successfully applied to controlled atmosphere storage (60). It has been shown that Ag^+ is an effective anti-
ethylene agent (61–63). In addition to the modification of ethylene action, manipulation of the level of ethylene in plants either by stimulation or inhibition of ethylene

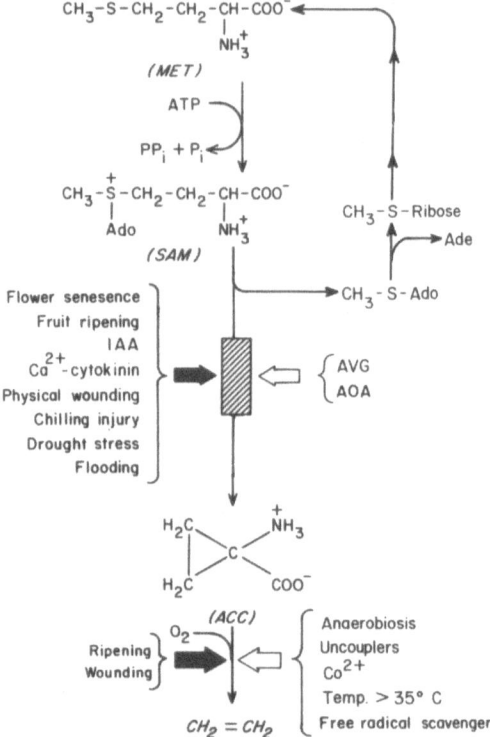

Fig. 9. Regulation of ethylene biosynthesis.———>, catalytic reaction ; ▨▨▨ , this reaction is normally suppressed and is the rate-limiting step in the pathway ;■■■▶, induction of synthesis of the enzyme ;▭▷, inhibition of the reaction. MET, Ado, and
Ade stand for methionine, adenosine, and adenine, respectively.

production is of great potential value in agriculture. Such manipulation is possible through the application of knowledge obtained from studies of the biosynthesis of ethylene and its regulation. The recent dicovery of ACC as the key intermediate in the pathway of ethylene biosynthesis (22) has opened the door for many fundamental as well as applied studies. The detailed pathway of ethylene biosynthesis has been defined, and our understanding of the regulation of ethylene production has been greatly clarified. The current status of knowledge with respect to the pathway and regulation of ethylene biosynthesis is summarized in Fig. 9. An important conclusion emerging from these studies is that an extensive synthesis of ACC synthase is induced by certain internal (developmental) or external factors, including fruit ripening, flower senescence, application of auxins and ethylene, physical wounding, chemical injury, root anaerobiosis, drought, and chilling injury (64), and that this increase is responsible for the onset of massive ethylene production. Thus, the new synthesis of ACC synthase is the first event leading to the rise in ethylene synthesis. The biochemical mechanism by which these quite diverse stimuli result in synthesis of the same rate-controlling enzyme, ACC synthase, remains to be elucidated. This onset of ethylene production can be suppressed by application of inhibitors of ACC synthase, such as AVG and AOA, which are inhibitors of pyridoxal phosphate-dependent enzymes. Once ACC synthesis is induced the overall ethylene production rate can be further regulated at the step at which ACC is converted to ethylene.

Acknowledgements
Our work cited herein has been supported by Research Grants from the National Science Foundation (PCM-7809278 and PCM-8114933) and U.S.-Israel Agricultural Research and Development Fund (I-145-79). I thank Gordon Von Abrams for reading this manuscript.

REFERENCES

1. Abeles, F.B. : 1973. Ethylene in Plant Biology, Academic Press, New York.
2. Grabham, M. : 1903. *Gard. Chron.*, 36.
3. Girardin, J.P.L. : 1864. *Jahresber. Agr. Chem.* 7, 199–200.
4. Neljubow, D. : 1901. *Bot. Zentralbl.* 10, 128–138.
5. Cousins, H.H. : 1910. Annu. Rep. Dept. Agric. Jamaica, 7–15.
6. Biale, J.B. : 1940. *Science* 91, 458.
7. Miller, E.V., Winston, J.R. and Fisher, D.F. : 1940. *J. Agric. Res.* 60, 269–278.
8. Elmer, O.H. : 1932. *Science* 75, 193.
9. Kidd, F. and West, C. : 1932. Gt. Br. Dept. Sci. Ind. Res. Food Invest. Board Rep., 55–58.
10. Gane, R. : 1934. *Nature* 134, 1008.
11. Burg, S.P. : 1962. *Annu. Rev. Plant Physiol.* 13, 265–302.
12. Pratt, H.K. and Goeschl, J.D. : 1969. *Annu. Rev. Plant Physiol.* 20, 541–581.
13. Yang, S.F. and Pratt, H.K. : 1978. Biochemistry of Wounded Plant Storage Tissues (Kahl, G., ed.), pp. 595–622. de Gruyter, Berlin.
14. Lieberman, M. and Mapson, L.W. : 1964. *Nature* 204, 343–345.
15. Yang, S.F. : 1974. *Recent Adv. Phytochem.* 7, 131–164.

16. Baur, A.H. and Yang, S.F. : 1972. *Phytochemistry* 11, 3207–3214.
17. Adams, D.O. and Yang, S.F. : 1977. *Plant Physiol.* 60, 892–896.
18. Shapiro, S.K. and Schlenk, F. : 1980. *Biochim. Biophys. Acta* 633, 176–180.
19. Backlund, P.S. and Smith, R.A. : 1981. *J. Biol. Chem.* 256, 1533–1535.
20. Shapiro, S.K. and Barrett, A. : 1981. *Biochem. Biophys. Res. Commun.* 102, 302–307.
21. Yung, K.H., Yang, S.F. and Schlenk, F. : 1982. *Biochem. Biophys. Res. Commun.*, 104, 771–777.
22. Adams, D.O. and Yang, S.F. : 1979. *Proc. Natl. Acad. Sci. U.S.A.* 76, 170–174.
23. Boller, T., Herner, R.C. and Kende, H. : 1979. *Planta* 145, 293–303.
24. Yu, Y.B., Adams, D.O. and Yang, S.F. : 1979. *Arch. Biochem. Biophys.* 198, 280–286.
25. Davis, L. and Metzler, D.E. : 1972. The Enzymes (Boyer, P.D., ed.), Vol. 7, pp.33–74, Academic Press, New York and London.
26. Konze, J.R. and Kende, H. : 1979. *Planta* 146, 293–302.
27. Yu, Y.B., Adams, D.O. and Yang, S.F. : 1980. *Plant Physiol.* 66, 286–290.
28. Yu, Y.B. and Yang, S.F. : 1979. *Plant Physiol.* 64, 1074–1077.
29. Imaseki, H. and Watanabe, A. : 1978. *Plant Cell Physiol.* 19, 345–348.
30. Hoffman, N.E., Yang, S.F., Ichihara, A. and Sakamura, S. : 1982. *Plant Physiol.* 70, 195–199.
31. Hiyama, T., Koide, H. and Nozaki, H. : 1975. *Bull. Chem. Soc. Jpn.*, 48, 2918–2921.
32. Cameron, A.C., Fenton, C.A.L., Yu, Y., Adams, D.O. and Yang, S.F. : 1979. *Hortscience* 14, 178–180.
33. Lürssen, K., Naumann, K. and Schröder, R. : 1979. *Z. Pflanzensphysiol.* 92, 285–294.
34. Lizada, M.C.C. and Yang, S.F. : 1979. *Anal. Biochem.* 100, 140–145.
35. Burroughs, L.F. : 1957. *Nature* 179, 360–361.
36. Vahatalo, M.L. and Virtanen, A.I. : 1957. *Acta Chem. Scand.* 11, 741–743.
37. Burroughs, L.F. : 1960. *J. Sci. Food Agric.* 11, 14–18.
38. Hoffman, N.E. and Yang, S.F. : 1980. *J. Am. Soc. Hort. Sci.* 105, 492–495.
39. Lieberman, M. : 1979. *Annu. Rev. Plant Physiol.* 30, 533–591.
40. Hoffman, N.E. and Yang, S.F. : 1982. *Plant Physiol.* 69, 317–322.
41. Burg, S.P. and Burg, E.A. : 1965. *Science* 148, 1190–1196.
42. Gazit, S. and Blumenfeld, A. : 1970. *J. Am. Soc. Hort. Sci.* 95, 229–231.
43. Yu, Y.B., Adams, D.O. and Yang, S.F. : 1979. *Plant Physiol.* 63, 589–590.
44. Jones, J.F. and Kende, H. : 1979. *Planta* 146, 649–656.
45. Yoshii, H. and Imaseki, H. : 1981. *Plant Cell Physiol.* 22, 369–379.
46. Bufler, G., Mor, Y., Reid, M.S. and Yang, S.F. : 1980. *Planta* 150, 439–442.
47. Yu, Y.B. and Yang, S.F. : 1980. *Plant Physiol.* 66, 281–285.
48. Cooper, W.C. and Henry, W.H. : 1971. *Agric. Food Chem.* 19, 559–563.
49. Riov, J. and Yang, S.F. : 1982. *J. Plant Growth Regul.*, submitted.
50. McMichael, B.L., Jordan, W.R. and Powell, R.D. : 1972. *Plant Physiol.* 49, 658–660.
51. El-Beltagy, A.S. and Hall, M.A. : 1974. *New Phytol.* 73, 47–59.
52. Apelbaum, A. and Yang, S.F. : 1981. *Plant Physiol.* 68, 594–596.
53. Hoffman, N.E., Yang, S.F. and McKeon, T. : 1982. *Biochem. Biophys. Res. Commun.* 104, 765–770.
54. Amrhein, N., Schneebeck, D., Skorupha, H., Tophof, S. and Stockigt, J. : 1981. *Naturwissenschaften*, 67, S.619–620.
55. Jackson, M.B. and Campbell, D.J. : 1976. *New Phytol.* 76, 21–29.
56. Zeroni, M., Jerie, P.H. and Hall, M.A. : 1977. *Planta* 134, 119–125.
57. Bradford, K.J. and Yang, S.F. : 1980. *Plant Physiol.* 65, 322–326.
58. Riov, J. and Yang, S.F. : 1982. *Plant Physiol.*, 69, 687–690.
59. Riov, J. and Yang, S.F. : 1982. *Plant Physiol.*, 70, 136–141.
60. Burg, S.P. and Burg, E.A. : 1969. *Qual. Plant Mater. Veg.* 19, 185–200.
61. Beyer, E.M. : 1976a. *Plant Physiol.* 58, 268–271.

62. Beyer, E.M. : 1976b. *HortScience* **11**, 195–196.
63. Veen, H. : 1979. *Planta* **145**, 467–470.
64. Wang, C.Y. and Adams, D.O. : 1980. *Plant Physiol.* **66**, 841–843.

Regulation of Ethylene Biosynthesis in Auxin-Treated Plant Tissues

Hidemasa Imaseki

INTRODUCTION

Physiological significance of ethylene produced by plants in minute amounts has been well established (1). It is surprising that such a small molecule as ethylene regulates a number of physiological as well as biochemical processes occurring at different times in the life cycle of plants. To regulate such diverse processes at different stages of plant development, ethylene content in tissues must be precisely controlled. In fact, the rate of ethylene production by plant tissues varies remarkably with changes in the physiological state of tissues and by the chemical and physical stimuli imposed on tissues.

In dark-grown dicot seedlings, ethylene produced in apical plumule and hook portions keeps plumule and hook closed, but when seedlings are irradiated by light the rate of ethylene production decreases leading to their opening (2,3). Shortly before petiole abscission occurs, the rate of ethylene production in the abscission zone increases transiently and if ethylene is positively removed from the ambient air to lower the endogenous content of the gas, abscission is delayed (4). Similarly, the ripening of many fruits starts after a rise in the ethylene production rate. It is well known that exogenously applied ethylene in amounts as low as one microliter per liter of air causes abscission and fruit ripening. Flower senescence of the morning glory follows an increase in the rate of production (5). A transient increase in the ethylene production rate is also observed when plant tissues are mechanically injured (6), although its significance is not yet fully known.

Thus, one of the most notable characteristics of ethylene production is the rapid increase and decrease in the rate of formation in tissues. Since ethylene formed in tissues is released into ambient air (or water) by physical diffusion depending on a concentration gradient between the inside and outside of the tissues, the endogenous concentration of the gas will be rapidly decreased if gas formation in cells ceases. Therefore, the rate of ethylene biosynthesis in cells controls the cellular concentration of the gas and is physiologically of significance.

For some time, auxin has been known to increase the production rate of ethylene in vegetative tissues (1,7). Auxin supplied to stem or leaf tissues increases the rate of ethylene production by the tissues after a lag period of about 1 hr (8,9), but a detailed kinetic study with hypocotyl segments excised from etiolated mung bean seedlings revealed that the lag period is as short as 15 min (10), a comparable value to that of auxin action on stem elongation. After the lag period, the rate of ethylene production increases linearly with time for a few hours. Degree of the increase in the rate after the lag period is a function of concentration of auxin supplied to tissues. Concentration of auxin supplied does not affect length of the lag period. In mung bean hypocotyls, indole-3-acetic acid (IAA) at 0.3 μM is the lowest concentration to detect an increase in ethylene production and the maximum rate increase is obtained at 0.5 to 1.0 mM IAA (10). Removal of exogenous auxin from the treated tissues results in a rapid decline in the rate of ethylene production indicating that continuous action of auxin is necessary to maintain the increased production rate by the tissues (8,9).

The increase in ethylene production by auxin action is coupled with synthesis of protein and RNA, because it is inhibited by various inhibitors of protein and RNA synthesis (8,9,11). When cycloheximide was added to the auxin-activated tissues to inhibit further synthesis of protein, the ethylene-producing activity present decreased rapidly with a half-life of about 30 min (9). These results strongly suggest that auxin induces the continuous production of protein(s) which is essential to form ethylene and that the protein is rapidly inactivated in cells.

Auxin-induced ethylene production also is greatly affected by other growth regulators which occur in nature; cytokinins stimulate and abscisic acid (ABA) suppresses the auxin action (10,12,13).

Although methionine has long been known to be a good precursor of ethylene (14,15), the biochemical pathway of conversion of methionine to ethylene has not been fully elucidated until recently. One breakthrough was a finding by Adams and Yang (16) that a methionine metabolite accumulated in apple fruit tissue placed in an anaerobic condition was an immediate precursor of ethylene and this was 1-aminocyclopropane-1-carboxylic acid (ACC). Lürssen et al. (17) also independently found that ACC supplied to leaf segments of soybean and other species was efficiently converted to ethylene and proposed that this was a precursor. Yang and his associates established that many plant tissues form ethylene from ACC supplied to tissues (18) and produce ACC (19).

The activated methionine, S-adenosylmethionine (SAM) has been proposed as an intermediate of ethylene biosynthesis (20,21) and formation of ACC from SAM was suggested (16,17). Proof was presented by Yu et al. (22) and Boller et al. (23)

who found in the extract of ripe tomato fruits an enzyme activity catalyzing ACC formation from SAM in the presence of pyridoxalphosphate: ACC synthase. In auxin-treated tissues, it was also reported that auxin induced production of an enzyme system responsible for ACC formation and that an enzyme which catalyzed ethylene synthesis from ACC was already present in tissues not treated with auxin (24,25).

Parallel to these studies, we have been studying the regulatory mechanism of auxin-induced ethylene production using IAA-treated mung bean (*Vigna radiata* Wilczek) hypocotyl segments, and also found ACC to be a methionine metabolite which was formed in detectable amounts only when the hypocotyl segments were treated with IAA and which converted to ethylene much more efficiently than methionine (26). In this article, we will summarize our recent results on the regulatory mechanism of auxin-induced ethylene production.

REGULATION OF ETHYLENE PRODUCTION BY THE ENDOGENOUS CONTENT OF ACC

We have shown that ACC accumulated when mung bean hypocotyl segments were incubated with IAA at concentrations which increased the rate of ethylene production. When ACC at various concentrations was supplied to tissue segments in the absence of IAA, the rate of ethylene production was almost proportional to the

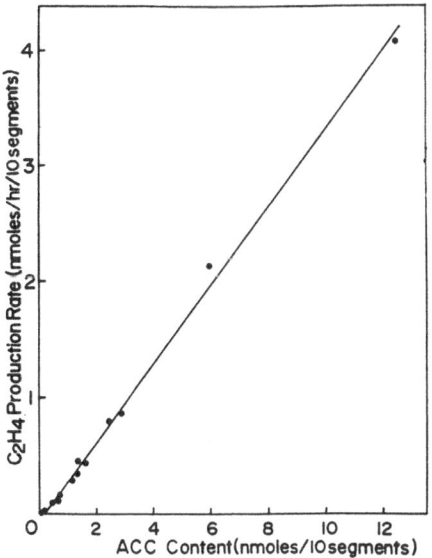

Fig. 1. Relationship between the rate of IAA-induced ethylene production and the endogenous ACC content in mung bean hypocotyl segments. Hypocotyl segments were incubated with various concentrations of IAA for various lengths of time. The rate of ethylene production was measured during one hour immediately before the tissues were extracted for ACC assay.

concentration of ACC in the incubation medium below 1 mM; thus, it was presumed that in the presence of IAA, the endogenous content of ACC which was increased by IAA action might regulate the rate of ethylene production.

Hypocotyl segments can be variously activated in terms of their ethylene producing activity by treating them with various concentrations of IAA for various lengths of time. Using these variously activated tissue segments, we measured both rates of ethylene production and the ACC contents. When the endogenous ACC contents of the tissues were plotted against the rates of ethylene production, a linear relationship was obtained (27) (Fig. 1).

The ethylene producing activity of the IAA-treated tissue segments is further modified by the presence of cytokinin and ABA. At any concentration of IAA between 0.3 μM and 0.1 mM, 5 μM benzyladenine (BA) or kinetin nearly doubled the producing activity and 0.1 mM ABA reduced it by half (10,12). Under these conditions, changes in the ethylene-producing activity of the IAA-treated hypocotyl segments also correlated with changes in the ACC contents (Table I). Since extents of the effects of BA and ABA in the absence of IAA on both the ethylene-producing activity and the increase in the ACC content were very small as compared with those in the presence of IAA, the action of BA and ABA in the presence of IAA is to modify the action of auxin.

These results clearly indicate that auxin increases the synthetic rate of ACC and the endogenous content of ACC primarily controls the rate of ethylene production. It could be argued, however, that auxin might have suppressed degradation and/or inactivation of ACC without affecting the synthetic rate of ACC which might be constitutively high. This is unlikely because radioactive ACC was not detected in the tissue segments supplied with U-[14]C- or 3,4-^{14}C-methionine if the tissue was not treated with IAA (26), indicating that the ACC synthetic activity in the tissue without IAA treatment is exceedingly low.

TABLE I

Effects of Indoleacetic Acid, Benzyladenine and Abscisic Acid on Rate of Ethylene Production and the Endogenous ACC Content in Mung Bean Hypocotyl Segments (27)

Additions[a]	C_2H_4 production rate[b] (nmol/hr · 10 segments)	ACC content[b] (nmol/hr · 10 segments)
+IAA	2.54 ± 0.24 (100)	7.88 ± 0.35 (100)
+IAA+BA	4.42 ± 0.41 (174)	17.50 ± 0.38 (222)
+IAA+ABA	1.14 ± 0.14 (45)	3.94 ± 0.20 (50)
−IAA	0.010 ± 0.001 (100)	0.16 ± 0.03 (100)
−IAA+BA	0.075 ± 0.002 (750)	0.15 ± 0.02 (94)
−IAA+ABA	0.11 ± 0.002 (110)	0.14 ± 0.02 (88)

[a] IAA, 0.5mM ; BA, 5μM ; ABA, 0.1mM
[b] Ten hypocotyl segments were incubated with a medium containing the indicated additions. Ethylene production between the 3rd and 4th hours after start of incubation was determined, then the tissues were extracted for ACC determination.

TABLE II
Effects of Indoleacetic Acid, Benzyladenine and Abscisic Acid on Rate of Ethylene Production
and ACC Synthase Activity in Mung Bean Hypocotyls (27)

Additions[a]	C_2H_4 production rate[b] (nmol/hr·50 segments)		ACC synthase activity[b] (nmol/hr·mg protein)	
+IAA	14.34 ± 0.52	(100)	0.190 ± 0.011	(100)
+IAA+BA	25.55 ± 0.81	(178)	0.383 ± 0.02	(202)
+IAA+ABA	4.68 ± 0.56	(33)	0.129 ± 0.006	(68)
−IAA	0.089 ± 0.005	(100)	0.007 ± 0.002	(100)
−IAA+BA	0.375 ± 0.027	(421)	0.018 ± 0.009	(257)
−IAA+ABA	0.071 ± 0.009	(80)	0.004 ± 0.001	(57)

[a] IAA, 0.5mM ; BA, 5μM ; ABA, 0.1mM
[b] Triplicate lots of 50 segments were incubated with media containing the indicated additions.
Ethylene production was determined between the 3rd and 4th hours after start of incubation,
then at the 4th hour the tissues were extracted for ACC synthase assay. The data of ethylene
production are mean values of the triplicate experiments with standard errors. ACC synthase
was assayed in duplicate.

In the absence of IAA, ABA slightly decreased the ACC content but BA did not
alter significantly it, although BA increased the ethylene-producing activity several
fold. This will occur if cytokinin stimulates both the activity of ethylene formation
from ACC and the increase of synthetic rate of ACC. Ethylene formation from ACC
by hypocotyl segments in the absence of IAA was always slightly stimulated by BA,
and the effectiveness of stimulation was greater as ACC concentrations supplied to
the tissue were lower. Moreover, BA indeed increased the ACC-forming activity even
in the absence of IAA (Table II). Cytokinin also was reported to increase greatly the
ethylene producing activity of mung bean hypocotyls in the presence of Ca^{2+} after a
long lag period (28). Therefore, when the ethylene-producing activity is low as in the
case of the absence of IAA, cytokinin could stimulate significantly ethylene production without affecting the ACC content by stimulating both ethylene formation
from ACC and an increase in ACC synthetic activity.

REGULATION OF THE ACC CONTENT BY ACC SYNTHASE ACTIVITY

ACC is formed from SAM by ACC synthase (22,23), and IAA treatment of mung
bean hypocotyl segments greatly increased ACC synthase activity in the tissue
(Table II). As in the case of the ACC content, the increase of ACC synthase activity
is stimulated by BA and reduced by ABA. Magnitude of the increase in synthase
activity in the presence of BA was comparable to that in ACC content under the
same conditions (Tables I and II). The decrease by ABA of auxin-induced ACC
synthase activity was smaller than that expected from the ACC content, but ABA
added at 0.1 mM to the reaction mixture of ACC synthase inhibited enzyme activity
by about 14%.

These results show that the increase in ACC content by IAA treatment with or without BA or ABA is supported by an increase of ACC synthase activity in the tissue. Thus the major mechanism of hormonal control of ethylene production in vegetative tissues is to regulate the endogenous activity of ACC synthase, which controls the endogenous content of ACC.

IDENTITY OF THE LABILE PROTEIN FORMED BY AUXIN ACTION

The increase in the ethylene-producing activity by auxin is maintained by continuous production of protein(s) which is quite labile in the tissue (8,9). Physiological properties of the protein are similar to those of ACC synthase, but identity of the labile protein must be determined.

Whether an increase in ACC synthase activity by IAA treatment is due to de novo synthesis of the enzyme is one criterion to know if the labile protein is ACC synthase. The increase in ACC synthase activity was severely inhibited by cyclo-heximide and 2-(4-methyl-2,6-dinitroanilino)-N-methylpropionamide (MDMP), inhibitors of mRNA translation, and actinomycin D and α-amanitin, inhibitors of gene transcription (Table III). These inhibitors inhibited neither ethylene formation from ACC by hypocotyl segments nor ACC synthase reaction in vitro. Therefore, the increase in ACC synthase activity by auxin action is a result of de novo synthesis of the enzyme molecules through activation of genes.

Another criterion is whether ACC synthase is labile in the tissue as the labile protein is. Cycloheximide was added to the tissue segments treated with IAA for 4 hr to inhibit further synthesis of ACC synthase, and changes with time in the enzyme activity were measured. If ACC synthase is stable in the tissue, its activity present at the time of cycloheximide addition should be maintained for some time thereafter. The ACC synthase activity decayed by the first order reaction and its half-life was estimated as about 25 min (Fig. 2), thus the enzyme is labile in the tissue being

TABLE III
Effects of Inhibitors of Protein and RNA Synthesis on IAA-induced Ethylene Production and IAA-induced Increase in ACC Synthase Activity

Inhibitors	C_2H_4 production rate[a] (nmol/hr·50 segments)	ACC synthase activity[a] (nmol/hr·mg protein)
Control	17.61 ± 0.04 (100)	0.397 ± 0.003 (100)
CHI 14μg/ml	1.50 ± 0.08 (8.5)	0.022 ± 0.007 (5.5)
CHI 28μg/ml	0.84 ± 0.13 (4.8)	0.016 ± 0.002 (4.0)
MDMP 15μg/ml	1.08 ± 0.05 (6.1)	0.010 ± 0.001 (2.5)
Act. D 25μg/ml	8.84 ± 0.23 (50.2)	0.070 ± 0.003 (17.6)
α-Amanitin 5μg/ml	6.40 ± 0.24 (36.3)	0.042 ± 0.002 (10.6)

Duplicate lots of 50 segments were incubated with 0.5mM IAA+5μM BA media containing the indicated inhibitors. Ethylene production rate was determined between the 3rd and 4th hours after start of incubation, then at the 4th hour the tissues were extracted for ACC synthase assay. CHI, cycloheximide ; Act. D, actinomycin D

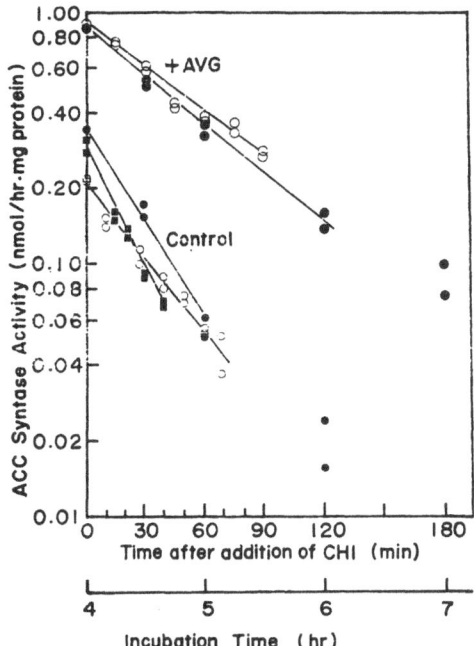

Fig. 2. Decay of ACC synthase activity after cycloheximide was added to the tissues. Hypocotyl segments were incubated with 0.5mM IAA plus 5μM BA in the presence or absence of 0.1mM AVG for four hours, then cycloheximide was added to the incubation medium at a final concentration of 0.1mM. The tissues were extracted at various times for assay of ACC synthase activity. The results of three (for control) and two (for AVG-treated) separate experiments were presented.

inactivated rapidly. The half-life of the activity decay is comparable to that of the ethylene-producing activity (9), and lability of the ethylene-producing activity can be fully attributed to the in vivo lability of ACC synthase.

Mechanism of the inactivation of ACC synthase is not known. Proteolytic degradation of the enzyme is one of the possible mechanisms. Since synthesis and inactivation of ACC synthase are quite rapid, inhibition of inactivation would cause more accumulation of ACC synthase than the control leading to more enhanced ethylene production. However, agents which inhibit proteolytic enzymes such as phenylmethylsulfonyl fluoride and p-chloromercuribenzoate did not affect the rate of auxin-induced ethylene production suggesting that a simple proteolysis may not be the cause.

REGULATION OF SYNTHESIS OF ACC SYNTHASE BY ETHYLENE

Recently we found that aminoethoxyvinylglycine (AVG), an inhibitor of ACC synthase reaction, remarkably increase levels of ACC synthase activity in the IAA-

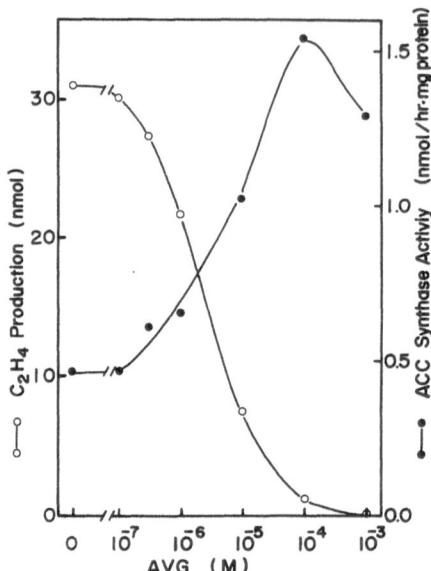

Fig. 3. Relationship between ethylene production and ACC synthase activity of IAA-treated hypocotyl segments as affected by various concentrations of AVG. The tissues were incubated with IAA (0.5 mM) plus BA (5 μM) in the presence of AVG as indicated for four hours. Amounts of ethylene produced during the four-hour period and ACC synthase activity at the 4th hour were plotted against concentrations of AVG.

treated tissue, while it severely inhibited ethylene production (27). There is a clear inverse relationship between inhibition of ethylene production and the increase in ACC synthase activity (Fig. 3), indicating that inhibition of the production is closely related to the increase in enzyme activity. A concentration of AVG which inhibited almost completely ethylene production (0.1 mM) caused 3 to 4-fold increase in ACC synthase activity as compared to the control. As the AVG-induced increase in ACC synthase activity was also inhibited by inhibitors of translation and transcription, the increase was due to accumulation by *de novo* synthesis of ACC synthase molecules.

AVG is a potent inhibitor of ACC synthase reaction and its presence abolishes formation of the reaction products, ACC and methylthioadenosine (MTA). Therefore, the AVG action may be related to release from the negative feedback control of synthesis of ACC synthase by the reaction products. Exogenous supply of 1 mM ACC to mung bean hypocotyl segments treated with both IAA and AVG suppressed the activity increase caused by AVG by about 50%, but MTA had no effect at all (Table IV). Ethylene (10 μl/liter) also suppressed the increase induced by AVG to a similar degree to ACC. Suppression of the AVG-induced increase in ACC synthase activity by ACC or ethylene was not complete, and the maximum suppression was 50 to 60% at higher concentrations of applied ACC or ethylene.

Since ethylene formation from ACC is not inhibited by AVG, the suppressive effect of ACC may be attributed to the effect of ethylene formed from applied

TABLE IV
Effects of AVG, ACC and Ethylene on IAA-induced Increase in ACC Synthase Activity

Additions	ACC synthase activity (nmol/hr·mg protein)	Activity increment by AVG (nmol/hr·mg protein)
Control	0.585 ± 0.008 (100)	
+AVG	1.990 ± 0.014 (340)	1.405 (100)
+AVG+ACC	1.276 ± 0.024 (218)	0.691 (49)
+AVG+C_2H_4	1.107 ± 0.031 (189)	0.522 (37)
+AVG+ACC+C_2H_4	1.150 ± 0.015 (197)	0.565 (40)

Control tissues were incubated with medium containing 0.5mM IAA and 5μM BA. AVG, 0.1 mM ; ACC, 1mM ; ethylene, 10μl/liter. The tissue segments were extracted for ACC assay at the 4th hour after start of incubation.

TABLE V
Effects of AVG on Ethylene Production and ACC Synthase Activity in Mung Bean Hypocotyls in the Absence of IAA

Additions	ACC synthase activity (nmol/hr·mg protein)	C_2H_4 production rate (nmol/hr·50 segments)
Control	0.007 ± 0.003 (100)	0.089
+AVG 0.1mM	0.039 ± 0.005 (557)	0.009
+BA 5μM	0.018 ± 0.009 (100)	0.375
+BA+AVG	0.092 ± 0.002 (511)	0.013
+ABA 0.1mM	0.004 ± 0.001 (100)	0.071
+ABA+AVG	0.024 ± 0.001 (600)	0.009

Ethylene production rate was determined between the 3rd and 4th hours after start of incubation, and ACC synthase activity was determined at the 4th hour.

ACC. If the action of ACC is independent of the action of ethylene, simultaneous application of ACC and ethylene to the IAA-AVG-treated tissue should produce greater effect than individual treatment. Since this was not the case (Table IV), the suppressive effect of ACC is considered to be due to the effect of ethylene.

Suppression by ethylene of the AVG-induced increase of ACC synthase activity was observed at and above 0.3 μl/liter and reached maximum at about 10 μl/liter. When ethylene in air at 0.3 μl/liter is equilibrated with the aqueous phase of tissue, the total amounts of ethylene present in hypocotyl segments is roughly 7.5 pl per 10 segments, and this amount of ethylene is produced by the tissue segments without IAA treatment in about 20 min. Therefore, it is highly likely that the amount of ethylene produced by the tissue in the absence of IAA is enough to repress partially synthesis of ACC synthase, and that produced by the tissue treated with 1 μM IAA for 1 hr results in maximum repression. In fact, when low ethylene production by the tissue in the absence of IAA was inhibited by AVG, ACC synthase activity increased several fold regardless of the presence or absence of BA or ABA (Table V). Therefore, in plant tissues under natural conditions where ethylene production is normally very low, the produced ethylene regulates its own synthesis through controlling the synthesis of ACC synthase.

POSSIBLE REGULATION OF INACTIVATION OF ACC SYNTHASE

A remarkable increase of synthesis of ACC synthase by inhibition of ethylene formation by AVG cannot be fully understood only by release from the feedback repression by ethylene, because the effect of AVG was only partially reversed by ethylene. As already discussed, ACC synthase in tissues is rapidly inactivated, and any factors which delay inactivation of the enzyme would cause increased enzyme accumulation. ACC synthase activity in the AVG-treated tissue had a longer half-life than that in the control tissue (Fig.2). The enzyme once mixed with AVG, then freed of AVG by gel filtration did not change its activity before and after the AVG treatment. Therefore, the lengthened half-life of the enzyme in the AVG-treated tissue is considered to be an effect of AVG on *in vivo* inactivation of the enzyme. This may indicate presence of a control site for inactivation of ACC synthase. Although we assumed in a previous section that a simple proteolysis might not be the cause of inactivation, validity of the assumption must be seriously examined by further experiments.

POSSIBLE REGULATION OF ETHYLENE FORMATION FROM ACC

Many tissues of various species produce ethylene without a significant lag period when ACC is supplied to tissues indicating that the ethylene-forming activity is constitutive in cells (*18,24–26*). There is little evidence that ethylene formation from ACC is regulated by endogenous factors. In postclimacteric Avocado fruits (*29*) and senescent carnation flowers (*30*), a large amount of ACC was present while ethylene production was quite low. In these cases, however, it is presumed that the ethylene-forming enzyme is inactivated in deteriorating cells whereas ACC synthase is still active. Most of the endogenous factors and environmental stresses (*31*) that alter the rate of ethylene production regulate ACC synthesis, a step before the final step of the biosynthetic sequence.

   In 1973, Sakai and Imaseki (*32*) found a protein in mung bean hypocotyls which inhibited ethylene production of stem segments of IAA-treated mung bean and pea seedlings. The protein was isolated and found to have a molecular weight of 112,000, consisting of two subunits of identical size. If hypocotyl segments are incubated with a medium containing IAA and the protein or if the protein is added to the tissue segments previously treated with IAA, the ethylene-producing activity is severely inhibited. But if the inhibited tissue segments are transferred to a fresh medium without the protein, the rate of ethylene production is restored (*33*). The production rate after removal of the protein from the inhibited tissue was often greater than the control. Presence of similar proteins has been known for a variety of legume seeds (*34*) and for a seaweed (*35*). The protein does not degrade, inactivate or bind irreversibly IAA, which is essential to the induced ethylene production. Although inhibitory effect of the protein is greatest on ethylene production, the protein also inhibits, to a lesser degree, many other metabolisms including protein and RNA synthesis (*36*). The inhibitory effect of the protein, however, is not mediated through its effects on protein and RNA synthesis, because induction by auxin of the ethylene-producing system is not inhibited by the protein, rather it inhibits only synthesis of ethylene catalyzed by the ethylene-producing system. This was finally proved by the results that the protein did not inhibit auxin-induced ACC formation, but inhibited ethylene formation from ACC (*26*).

We also found that under the conditions where IAA-induced ethylene production was inhibited more than 80%, the applied inhibitory protein was present only in the epidermis of hypocotyl segments (*33*). To know possible localization in the stem segments of the ethylene-forming enzyme, hypocotyl segments were peeled to separate into epidermal strips and peeled segments consisting of cortex, vascular systems and pith, and ACC was supplied to them separately. The peeled segments produced ethylene at a rate similar to or a little higher than the unpeeled segments, whereas the epidermal strips did not produce ethylene from ACC at all. A striking result was that the inhibitory protein inhibited ethylene formation from ACC by the unpeeled segments, but not at all from that by the peeled segments (Table VI). Moreover, ethylene formation occurring in a combined incubation of once separated epidermal strips and peeled segments with ACC was not inhibited by the protein. The peeled segments exposing cortex cells did not inactivate the protein. Thus, epidermis is essential for the inhibitory protein to inhibit ethylene formation from ACC, occurring in the inner cells of a segment.

At present, this result is hard to explain. It seems that the inhibitory protein interacts with epidermal cells and the affected epidermal cells transmit some signals to the inner cells to reduce the ethylene-forming activity in the latter cells. This is an attractive hypothesis, but is purely speculative. Ethylene production in many plant tissues is known to be affected by treatments of tissues which modify physical

TABLE VI
Effects of the Inhibitory Protein on ACC-dependent Ethylene Production of Unpeeled and Peeled (Epidermis Removed) Segments of Mung Bean Hypocotyls

	Ethylene production rate (nmol/hr · 10 segments)	
	Unpeeled	Peeled
Control	2.35	2.36
+Inhibitory Protein (44 μg/0.5ml)	0.82	2.32

Ten hypocotyl segments, either peeled or unpeeled, were incubated with 1mM ACC and ethylene production rate during the first hour was measured.

or biochemical properties of cellular membranes (37–39), and it has been considered
that the cellular membranes are associated with the controlling mechanism of ethylene production (7). It may be possible that signals generated by conformational
changes of the cellular membranes of one cell are transferred to other cells through
the membranes, and that in highly organized organs such as stem and leaf, epidermal
cells control biochemical reactions occurring in other cells. It was reported that leaf
senescence was retarded by removing the epidermal layers (40) and that wound
signals generated at the wounded site were transmitted to other parts to increase
ethylene production (41).

CONCLUDING REMARKS

Ethylene production in plant tissues is regulated at various steps, but the major step
to be regulated is ACC formation catalyzed by ACC synthase. In auxin-treated
tissues, auxin induces continuously *de novo* synthesis of ACC synthase and the increasing enzyme supplies an increasing amount of ACC, an immediate precursor of
ethylene, to the ethylene-forming enzyme which is a constitutive enzyme. The rate
of ethylene production is primarily determined by the endogenous content of ACC.
However, not all of ACC synthesized is converted to ethylene, but only a small
fraction is used for ethylene synthesis.

Nature of the ethylene-forming enzyme is not known, and its biochemical
characterization is an important problem to be studied in future. An enzyme activity
forming ethylene from ACC in an extract of etiolated pea seedlings was reported
(42). We also established with the extract of mung bean hypocotyls a cell-free system
producing ethylene from ACC in the presence of NADH, Mn^{2+} and unidentified
cofactors contained in an ethanol extract of the hypocotyls (Takakuwa, Suzuki and
Imaseki, unpublished data). However, horseradish peroxidase catalyzes the same reaction and the ethylene-forming activities in the mung bean extract were separated
on a DEAE-cellulose column concomitant with peroxidase activity (Suzuki and
Imaseki, unpublished data). Moreover, an enzyme extract from the epidermal strips
which had no activity to form ethylene from ACC also contained the ethylene-
forming activity in the above mixture. Therefore, our *in vitro* activity of ethylene
formation from ACC must be catalyzed by peroxidase. Although the possibility that
an isoenzyme of peroxidases functions as the ethylene-forming enzyme cannot be
ruled out, properties of ethylene-forming reaction by peroxidase or by the crude
extract are different from those of *in vivo* formation of ethylene from ACC in several
respects. An ethylene-forming reaction that reflects the *in vivo* reaction must be
sought before the regulatory mechanism of the final step of ethylene biosynthesis is
understood.

The auxin action to induce synthesis of ACC synthase is through activation of
mRNA synthesis. Since auxin is known to increase the content of mRNA specific
to cellulase in pea stem (43), it is likely that synthesis of mRNA specific to ACC
synthase is activated by auxin action. However, there is a slight possibility that auxin
induces other mRNAs coded for protein factors which regulate translation of mRNA

for ACC synthase, since it is apparent that auxin induces compositional changes of several mRNAs in soybean tissues (44). Thus, whether or not auxin indeed increases mRNA specific for ACC synthase should be conclusively studied by determining the mRNA content.

In vivo inactivation of ACC synthase occurs in the first order reaction with a half-life of activity decay of about 25 min. Therefore, the content of ACC synthase in tissues at a certain time is determined by the rate of enzyme synthesis at that time. If the rate of enzyme synthesis is decreased for any reason, for example, by lowered auxin content or increased ABA content, the ACC synthase activity will decrease quickly leading to rapid reduction in the ethylene production rate. Since the induction of ACC synthase by auxin also proceeds rapidly (the lag period is about 15 min), if the endogenous content of auxin or cytokinin increases, the rate of ethylene production increases rapidly. These properties of the ethylene producing enzyme system well account for rapid changes in the ethylene production rate normally observed in nature.

Acknowledgement
Studies described in this article were carried out in collaboration with Isao Todaka and Hiroshi Yoshii, and were supported in part by grants from the Ministry of Education, Science and Culture, Japan.

REFERENCES

1. Abeles, F.B. : 1973. Ethylene in Plant Biology, Academic Press, New York and London.
2. Kang, B.G., Yocum, C.S., Burg, S.P. and Ray, P.M. : 1967. Science 156, 958–959.
3. Goeschl, J.D., Pratt, H.K. and Bonner, B.A. : 1967. Plant Physiol. 42, 1077–1080.
4. Jackson, M.B. and Osborne, D.J. : 1970. Nature 225, 1019–1022.
5. Kende, H. and Hanson, A.D. : 1976. Plant Physiol. 57, 523–527.
6. Saltveit, Jr., M.E. and Dilley, D.R. : 1978. Plant Physiol. 61, 447–450.
7. Lieberman, M. . 1979. Annu. Rev. Plant Physiol. 30, 533–591.
8. Kang, B.G., Newcomb, W. and Burg, S.P. : 1971. Plant Physiol. 47, 504–509.
9. Sakai, S. and Imaseki, H. : 1971. Plant Cell Physiol. 12, 449–459.
10. Imaseki, H., Kondo, K. and Watanabe, A. : 1975. Plant Cell Physiol. 16, 777–787.
11. Abeles, F.B. : 1966. Plant Physiol. 41, 585–588.
12. Kondo, K., Watanabe, A. and Imaseki, H. : 1975. Plant Cell Physiol. 16, 1001–1007.
13. Fuchs, Y. and Lieberman, M. : 1968. Plant Physiol. 43, 2029–2036.
14. Lieberman, M., Kunishi, A.T., Mapson, L.W. and Wardale, D.A. : 1966. Plant Physiol. 41, 376–382.
15. Burg, S.P. and Clagett, C.O. : 1967. Biochem. Biophys. Res. Commun. 27, 125–130.
16. Adams, D.O. and Yang, S.F. : 1979. Proc. Natl. Acad. Sci. U.S.A. 76, 170–176.
17. Lürssen, K., Naumann, K. and Schröder, R. : 1979. Z. Pflanzenphysiol. 92, 285–294.
18. Cameron, A.C., Fenton, C.A.L., Yu, Y., Adams, D.O. and Yang, S.F. : 1979. Hort Science 14, 178–180.
19. Bradford, K.J. and Yang, S.F. : 1980. Plant Physiol. 65, 322–326.
20. Burg, S.P. : 1973. Proc. Natl. Acad. Sci. U.S.A. 70, 591–597.
21. Adams, D.O. and Yang, S.F. : 1977. Plant Physiol. 60, 892–896.
22. Yu, Y., Adams, D.O. and Yang, S.F.: 1979. Arch. Biochem. Biophys. 198, 280–286.
23. Boller, T., Herner, R.C. and Kende, H. : 1979. Planta 145, 293–303.

24. Jones, J.F. amd Kende, H. : 1979. *Planta* 146, 649–656.
25. Yu, Y., Adams, D.O. and Yang, S.F. : 1979. *Plant Physiol.* 63, 589–590.
26. Yoshii, H., Watanabe, A. and Imaseki, H. : 1980. *Plant Cell Physiol.* 21, 279–291.
27. Yoshii, H. and Imaseki, H. : 1981. *Plant Cell Physiol.* 22, 369–379.
28. Lau, O. and Yang, S.F. : 1974. *Planta* 118, 1–6.
29. Hoffman, N.E. and Yang, S.F. : 1980. *J. Amer. Soc. Hort. Sci.* 105, 492–495.
30. Bufler, G., Mor, Y., Reid, M.S. and Yang, S.F. : 1980. *Planta* 150, 439–442.
31. Apelbaum, A. and Yang, S.F. : 1981. *Plant Physiol.* 68, 594–596.
32. Sakai, S. and Imaseki, H. : 1973. *Planta* 113, 115–125.
33. Sakai, S. and Imaseki, H. : 1973. *Plant Cell Physiol.* 14, 881–892.
34. Sakai, S. : 1975. *Plant Cell Physiol.* 16, 529–532.
35. Watanabe, T. and Kondo, N. : 1976. *Agric. Biol. Chem.* 40, 1877–1878.
36. Todaka, I., Watanabe, A. and Imaseki, H. : 1978. *Plant Cell Physiol.* 19, 561–571.
37. Mattoo, A.K., Baker, J.E., Chalutz, E. and Lieberman, M. : *Plant Cell Physiol.* 18, 715–719.
38. Imaseki, H. and Watanabe, A. : 1978. *Plant Cell Physiol.* 19, 345–348.
39. Odawara, S., Watanabe, A. and Imaseki, H. : 1977. *Plant Cell Physiol.* 18, 569–576.
40. Kasamo, K. : 1976. *Plant Cell Physiol.* 17, 1297–1307.
41. Saltveit, Jr., M.E. and Dilley, D.R. : 1979. *Plant Physiol.* 64, 417–420.
42. Konze, J.R. and Kende, H. : 1979. *Plant* 146, 293–301.
43. Verma, D.P.S., Maclachlan, G.A., Byrne, H. and Ewings, D. : 1975. *J. Biol. Chem.* 250, 1019–1026.
44. Zurfluh, L.L. and Guilfoyl, T.J. : 1982. *Plant Physiol.* 69, 332–337.

Natural Abundance Ratios of Isotopes of Hydrogen, Carbon, Nitrogen, Oxygen, and Sulfur in Metabolic Kinetics

Bruce N. Smith

INTRODUCTION

An element is uniquely characterized by the number of protons in its nucleus, but the number of neutrons associated with these protons can vary. As a result an element may have several isotopes differing in atomic weight and stability but not appreciably in chemical properties. Relatively few of the possible isotopes of any element are stable; of the 1,500 or so known isotopes, only about 280 are non-radioactive (1). The isotopes of an element have similar chemical properties but the difference in mass will affect the vibration frequency leading to differences in reaction rates between isotopes.

Among elements of great importance to living organisms are hydrogen, carbon, nitrogen, oxygen and sulfur. We will concentrate in this paper on these five elements. Most of the research done to date on isotopic fractionation in organisms has been done on these elements. The technique has so far been applied to relatively few systems and thus would seem to represent a new frontier in plant biochemistry.

ISOTOPE RATIOS IN METABOLISM

Let us consider an equilibrium process:
$$a\ A_1 + b\ B_2 \rightleftharpoons a\ A_2 + b\ B_1$$
where A and B are molecules having an element as a common constituent. The subscripts 1 and 2 indicate that the molecules contain only the light or heavy molecule, respectively. The equilibrium constant, K, can be expressed in terms of the partition

functions, Q, as :

$$K = \frac{Q\,(A_2)}{Q\,(A_1)} \Bigg/ \frac{Q\,(B_2)}{Q\,(B_1)}$$

This means that the equilibrium constant K is the quotient of two partition function ratios, one for the two molecules of A, and one for B. Thus a calculation of the partition functions allows the calculations of equilibrium constants for chemical reactions. The partition function Q of a molecule is defined by

$$Q = \sum_n g_n^{-E_n/kT}$$

where the summation extends over all the allowed energy levels E_n of the molecules and g_n is the statistical weight of the nth level of E_n (2). Calculation of partition function ratios of isotopic molecules demonstrates the relationship between the equilibrium constant and the vibration frequency and temperature (3,4). As temperatures increase to very high values, K approaches unity and there is no fractionation (5).

Usually we are interested in the fractionation factor, rather than in the equilibrium constant. The fractionation factor, α, is defined as the ratio of the numbers of any two isotopes in one chemical compound divided by the corresponding ratio for the other chemical species. At equilibrium, in general $\alpha = K^{1/n}$, where n is the maximum number of exchangeable atoms (2,6).

The temperature-dependent exchange reaction ($^{18}O/^{16}O$) for calcite and water has been used to establish an isotopic thermometer which has become a powerful tool for measuring earth surface temperature fluctuations during geologic time (7,8).

It has been recognized for over thirty years that, for greatest precision, a comparative method of measurement using the null-point principle must be employed (9). As a consequence, the amplitude of each isotope is not measured individually, but a ratio is measured. This ratio is quickly compared to the isotope ratio of a standard gas. The two ratios are then compared to yield a value termed S and expressed as parts per thousand (‰). Table I lists the average isotopic ratio of light elements of biological importance.

Table II lists the 5 notations for the various elements and references the standards against which they are compared. In practice, the fractionation factor describing a process can be defined as:

$$\alpha = 1 + (\delta_1 - \delta_2)/1000$$

TABLE 1
Average Isotopic Ratio of Light Elements on Earth

Isotopes	Ratio
$^1H\,/\,^2H$	6,410
$^{12}C\,/\,^{13}C$	89.1
$^{14}N\,/\,^{15}N$	277
$^{16}O\,/\,^{18}O$	501
$^{32}S\,/\,^{34}S$	22.6

TABLE II
Definition of Notations and Standards Used for Each Element

---

(1) $\delta D \ \permil = \dfrac{D/H_{sample} - D/H_{std}}{D/H_{std}} \times 1000$

standard mean ocean water (s.m.o.w.)

(2) $\delta\,^{18}O \ \permil = \dfrac{^{18}O/^{16}O_{sample} - ^{18}O/^{16}O_{std}}{^{18}O/^{16}O_{std}} \times 1000$

standard s.m.o.w. or p.d.b. limestone

(3) $\delta\,^{13}C \ \permil = \dfrac{^{13}C/^{12}C_{sample} - ^{13}C/^{12}C_{std}}{^{13}C/^{12}C_{std}} \times 1000$

standard p.d.b. limestone

(4) $\delta\,^{14}N \ \permil = \dfrac{^{15}N/^{14}N_{sample} - ^{34}S/^{32}S_{std}}{^{15}N/^{14}N_{std}} \times 1000$

standard atmospheric N_2

(5) $\delta\,^{34}S \ \permil = \dfrac{^{34}S/^{32}S_{sample} - ^{34}S/^{32}S_{std}}{^{34}S/^{32}S_{std}} \times 1000$

standard meteorite troilite (FeS)

Most biological systems are not in chemical equilibrium with their environment although they do obey the laws of thermodynamics. Thus isotopic fractionation in living systems can best be explained on a kinetic basis rather than in terms of equilibrium. Thus rate constants are important. For a simple reaction where molecule A containing isotopes 1 and 2 is transformed to molecule B containing the same two isotopes, the rate constants K_1 and K_2 can be applied to the reactions:

$$A_1 \xrightarrow{\ K_1\ } B_1, \quad A_2 \xrightarrow{\ K_2\ } B_2$$

and the isotopic fractionation $\alpha = K_1/K_2$. Harrison and Thode (10) attempted to accurately describe a kinetic isotope affect:

$$\alpha = K_1 = \left[\frac{Q^1\ (A_2)}{Q^1\ (A_1)} \middle/ \frac{Q^1\ (A_2^*)}{Q^1\ (A_1^*)} \right] \left[\frac{M_2^*}{M_1^*} \right]^{\frac{1}{2}}$$

where A* is the activated complex, M* the reduced mass of the activated complex and Q^1, the reduced partition function. Because of the difficulty in identifying the active complex in a biological reaction, few attempts have been made to quantitatively define a unidirectional metabolic system and predict the isotope effects involved.

Hydrogen

Approximately one atom of hydrogen in 6,400 in seawater is ^{2}H. The large mass difference between ^{2}H and ^{1}H results in a large fractionation which can easily be measured despite the rather low abundance of the heavy isotope (Table I). Plants appear to concentrate ^{1}H preferentially over ^{2}H (*11,12*). The fractionation appears to be associated with photosynthesis (*13*). Apparently photosynthesis forms a pool of reductant (presumably NADPH) enriched in the light isotope (*14*). Carboxyl and hydroxyl hydrogen are exchanged very quickly but apparently hydrogen bound to carbon does not readily exchange and can represent a rather long-term isotopic record (*15*). In addition to photosynthetic fractionation, an additional fractionation occurs in lipid synthesis (*13,14*). In general the total ^{2}H/^{1}H ratio for plant tissue reflects meteoric water (*13*). This can be used to determine the relative elevation or distance inland along a storm track where a particular plant has been grown (*16*). Ziegler *et al.* (*17*) found isotopic differences between C_3, C_4 and CAM species grown together in a growth chamber. They attribute the differences to differnces in transpiration water loss. Much more work needs to be done on hydrogen fractionation in living systems.

Carbon

Carbon isotopes have proven to be a very effective means of distinguishing between plants with the C_3 and those with the C_4 photosynthetic pathway (*18–20*). The isotopic discrimination by C_4, CAM plants and aquatic plants is largely due to diffusion (*21,22*). C_3 species exhibit enzymatic fractionation of the isotopes (*23*). Plants exhibiting Crassulacean Acid Metabolism (CAM) may be either obligate or facultative. When they are in the CAM mode, the carbon fixed is, like that in C_4 plants, rather heavy. Whereas, when CAM plants are operating in the C_3 mode, they have isotopic values similar to C_3 plants. Thus obligate CAM plants have isotopic values similar to C_4 plants while facultative CAM plants may exhibit the entire range of δ^{13}C values from −8 to −36 ‰ (*24*).

In addition to photosynthetic fractionation, isotopic differences are noted between different chemical fractions (*25*) with the lipid fraction having the lowest ratio. The lipid fraction has a low ^{13}C/^{12}C ratio as a result of fractionation during the oxidation of pyruvate to acetyl coenzyme A (*26*).

Small isotopic differences exist between C_4 monocots and C_4 dicots (*20*). Small isotopic differences have been shown due to temperature and light intensity with minimal fractionation associated with conditions giving maximal growth (*27*).

The effects of increasing CO_2 concentration is to increase the isotopic discrimination (*23*). This is true for aquatic plants, C_3 and C_4 species (*21*). When the CO_2 level is high, diffusional resistance is much greater than carboxylation resistance. Of course any change in the isotopic composition of the source carbon will be reflected in the isotopic ratio of the plant material.

Small, but real, isotopic differences may prove very useful in the investigation of environmental influences on plants as well as changes in biochemistry. Specific differences in intramolecular distribution of isotopes may give specific information about metabolic events (28).

Nitrogen

Natural abundance ratios of $^{15}N/^{14}N$ allow interpretation of the nitrogen cycle. Terrestrial nitrogen fixing plants have a lower $^{15}N/^{14}N$ ratio than non-nitrogen fixing plants (29). Soil organic nitrogen and animal tissues are enriched in ^{15}N relative to atmospheric nitrogen. Nitrogen-fixing plants have values close to atmospheric nitrogen. Nodulating and non-nodulating strains of soybeans showed consistent $\delta^{15}N$ differences as expected (30). Fertilizer nitrogen fixed via the Haber process shows considerable isotopic fractionation and thus can be readily distinguished from biologically fixed nitrogen (30). Denitrification often results in isotopic discrimination of more than 20 ‰ (29). However the isotope affect associated with denitrification ranges from 0 ‰ at low NO_2^- concentration to about 20 ‰ at high NO_2^- concentration. The isotopic composition of animal nitrogen can be used to obtain information about an animals diet (e.g. ratio of legume to non-legume food sources) (31).

Oxygen

Oxygen isotopes have been used very effectively by geochemists to determine paleotemperatures (8). They have utilized invertebrate carbonate exoskeletons which are thought to be deposited in equilibrium with seawater. A linear relationship between carbonate $^{18}O/^{16}O$ ratios at seawater temperature was established (7) and has proven extremely useful in the study of past climatic variations (8). More recently attempts have been made to use oxygen isotopes from cellulose to reconstruct paleoclimates (32). Water oxygen is fractionated during transpiration (33).

Atmospheric oxygen is enriched in ^{18}O in comparison with ocean water oxygen. However, if oxygen is released from water by electrolysis or ultraviolet radiation, the oxygen is depleted in ^{18}O. Photosynthetic oxygen, on the other hand, is enriched in ^{18}O compared with the source water (34). Since respiration involves preferential uptake of ^{16}O, the combination of photosynthesis and respiration could account for the observed relationship between seawater and atmospheric oxygen (35). Metzner and Gerster (36) have suggested that an explanation for photosynthetic O_2 isotopes is that the sources of oxygen is CO_2 rather than water. Generally speaking, however, abundant evidence exists for water as the source of oxygen (37). Light oxygen released in photosynthesis must be shunted to a still-unknown acceptor.

Sulfur

The major fractionation step for sulfur isotopes results from the reduction of sulfate by bacteria and reduction of sulfite by yeast (38). In each case, ^{32}S is enriched in the product sulfide. Incorporation of cellular sulfur by uptake of sulfate resulted in a very small fractionation (<1.005) and release of H_2S by hydrolysis of the sulfhydryl bond (-C-SH) from cysteine yielded slight enrichment of ^{32}S (ca. 1.005) (39). During oxidation of H_2S, either chemosynthetically or photosynthetically, the fractionation is less than during the reduction step (5). No enrichment resulted from oxidation of elemental sulfur to sulfate (*38*). Similar results for *Chlorella* (*40*) and spinach chloroplasts (*41*) have recently been reported.

Study of isotopic discrimination in plants would seem to hold great promise for future research on the kinetics of metabolism and the effect of environmental parameters. I would strongly encourage others to consider the possibilities of using stable isotopes.

REFERENCES

1. Fowler, W.A. : 1967. Nuclear Astrophysics, Amer. Phil. Soc., Philadelphia.
2. Hoefs, J. : 1983. Stable Isotope Geochemistry, Springer-Verlag, Heidelberg.
3. Bigeleisen, J. and Mayer, M.G. : 1947. *J. Chem. Phys.* 15, 261–267.
4. Urey, H.C. : 1947. *J. Chem. Soc.* 562–581.
5. Kaplan, I.R. : 1975. *Proc. R. Soc. Lond. B.* 189, 183–211.
6. Rankama, K. : 1954. Isotope Geology, Pergamon, London.
7. Epstein, S., Buchsbaum, H.A. and Lowenstam, H.A. : 1953. *Bull. Geol. Soc. Amer.* 64, 1315–1325.
8. Bowen, R. : 1966. Paleotemperature Analysis, Elsevier, Amsterdam.
9. McKinney, C.R., McCrea, J.M., Epstein, S., Allen, H.A. and Urey, H.C. : 1950. *Rev. Scient. Instruct.* 21, 724–730.
10. Harrison, A.G. and Thode, H.C. : 1957. *Trans. Farad. Soc.* 53, 1–4.
11. Zborowski, G., Ponticorvo, L. and Rittenberg, D. : 1967. *Proc. Natl. Acad. Sci. U.S.A.* 58, 1660–1663.
12. Schiegl, W.L. and Vogel, J.C. : 1970. *Earth Planet Sci. Lett.* 7, 307–313.
13. Smith, B.N. and Epstein, S. : 1970. *Plant Physiol.* 46, 738–742.
14. Estep, M.F. and Hoering, T.C. : 1981. *Plant Physiol.* 67, 474–477.
15. Smith, B.N. and Jacobson, B.S. : 1976. *Plant Cell Physiol.* 17, 1089–1092.
16. Smith, B.N. : 1975. *Naturwissenschaften* 62, 390.
17. Ziegler, H., Osmond, C.B., Stichler, W. and Trimborn, P. : 1976. *Planta* 128, 85–92.
18. Smith, B.N. and Epstein, S. : 1971. *Plant Physiol.* 47, 380–384.
19. Bender, M.M. : 1971. *Phytochemistry* 10, 1239–1244.
20. Troughton, J.H., Card, K.A. and Hendy, C.H. : 1974. *Carnegie Inst. Washington Yearbook* 73, 768–780.
21. Vogel, J.C. : 1980. Fractionation of the Carbon Isotopes During Photosynthesis, Springer-Verlag Berlin.
22. O'Leary, M.H. : 1981. *Phytochemistry* 20, 553–567.
23. Park, R. and Epstein, S. : 1960. *Geochim. Cosmochim. Acta* 21, 110–126.
24. Osmond, C.B., Allaway, W.G., Sutton, B.G., Troughton, J.H., Queiroz, O., Luttge, U. and Winter, K. : 1973. *Nature* 246, 41–42.
25. Park, R. and Epstein, S. : 1961. *Plant Physiol.* 36, 133–138.
26. DeNiro, M.J. and Epstein, S. : 1977. *Science* 197, 261–263.
27. Smith, B.N., Oliver, J. and McMillan, C. : 1976. *Botan. Gaz.* 137, 99–104.
28. Monson, K.D. and Hayes, J.M. : 1982. *Geochim Cosmochim. Acta* 46, 139–149.
29. Wade, E., Kadonaga, T. and Matsuo, S. : 1975. *Geochem. J.* 9, 139–148.
30. Kohl, D.A., Shearer, G. and Harper, J.E. : 1980. *Plant Physiol* 66, 61–65.
31. DeNiro, M.J. and Epstein, S. : 1981. *Geochim. Cosmochim. Acta* 45, 341–351.
32. Epstein, S., Thompson, P. and Yapp, C.J. : 1977. *Science* 198, 1209–1215.
33. Forstel, H. : 1978. *In* Environmental Biogeochemistry and Geomicrobiology, Vol. 3 (Krumbein, W.E., ed.), pp.811–835, Ann Arbor Science, Ann Arbor.
34. Dole, M. and Jenks, G. : 1944. *Science* 100, 409.
35. Lane, G.a. and Dole, M. : 1956. *Science* 123, 574–576.

36. Metzner, H. and Gerster, R. : 1976. *Photosynthetica* 10, 302–306.
37. Stemler, A. and Radmer, R. : 1975. *Science* 190, 457–458.
38. Kaplan, I.R. and Rittenberg, S.C. : 1964. *J. Gen. Microbiol.* 34, 195–212.
39. Rees, C.E. : 1973. *Geochim. Cosmochim. Acta* 37, 1141–1162.
40. Spedding, D.J., Ziegler, I., Hampp, R. and Ziegler, H. : 1980. *Z. Pflanzenphysiol.* 97, 205–214.
41. Hampp, R., Spedding, D.J., Ziegler, I. and Ziegler, H. : 1980. *Z. Pflanzenphysiol.* 99, 113–119.

Chilling Stress in Plants and Their Products: Causes, Responses and Amelioration

Douglas Graham

INTRODUCTION

That certain plants are injured by exposure to temperatures a little above the freezing point of their tissues has been recognised for at least 200 years (1). The term chilling injury was introduced (2) to distinguish the effects from freezing injury. Chilling injury occurs in plants (and their fruits) of lowland tropical and sub-tropical origin when they are exposed to low temperatures ranging from just above the freezing point of the tissue (usually about $-1.5°C$) up to as high as $15-20°C$ for some sensitive species. Some examples of chilling-sensitive and chilling-resistant plants are shown Table I. In a few cases, *e.g.* peach, the fruit is somewhat chilling sensitive, whereas the tree is cold resistant. These are thus examples where a clear distinction cannot be drawn between chilling sensitivity and chilling resistance. The limiting temperature below which injury occurs varies with species or variety and, in most cases, follows a time-temperature relationship, that is, for a particular plant or fruit, injury will occur more quickly at lower temperatures and less quickly at higher temperatures. The injury becomes most apparent after return to higher, ambient temperatures and may be manifested in various symptoms. These include necrosis, shown by the browning of the tissues (*e.g.* in fruits of avocado and pineapple, tubers of sweet potato and leaves of many species), surface pitting, due to collapse of cells (*e.g.* fruit of papaya and leaves) wilting (*e.g.* in leaves) poor ripening (*e.g.* tomato fruit), poor or arrested seed germination, abnormal pollen formation and fruit setting and poor growth (for further examples, see 3). Plants of temperate zone origin are chilling-resistant and do not suffer chilling injury except in some cases of very

175

TABLE I

Some Chilling-sensitive and Chilling-resistant Plants and Their Products of Horticultural and Agricultural Importance

Chilling-sensitive		Chilling-resistant	
Avocado	Orange	Apple	Pea
Banana	Papaya	Apricot	Peach (tree)
Bean (*Phaseolus*)	Passionfruit	Barley	Pear
Cucumber (and	(some species)	Cabbage (and other	Plum
other cucurbits)	Peach fruit	*Brassica* spp.)	Potato
Egg plant	(partially)	Carrot	Wheat
Grapefruit	Pineapple	Cherry	
Lemon	Rice	Onion	
Lime	Sorghum	Leek	
Maize	Sugar-cane	Lettuce	
Mango	Sweet potato	Passionfruit	
Melon	Tomato	(some species)	

prolonged cold exposure, for example, during refrigerated storage of certain varieties of apple fruits. It should be borne in mind, however, that most chilling-resistant plants cannot develop reproductive cells near 0°C.

Chilling injury may occur as a result of refrigerated storage of chilling-sensitive produce (*e.g.* fruits and vegetables). The potential for chilling injury restricts the use of refrigeration for the storage and transport of chilling-sensitive, tropical produce. Practical storage life of a commodity depends on the degree of chilling sensitivity and the respiration rate of the commodity (*4*). Hence chilling-sensitive produce, which usually cannot be stored below 7–10°C, has a short storage life because at these temperatures respiration rate is generally quite high.

Chilling injury also occurs in the field, especially where fruits such as pineapples and papaya are grown near to their climatic limit, as for example, in coastal, southern Queensland and northern New South Wales in Australia. Field chilling is probably a widespread phenomenon which, however, is poorly documented. The risk of field chilling and freezing restricts the geographical regions and the seasons in which such plants may be grown. Furthermore, with the increase in energy costs it often becomes prohibitively expensive to grow horticultural produce in protected systems (*e.g.* glasshouses) which has been common practice in Europe, North America, Japan and elsewhere. Lowering the optimum growth temperature for such plants, as well as the minimum temperature at which chilling injury occurs, would alleviate, at least partly, such increased costs and conserve dimminishing energy reserves. It would, therefore, be a considerable economic benefit if chilling sensitivity could be eliminated or ameliorated.

The relationships between chilling, freezing and heat stress in plants are poorly understood. The current expansion of interest in temperature stress, and the related phenomenon of water stress, is particularly appropriate at the present time because of the potential changes in world weather patterns which seem likely as a consequence of Man's assault on the environment. The well-documented increase in atmospheric concentration of carbon dioxide, attributed largely to fossil fuel combustion, is predicted (largely by computer modelling) to lead to an increase in global annual surface mean temperature of 2° to 3°C (see, for example, 5), although others dispute this prediction (6). The consequences of such global temperature changes are likely to be major shifts in agriculture and horticulture over considerable regions of the world. Whether such changes would be beneficial or not is, of course, highly problematical and could, no doubt, vary from place to place. What is important is that biological research should develop basic understanding of temperature stress to cope with these potential changes in climate. Such knowledge could be used to adapt agriculture and horticulture to minimise deleterious effects of global temperature changes and even take advantage of them. In any case, the resultant information will be applicable to current cropping and postharvest strategies with consequent benefits to farmers and consumers.

CAUSES OF CHILLING INJURY

A prime aim of the work in our laboratory is to determine the causes of chilling injury at the cellular and molecular level with a view to developing strategies for overcoming or ameliorating chilling injury in plants.

Membranes and Lipids
It has been proposed by J.M. Lyons and J.K. Raison (3) that the primary cause of chilling injury is due to the physical response of the membrane lipids to low temperature. They postulated that the membrane lipids of chilling-sensitive species undergo a physical change, generally at a precise temperature around 10–15°C, which results in a change in the properties of the cell membranes. For some membranes, as a consequence of the changes in physical properties of the lipids at low temperature, the associated membrane proteins (particularly enzymes) show much decreased activity or function. By contrast, the lipids of chilling-resistant plants do not show these physical responses until much lower temperatures are reached, often below the freezing point of the tissue. A key feature of the Lyons-Raison hypothesis is the close correlation between the temperature at which the physical change in lipids occurs and that at which the biological function, such as enzymic activity, growth, ion transport *etc.* is impaired. This is illustrated in Fig. 1. Electron spin resonance of spin-labelled lipid probes has been used as a principal method for determining the physical changes with temperature (*e.g.* 7) but more recently fluorescent probes (8) and differential scanning calorimetry (*e.g.* 9) and other physical techniques have been used (for other examples, see ref. 10).

The 'break' temperatures in the Arrhenius plots (log rate against reciprocal absolute temperature ; see Fig. 1) are interpreted as the temperatures at which physical changes, either phase separations or phase transitions, in membrane lipid properties occur and result in changes the properties of the membrane, manifested in their biological functions. The 'break' around 10–15°C is considered to be the critical temper-

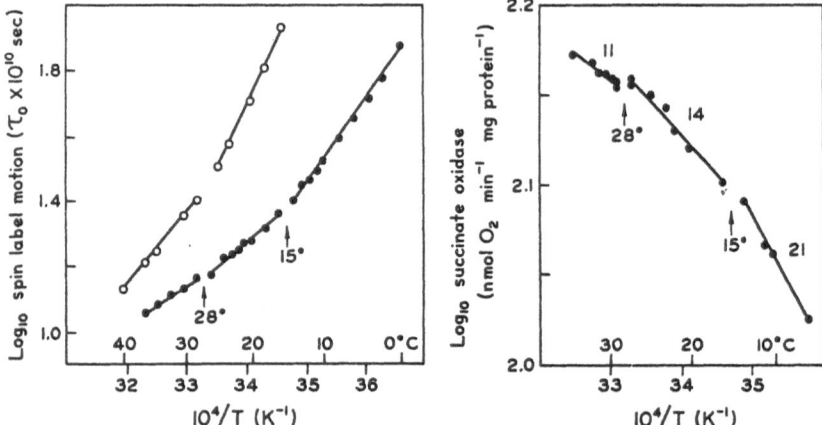

Fig. 1A. The change in logarithm of spin label motion as a function of the reciprocal of the absolute temperature for spin labels intercalated into the membranes of mitochondria from mung bean hypocotyl. For the spin label with the nitroxide substituent on carbon 16 of methyl stearate (16 nitroxide methyl stearate), τ_0 was calculated over the temperature range of $0-40°C$ (●). For the 12 nitroxide analogue the motion was more restricted and τ_0 was calculated over the range of $15-45°C$ (○). The changes in slope, indicated by the arrows, occur at $15°$ and $28°C$. From (7).

Fig. 1B. Changes in the logarithm of succinate oxidase activity of mung bean mitochondria as a function of the reciprocal of the absolute temperature. Measurements were made on two preparations of mitochondria from 4-day hypocotyl tissue. These preparations had specific activities of 144 and 149 moles O_2 min^{-1} mg $protein^{-1}$. The composite plot was made by normalizing the rates of the two preparations at $36°C$. The arrows at $15°C$ and at $28°C$ indicate the temperatures below which the E_a increased. From (7).

ature below which chilling injury is likely to occur. Chilling-resistant plants produce linear Arrhenius plots in the range of about $0-40°C$ (11). Some examples referring to the mitochondrial enzyme complex, succinoxidase, in chilling-sensitive and chilling-resistant fruits and vegetables are shown in Fig. 2. It should be noted that in this earlier work (12) only a single 'break' temperature, the lower one, is shown. Subsequent studies have shown that measurements made over an extended temperature range show a second 'break' temperature in the upper range in chilling-sensitive plants (see Fig. 1).

Controversy has surrounded this hypothesis recently (see, for example, 13-16), due partly to the difficulty in interpreting the presence or absence of 'break' temperatures in Arrhenius plots. This has been a long standing problem with Arrhenius plots and was much discussed in the nineteen twenties and thirties in relation to the Crozier hypothesis (see 17). An interesting reminder of this occurred recently when Professor Tang Pei-sung, Director of the Institute of Botany, Beijing, China visited our laboratory and recalled work he had done in the U.S.A. in the 1930s on temperature responses and the difficulties he had in interpreting results using Arrhenius

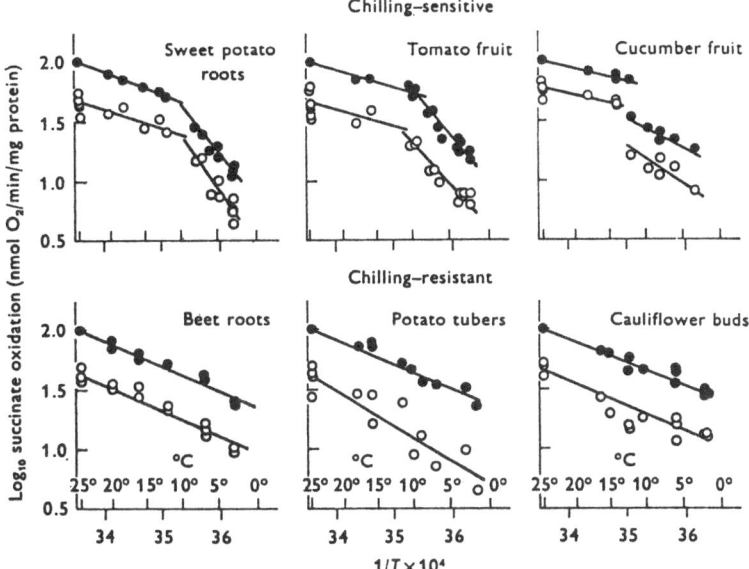

Fig. 2. Arrhenius plots of succinate oxidation by plant mitochondria obtained from chilling-sensitive and chilling-resistant tissues. Each plot showing state 3 (●) and state 4 (○) respiration represents data from three or more mitochondrial preparations. The log values were adjusted by a factor to a common value at $25°C$ in order to compensate for differences in the rates between the different preparations. From (12).

plots. Thus, for a brief moment; the two eras in this type of study came together, spanning some 50 years of research.

The question of the validity of Arrhenius plots (18) and their use in the interpretation of temperature responses came under discussion at a symposium on low temperature stress held in Hawaii in 1979 (10). A number of objective methods for the determination of 'breaks' or curves in such plots are reported in that volume (19–21). The use of statistical methods should minimise the frequently cavalier approach to fitting lines to Arrhenius plots.

Another approach is to use a diverse range of techniques, placing less reliance on a single technique such as e.s.r. This is now being done, especially in measuring the physical changes in membrane lipid properties. The use of fluorescent probes, such as parinaric acid (see Fig. 3) and differential scanning calorimetry (22) are providing confirmation of the observations made by the electron spin resonance method and substantiate the Lyons-Raison hypothesis.

A further controversial feature of the hypothesis concerns the lack of a consistent correlation between membrane lipid composition and chilling sensitivity (23, 24). In particular, attempts to show a correlation between the degree of unsaturation of fatty acids and cold sensitivity have produced conflicting results in higher organisms (23), although for many prokaryotes the lipid composition, especially the degree of fatty acid unsaturation, appears to be closely adapted to environmental temper-

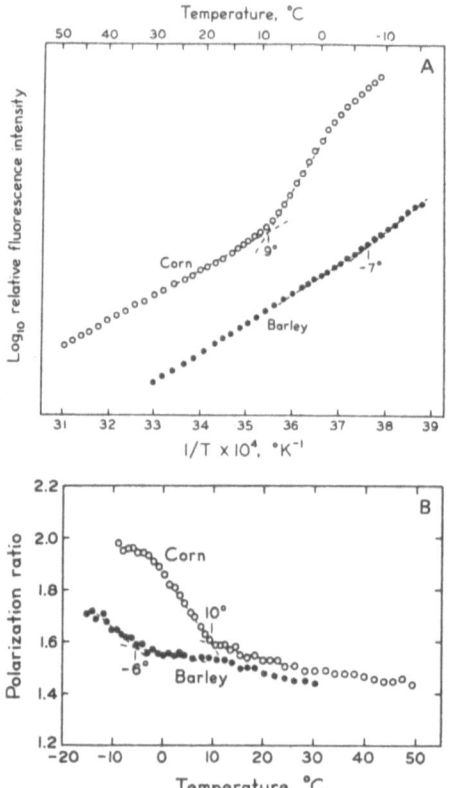

Fig. 3A. Logarithmic plots of *trans*-parinaric acid fluorescence intensity of corn and barley phospholipid vesicles. The phospholipids were dispersed by gentle sonication in 0.1M Tris-HCl buffer, pH 7.2, containing 5 mM Na_2EDTA and 25% or 33% (v/v) ethylene glycol. A suspension containing 400 μg lipid per 3 ml of buffer was labeled with 0.7 μg of *trans*-parinaric acid. Polarized light at 320 nm was used to excite the sample and fluorescence was measured at 420 nm, with the excitation polarizer parallel to the emission polarizer. From (8).

Fig. 3B. *Trans*-parinaric acid fluorescence polarization of corn and barley phospholipid vesicles. Fluorescence emission was measured with the polarizer parallel ($I_{\parallel}$) and perpendicular ($I_{\perp}$) to the orientation of the excitation polarizer. The polarization ratio is $I_{\parallel}/I_{\perp}$. From (8).

ature (25). Furthermore, phospholipid composition also does not show consistent differences between chilling-sensitive and chilling-resistant species (23) although there is a tendency for phospholipids to accumulate during the development of freezing resistance in some species (26).

Notwithstanding these results, further support for the Lyons-Raison hypothesis has recently come from the discovery (22) that certain disaturated phospholipids, present at about 1% the natural membrane lipid composition, can cause membrane preparations from chilling-resistant plants to exhibit the properties of chilling-sensitive membranes. That is, thermal transitions appear at around 8°C, when the disaturated components are added, compared with transitions below 0°C in their absence. Furthermore, chilling-sensitive plants appear to contain low but measurable amounts (in the region of 0.5–1.0% of membrane lipids) of the disaturated lipids but chilling-resistant plants contain only trace amounts of these components (22, N. Murata, personal communication). Thus the chemical identity of at least some of the components responsible for the chilling sensitivity of some plant species has been established which opens the way to understanding the mechanisms of cold sensitivity.

These recent observations also account for the previous inconsistency in the results with lipid composition. The minor compositional components of the membrane, such as the disaturated phospholipid fraction and sterols, have rarely been measured, due no doubt to the inherent difficulties in accurately measuring components representing perhaps only 1 or 2% of the membrane. There appears to be a fruitful field for those laboratories able to carry out such exacting analyses.

It is notable that the polarization technique with the fluorescent probe, parinaric acid, shows only a single 'break' temperature which appears equivalent to the lower temperature 'break' observed with the electron spin resonance technique. It is now proposed (27) that phase separation of membrane components begins below about 7°C in Nerium oleander (in warm-grown plants), corresponding to the lower 'break' temperature determined by e.s.r. The gel phase then expands as the temperature declines, complete solidification of the membrane occurring below 0°C. This would accord with the known physical behaviour of individual lipids, many of the more unsaturated species of fatty acids having freezing points well below 0°C, although it should be noted from the above discussion that the bulk lipid composition is not invariably the factor determining the phase transition temperature (22,27).

Proteins

While lipid composition of membranes is important in determining whether a plant is chilling-sensitive or not, it should not be overlooked that proteins also show responses to low temperatures (28). With membrane proteins a major obstacle to determining their temperature sensitivity has been the presence of the lipid components of the membrane. However, it has been proposed by Uritani and his colleagues that membrane proteins such as cytochrome c oxidase are intrinsically affected by low temperatures in the chilling range (29) since removal of the lipid component from purified preparations derived from sweet potato resulted in no drastic change in the 'break' temperature of the enzyme. This followed earlier studies on succinoxidase in mitochondria from sweet potato (30). On the other hand, studies with tomato fruit preparations (31) appear to show that depletion of lipids by detergent treatment results in loss of the 'break' in the Arrhenius plots evident in native preparations suggesting that it is the lipid component which is important in determining the 'break'. This matter needs resolving, especially the role of detergent molecules in these preparations.

There is also evidence that K^+-stimulated, Mg^{2+}-activated ATPase, another membrane bound protein, is intrinsically affected by temperature (32). No thermal transitions, measured by differential scanning calorimetry and electron spin resonance spectroscopy, could be found in plasma-membrane fractions from curd of cauliflower, a chilling-resistant plant, while biphasic Arrhenius plots were obtained for the enzymic activity. The enzyme thus appears to respond to cold by an increase in Arrhenius activation energy which is the result of low temperature effects on the protein itself rather than being mediated by bulk membrane lipids, as is the case with some other membrane-bound enzymes, e.g. succinate oxidase, of chilling-sensitive plants (7). The effect of low temperature on the ATPase protein is not likely to be physiologically important in this chilling-resistant plant, since presumably the curd of the cauliflower continues to grow and develop below the 'break' temperatures. The ATPase is probably not a rate-limiting enzyme, even at the depressed activities found at low temperatures.

The responses to cold of soluble, non-membrane enzymes are also being investigated (33). This is exemplified in Fig. 4 which shows results for the enzyme phosphoenol pyruvate (PEP) carboxylase which catalyses the reaction:

Phosphoenol pyruvate + CO_2 → oxaloacetate + inorganic phosphate

This enzyme has an anaplerotic function in supplying the tricarboxylic acid

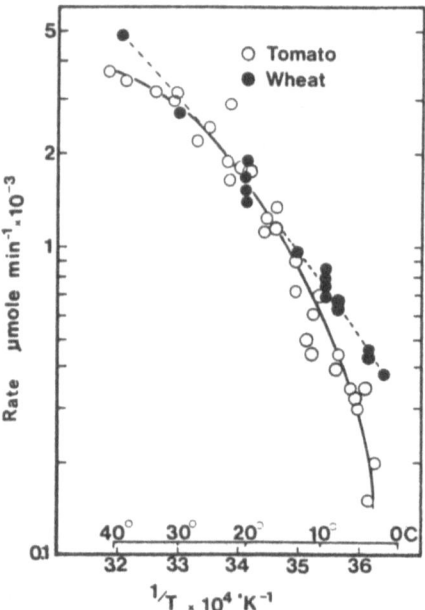

Fig. 4. Arrhenius plots of wheat and tomato leaf PEP-carboxylase activities from crude preparations after Sephadex G-25 treatment. Results have been normalized to 20°C to correct for different extractions of the enzyme. Both were measure at 2 mM PEP. Enzymic rates are μmoles PEP (or NADH oxidized) min^{-1}. From (33).

(TCA) cycle with C_4 carbon skeletons. This synthetic function is required to maintain turnover of the cycle when carbon skeletons are withdrawn in the conversion of α-oxoglutarate to glutamate, succinate to form tetrapyrole ring compounds (e.g. chlorophyll) and the sequestering of organic acids such as malate in the cell vacuole. Because of its very low activity relative to other enzymes and the key position it holds in linking the glycolytic and tricarboxylic acid cycle metabolic pathways, it is likely that this enzyme occupies a rate-limiting position in metabolism.

For PEP carboxylase from C_3 plants there is a clear distinction in response to chilling temperatures between chilling-resistant plants of temperate zone origin e.g. wheat, and chilling-sensitive species originating in the lowland tropics, e.g. domestic tomato (33 ; Fig. 4). However, further studies have shown that a number of tropical, chilling-resistant species including wild tomato (Lycopersicon hirsutum) and Passiflora species, mainly originating in cool, highland regions also have chilling-sensitive PEP carboxylases (34). Since these tropical plants vary in their chilling response from extremely chilling-sensitive (lowland ecotypes or species) to moderately chilling-resistant (highland ecotypes or species), as measured by a range of responses including germination (35), survival of chilling (36,37), ion leakage (38), amino acid uptake (39) and so on, it seems unlikely that the cold sensitivity of this enzyme found in both highland and lowland tropical plants is directly related to chilling injury. It may be, however, that this enzyme governs the growth of the plant type at low temperature because a major difference between temperate and tropical species is in their minimum temperature for growth (40). Both highland and lowland tropical species, such as wild tomato, have relatively high minimum growth temperatures. This correlates with the cold sensitivity of their PEP carboxylases. Further detailed studies of the behaviour of this and other enzymes at low temperatures are required to clarify the situation.

The kinetic properties of PEP carboxylases from C_3 plants change at low temperature (33). In particular, there is a sharp increase in K_m PEP at temperatures below about 8°C, i.e. there is a decrease in affinity of the enzyme for one of its principal substrates. This will presumably alter metabolism at low temperatures. However, it remains to be shown what the situation is in vivo since it is known that large changes occur in some metabolites, e.g. glucose 6-phosphate, alanine and glutamate, related to the glycolytic and tricarboxylic acid respiratory pathways on chilling (41,42). The possible activation of the enzyme by metabolites, such as glucose 6-phosphate, may overcome the apparent decline in enzymic activity in the cold. That the apparent cold sensitivity of PEP carboxylase from both chilling-sensitive (lowland) and chilling-resistant (highland) tropical species is not artefactual is supported by evidence from $^{14}CO_2$ fixation experiments in the dark in which marked decreases in fixation were shown at low temperatures in both types of plant (43). These would be consistent with the decline in activity of the enzyme observed in vitro.

A number of other soluble enzymes have been examined for their responses to cold. Extensive studies have been made on the photosynthetic enzyme, ribulose bisphosphate carboxylase, in cold-resistant plants such as rye (Secale cereale) (44) and potato (45). The results show changes in structural and kinetic properties at low temperatures which appear to be advantageous to cold-hardy species. The native enzyme from the cold-hardy potato, *Solanum commersonii,* had fewer readily titratable SH groups than in the enzyme from the less cold-hardy *S. tuberosum* (domestic potato) which seems to confer greater stability at low temperatures and suggests differences in protein conformation. Furthermore, the V_{max} at 5°C was greater in the cold-hardy species but this situation was reversed at 25°C, while there were no significant differences in K_m CO_2 between the two species at either temperature. The cold sensitivity of this enzyme from rye is decreased on cold hardening of the plant with changes similar to those shown for potato.

The properties of the enzyme malate dehydrogenase have been examined in latitudinal populations of the cold-resistant legume, *Lathyrus japonicus* from different climates in North America, ranging from cool to warm summer (46). The enzyme from the colder climate plants had lower heat stability, lower Arrhenius activation energy and lower specific activity compared with the enzyme from the warmer climate plants. The differences in thermal stability were largely associated with the mitochondrial isoenzymes.

Cold-lability of enzymes has been shown for two highly purified enzymes, pyruvate, phosphate dikinase from plants with the C_4 photosynthetic pathway (47), and phosphofructokinase (PFK) from potato tubers (48). In both cases, the tetrameric form of the enzyme dissociates into inactive dimers at low temperatures. The effect of cold on the potato PFK may account for the sweetening found on low temperature storage of potatoes (48,49). However, recent work indicates that, for a large number of potato varieties, no change in PFK activity in crude extracts could be found after storage of the potatoes at 8°C (50). This raises the interesting question of whether the observations made with highly purified preparations necessarily reflect the *in vivo* response of these soluble enzymes.

Microtubules, which are composed of the proteins α and β tubulin, form part of the cytoskeleton and are cold-sensitive. The tubulin dissociates into dimers in the cold both *in vivo* (51) and *in vitro* (52). The severity of chilling injury can be increased in chilling-sensitive cotton plants by the antimicrotubular drug colchicine which combines with tubulin dimers (53). It would be interesting to compare the properties of tubulin from chilling-sensitive and chilling-resistant species. Presumably in the latter, tubulin must be less cold-sensitive or in some way be protected from cold-induced dissociation.

It is apparent that the responses of proteins to cold in chilling-sensitive and chilling-resistant plants are complex and much more information is required in order to determine the significance of proteins in the causation of chilling injury and the ability to resist cold.

Metabolism

Little is known at present about the metabolic consequences of chilling on plants. One of the earlier hypotheses to account for chilling injury was the toxin hypothesis (for details, see 3). It was proposed that disturbance in cellular metabolism as a consequence of low temperature resulted in an imbalanced metabolism with the accumulation of compounds which might be toxic to the plant. Only limited evidence is so far available. Murata (*54*) has shown in banana fruit that ethanol and compounds such as those usually associated with anaerobic metabolism accumulated. Some of these compounds could adversely affect the plant and perhaps result in membrane damage. Recent work has shown that alanine (*41*) and glucose 6-phosphate (*42*) accumulate in the early stages of exposure of chilling-sensitive *Passiflora* leaves to chilling temperatures. This suggests impairment of the glycolytic pathway and inability to transfer the 3 carbon product of glycolysis, pyruvate, to the tricarboxylic acid cycle which is located in the mitochondrion, probably due to impairment of mitochondrial function in the cold. Alanine, but not ethanol, accumulates in consequence, which is perhaps surprising since it would be anticipated that mitochondrial dysfunction might lead to products similar to those of anaerobic metabolism, as was found in Murata's experiments (*54*). This aspect of metabolism in the cold needs further investigation. Clearly much remains to be done to elucidate the metabolism of plants in the cold.

RESPONSES TO CHILLING TEMPERATURES

Measurement of Physiological Responses to Chilling
While the symptoms of chilling injury are usually quite obvious, they are not easily quantified. Furthermore, they are late events, usually occurring when the plant or tissue may be approaching death. Consequently, it is desirable to have quantifiable measurements of chilling response, preferably during the early stages of chilling. A complicating factor in measuring chilling response is the large diurnal variation in chilling sensitivity so that manifestation of chilling injury can depend on the time of day that chilling treatment begins (*35,55*). Other factors which can complicate the measurements and their interpretation are that comparisons must often be made between quite disparate types of plant or commodity, an extreme example being comparison between, say, the chilling-resistant cauliflower, which is a flowering organ and the chilling-sensitive tomato which is a fruit. In order to overcome these genetic

TABLE II
Species of *Passiflora* (Passionfruit) Differing in Chilling Sensitivity

P. quadrangularis	Very chilling-sensitive
P. maliformis	Very chilling-sensitive
P. edulis forma *flavicarpa*	Very chilling-sensitive
P. edulis forma *edulis*	Chilling-resistant
P. incarnata	Chilling-resistant
P. cincinnata	Chilling-resistant
P. caerulea	Very chilling sensitivity
P. (flavicarpa × *cincinnata)*	Intermediate chilling sensitivity

differences and simplify interpretation of results, workers in our laboratory (particularly B.D. Patterson) have made use of plants which are closely related but which differ in their chilling sensitivity. In particular, two groups of plants have been studied. The first is a series of ecotypes of wild tomato, *Lycopersicon hirsutum*, collected by Professor C.M. Rick and his colleagues of the University of California, Davis, over an altitudinal range from near sea-level to over 3,000 m in the South American Andean Mountains. Thus within a species, a range of chilling-sensitive (lowland) and chilling-resistant (highland) types are available. The second group is a series of species of the genus *Passiflora* also differing in degree of chilling sensitivity (Table II). Since several of these species can be hybridised, it is possible to examine the chilling responses of the F_1 progeny (see below).

Physiological Responses

A wide diversity of measurements can be made to determine response to chilling temperatures. These range from integrated biological functions, such as growth, through organelle functions *e.g.* mitochondrial respiration and chloroplast photosynthetic activities, to more specific activities such as ion leakage or enzymic activities, as well as the physical responses such as those of membrane lipids discussed previously. Only a few examples of biological responses will be discussed here.

Seed germination is a complex phenomenon involving initial imbibition, during which membrane systems are re-organised from the disorganised state of the dry seed at the advent of water, followed by cell expansion and subsequent cell division. Some remarkably sharp responses of germination to low temperature have been shown and some surprisingly high temperatures are required to elicit germination in species which are normally considered to be cold-resistant. Two factors may be important, the first being the need to break dormancy and the second the sensitive initial stage of germination, imbibition, during which chilling-sensitive species are likely to be damaged in the process of membrane re-organisation. It has been suggested that membrane lipid properties are not the only factor in the sensitivity of seed germination to chilling but that some other factor, such as protein denaturation, is responsible for failure to germinate in the cold (*56*). The evidence on this point is at present incomplete and confusing. The observations on proteins in response to cold, mentioned previously, could have some relation to the germination responses in the cold, particularly as the cold sensitivity of proteins does not necessarily seem to be related to the chilling sensitivity or resistance of the plant from which they are derived.

Growth of the plant or its parts, measured as an increase in dry weight, or leaf area, or extension of mesocotyl or hypocotyl plus radicle show marked responses to low temperature in chilling-sensitive species (*7,14,57* and *58*). Chilling-sensitive plants such as *Sorghum bicolor* and mung bean (*Vigna radiata*) show more or less sharp decreases in growth below temperatures around 10–15°C. Such results are usually displayed as Arrhenius plots showing 'break' temperatures, although this approach has recently been criticised (*16*).

Ion leakage can be a useful measure of chilling sensitivity (*38,59*). The results shown in Fig. 5 indicate the wide differences in response between various species of

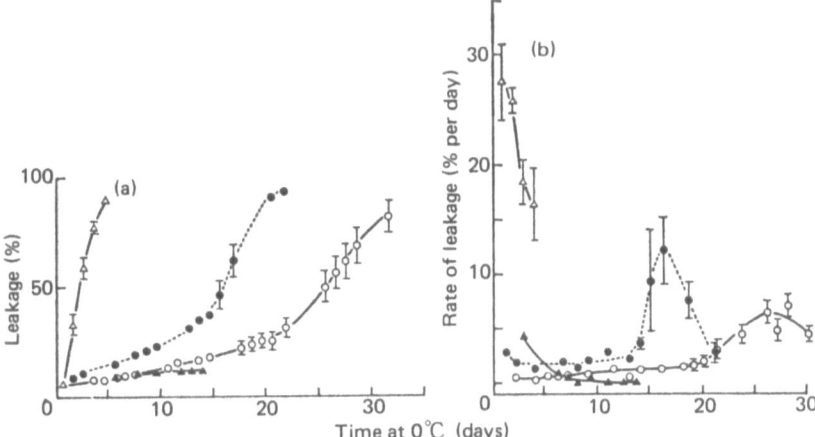

Fig. 5. (a) Time courses of leakage of electrolyte at 0°C from *P. flavicarpa* (△), *P. (flavicarpa × cincinnata)* (●) and *P. edulis* (○). (b) The rates of leakage derived from the time courses given in (a). Also included in both figures are data for *P. flavicarpa* at 10°C (▲). Rates were calculated from the increments of leakage between successive readings and plotted as percentage leakage per day. The bars show standard errors of the mean (n = 10). From (38).

Passiflora. The ionic contents (amino and organic acids, salts) leak immediately the chilling-sensitive *P. flavicarpa* is placed under chilling stress at 0°C and continue to do so for several days until the tissue is dead, whereas the very cold-resistant *P. caerulea* does not show appreciable leakage for at least 20 days (38). The hybrid between the chilling-sensitive *P. flavicarpa* and the chilling-resistant *P. cincinnata* shows intermediate behaviour (Fig. 5).

Other types of chilling responses which can be measured include protoplasmic streaming (60), pollen germination (35), tissue respiration (3,61), mitochondrial oxidation (3,7,11,12,61) and chloroplast (photosynthetic) activities including photoreduction and chlorophyll fluorescence (62).

A method for the objective measurement of the browning reaction in avocado fruit has recently been developed (62a).

Where measurements are made on subcellular systems e.g. mitochondria, it is important to use healthy tissue which has not been exposed to cold, for the isolation of the organelles. This was a prime aspect of the early work of Raison and Lyons (12, 63). Otherwise, comparisons may reflect differences in extraction or other aspects of 'survival' of the organelle rather than its response to temperature *in vitro*. They also found that mitochondria isolated from chilling-sensitive species could be stored in ice at around for 0°C for several hours with surprisingly little deterioration of their oxidative capacities, measured as oxygen evolution linked to succinoxidase activity at higher temperatures e.g. 25°C. This indicates the ready reversibility of the low temperature induced membrane effects described above, at least for several hours after the beginning of chilling.

The development of chilling injury is generally considered to follow a time-temperature relationship, the lower the temperature the greater and quicker the response, especially if the plant is subsequently returned to warm conditions. Retention at low temperature usually prevents manifestation of chilling injury. Reversibility of a chilling response is likewise time-temperature dependent. The primary events are considered to be instantaneous and immediately reversible. The time and degree of reversibility will obviously depend on the inherent sensitivity of the particular species, ranging from *Episcia reptans* which is killed by a few hours of chilling at 5°C through a continuum of moderately chilling-sensitive species to species that are generally considered to be chilling resistant but which nevertheless show cold injury *e.g.* certain varieties of apple fruits after prolonged (months) storage at around 0°C.

AMELIORATION OF CHILLING INJURY IN FIELD CROPS AND POSTHARVEST SITUATIONS

The widespread economic importance of chilling injury in tropical field crops and in tropical fruits and vegetables demands that ways of ameliorating or avoiding chilling damage be found. Three principal avenues to this end appear possible. These are a) breeding new, more cold-resistant cultivars, b) avoidance of chilling temperatures, involving storage of commodities under non-refrigerated conditions and c) treatments, whether before or during refrigerated storage, which may reduce or delay chilling injury. Each of these will be discussed in turn.

Plant Breeding
    It should be possible to select from existing cultivars those which are more cold-

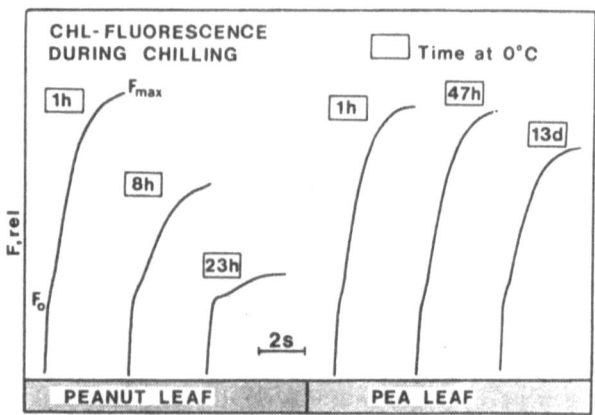

Fig. 6. Chlorophyll fluorescence rise at 0°C in dark-adapted peanut and pea leaves held at 0°C for various periods of time. F_0 indicates the initial fluorescence level, F_{max} the maximum reached. h, hours ; d, days. From (62).

resistant or introduce genes for cold-resistance from wild relatives of cultivated plants, such as the wild tomato (*L. hirsutum*), which can be hybridised with the cultivated tomato *L. esculentum*. In order to do this it is necessary for plant breeders to have rapid, and relatively simple, methods of screening large numbers of plants for cold-resistance (or sensitivity). Of the methods of measuring chilling injury so far tested in our laboratory, the most practical for use in plant breeding programmes is that using the decline in photosynthetic fluorescence as a result of chilling treatment (Fig. 6) (*62*). This response can be detected by a simple, inexpensive fluorometer. The method is applicable to any green tissue, including green fruits, and can rank the order of chilling sensitivity among varieties or very closely related species. It will also distinguish between chilling-sensitive and chilling-resistant species.

Avoidance of Refrigerated Storage

Avoidance of refrigeration would, of course, be ideal but unfortunately is not widely applicable. The banana fruit is the only fully documented example in which storage of mature green fruit is possible at ambient temperature. This is achieved by containment of the fruit in a modified atmosphere (elevated CO_2 and diminished O_2) with an ethylene absorbent (potassium permanganate) using sealed, polyethylene bags. The method may have application with other fruits such as avocado but at present has adverse effects when used on certain other fruits such as mangos (G.R. Chaplin, personal communication). Further research is required to apply the method to other fruits. It is unfortunately an axiom of postharvest management of most temperate commodities that low temperature is the most effective way to slow deterioration. Chilling injury of tropical commodities intervenes in this strategy for such products and consequently the quantities of such commodities lost postharvest can be very high indeed.

Storage Treatments

Treatments which delay or minimise chilling injury warrant further investigation. The postharvest application of calcium has been shown to prevent some physiological disorders of apples, particularly bitter pit and low temperature breakdown, as well as to delay senescence of the fruit (*65,66*). Calcium concentration also appears to be important in preventing or delaying the development of chilling injury in stored avocados.

Intermittent warming during the cool-storage period can have beneficial effects on cold-induced storage problems, especially for stone fruits (plums) (*68*). The effect may be due to removal of an accumulated 'toxic' metabolite. Maintenance of high humidity during chilling may prevent or delay the development of symptoms of chilling injury (*69*) and could be important in the application of the polyethylene bag method to tropical fruit storage at low temperatures. There are also indications that modified atmospheres minimise chilling injury in some fruits *e.g.* avocados (*70*). In the peach fruit the ability to ripen normally is lost after storage in air at 1°C for 21 to 28 days whereas addition of 20% CO_2 protects the fruit from this disability (*68*). The mechanism of this effect is unknown. There is thus a number of storage strategies which may minimise chilling injury but for which we presently have no explanation.

CONCLUSION

Widespread economic losses occur in tropical and subtropical crops either due to cold in the field or to the inability to use refrigeration during the postharvest life of the commodity because of the risk of chilling injury. While progress has been made in identifying the cellular causes and responses which result in chilling injury, much remains to be done to determine the means of preventing or ameliorating chilling injury. Some promising leads include breeding for cold resistance by introducing genes from cold resistant, wild relatives of the domesticated cultivars (e.g. in the tomato) and the manipulation of storage strategies using intermittent warming, modified atmospheres, humidity control, increased calcium, or a combination of these. In this brief review only a few salient topics have been discussed and it is clear that further research is required to clarify the role of certain cold-sensitive enzymes and other proteins, e.g. tubulin, the metabolic consequences of chilling and the mechanism by which certain minor lipid components can regulate the properties of plant membranes, to mention just a few topics of current interest.

Acknowledgement
    I am grateful to my colleagues, Dr. J.K. Raison and Dr. B.D. Patterson, for their critical reading of the manuscript.

REFERENCES

1. Molisch, H. : 1896. *Sitzungsber. Kaiserl. Akad. Wiss. Wien., Math. Naturwiss. Kl. Abt. 1* **105**, 82–95.
2. Molisch, H. : 1897. Untersuchungen über das Erfrieren der Pflazen, Fischer, Jena.
3. Lyons, J.M. : 1973. *Annu. Rev. Plant Physiol.* **24**, 445–466.
4. Wills, R.B.B., Lee, T.H., Graham, D., McGlasson, W.B. and Hall, E.G. : 1981. Postharvest : An Introduction to the Physiology and Handling of Fruit and Vegetables, New South Wales University Press, Kensington, Australia.
5. Wigley, T.M.L. and Jones, P.D. : 1981. *Nature* **292**, 205–208.
6. Idso, S.B. : 1981. *New Sci.* **92**, 444–446.
7. Raison, J.K. and Chapman, E.A. : 1976. *Aust. J. Plant Physiol.* **3**, 291–299.
8. Pike, C.S., Berry, J.A. and Raison, J.K. : 1979. in reference (*10*), pp.305–318.
9. McMurchie, E.J. : 1979. in reference (*10*), pp.163–176.
10. Lyons, J.M., Graham, D. and Raison, J.K. (eds.) : 1979. Low Temperature Stress in Crop Plants : The Role of the Membrane, Academic Press, New York.
11. Raison, J.K. : 1973. *Symp. Soc. Exp. Biol.* **27**, 485–512.
12. Lyons, J.M. and Raison, J.K. : 1970. *Plant Physiol.* **45**, 386–389.
13. Silvius, J.B., Read, B.D. and McElhaney, R.M. : 1978. *Science* **199**, 902–904.
14. Bagnall, D.J. and Wolfe, J.A. : 1978. *J. Exp. Bot.* **29**, 1231–1242.
15. Steponkus, P.L. : 1981. Encyclopedia of Plant Physiology, Vol 12A (Lange, O.L., Nobel, P.S., Osmond, C.B. and Ziegler, H., eds.), pp.371–402, Springer-Verlag, Berlin.
16. Bagnall, D.J. and Wolfe, J. : 1982. *Cryoletters* **3**, 7–16.

17. Johnson, F.H., Eyring, H. and Polissar, M.J. : 1954. The Kinetic Basis of Molecular Biology, Wiley, New York.
18. Wolfe, J. : 1979. in reference (10), pp.327–335.
19. Wolfe, J. and Bagnall, D. : 1979. in reference (10), pp.527–533.
20. Potter, J.F. and Ross, G.J.S. : 1979. in reference (10), pp.535–542.
21. Willcox, M.E. and Patterson, B.D. : 1979. in reference (10), pp.523–525.
22. Wright, L.C. and Raison, J.K. : 1981. Proc. XIIIth Int. Bot. Congr. Sydney 90.
23. Quinn, P.J. and Williams, W.P. : 1978. Prog. Biophys. Mol. Biol. 34, 109–173.
24. Bishop, D.G., Kenrick, J.R., Bayston, J.H., Macpherson, A.S., Johns, S.R. and Willing, R.I. : 1979. in reference (10), pp.375–389.
25. Inniss, W.E. and Ingraham, J.L. : 1978. Microbial Life in Extreme Environments (Kushner, D.J., ed.), pp.73–104, Academic Press, London.
26. Yoshida, S. and Sakai, A. : 1973. Plant Cell Physiol. 14, 353–359.
27. Raison, J.K., Pike, C.S. and Berry, J.A. : 1982. Plant Physiol. 70, 215–218.
28. Graham, D. and Patterson, B.D. : 1982. Annu. Rev. Plant Physiol. 33, 347–372.
29. Maeshima, M., Asahi, T. and Uritani, I. : 1980. Agric. Biol. Chem. 44, 2351–2356.
30. Yamaki, S. and Uritani, I. : 1974. Plant Cell Physiol. 15, 669–680.
31. Waring, A. and Glatz, P. : 1979. in reference (10), pp.365–374.
32. Wright, L.C., McMurchie, E.J., Pomeroy, M.K. and Raison, J.K. : 1982. Plant Physiol. 69, 1356–1360.
33. Graham, D., Hockley, D.G. and Patterson, B.D. : 1979. in reference (10), pp.453–461.
34. Graham, D., Hockley, D.G. and Patterson, B.D. : in preparation.
35. Patterson, B.D., Graham, D. and Paull, R. : 1979. in reference (10), pp.25–35.
36. Paull, R.E., Patterson, B.D. and Graham, D. : 1979. in reference (10), pp.507–519.
37. Patterson, B.D., Paull, R. and Smillie, R.M. : 1978. Aust. J. Plant Physiol. 5, 609–617.
38. Patterson, B.D., Murata, T. and Graham, D. : 1976. Aust. J. Plant Physiol. 3, 435–442.
39. Paull, R.E., Patterson, B.D. and Graham, D. : 1979. Aust. J. Plant Physiol. 6, 475–484.
40. Went, F.W. : 1957. The Experimental Control of Plant Growth, Ronald Press, New York.
41. Patterson, B.D., Pearson, J.A., Payne, L.A. and Ferguson, I.B. : 1981. Aust. J. Plant Physiol. 8, 395–403.
42. Pearson, J.A., Patterson, B.D. and Payne, L.A. : 1981. Proc. XIIIth Int. Bot. Congr. Sydney 260.
43. Graham, D. and ap Rees, T. : in preparation.
44. Huner, N.P.A. and MacDowall, F.D.H. : 1979. Can. J. Biochem. 57, 1036–1041.
45. Huner, N.P A., Palta, J.P., Li, P.H. and Carter, J.V. : 1981. Can. J. Biochem. 59, 280–289.
46. Simon, J-P. : 1979. Plant Cell Environ. 2, 23–33.
47. Sugiyama, T., Schmitt, M.R., Ku, S.B. and Edwards, G.E. : 1979. Plant Cell Physiol. 20, 965–971.
48. Dixon, W.L., Franks, F. and ap Ress, T. : 1981. Phytochemistry 20, 969–972.
49. Pollock, C.J. and ap Rees, T. : 1975. Phytochemistry 14, 613–617.
50. Walcott, P.J. : 1982. M.Sc. Thesis, Macquarie University, Sydney.
51. Hart, J.W. and Sabnis, D.D. : 1977. The Molecular Biology of Plant Cells (Smith, H., ed.), pp.160–180, Blackwell Scientific, Oxford.
52. Timasheff, S.N. and Grisham, L.M. : 1980. Annu. Rev. Biochem. 49, 565–591.
53. Rikin, A., Atsmon, D. and Gitler, C. : 1980. Plant Cell Physiol. 21, 829–837.
54. Murata, T. : 1969. Physiol. Plant. 22, 401–411.
55. King, A.I., Reid, M.S. and Patterson, B.D. : 1982. Plant Physiol. 70, 211–214.
56. Simon, E.W. : 1979. in reference (10), pp.37–45.
57. Bagnall, D. : 1979. in reference (10), pp.67–80.
58. McWilliam, J.R., Manokaran, W. and Kipnis, T. : 1979. in reference (10), pp.491–505.
59. Tatsumi, Y. and Murata, T. : 1978. J. Jpn. Soc. Horti. Sci. 47, 105–110.
60. Patterson, B.D. and Graham, D. : 1977. J. Exp. Bot. 28, 736–743.
61. Raison, J.K. : 1980. The Biochemistry of Plants : A Comprehensive Treatise, Vol. 2 (Davies,

D.D., ed.), pp.613–626, Academic Press, New York.
62. Smillie, R.M. : 1979. in reference (10), pp.187–202.
62a. Chaplin G.R., Wills, R.B.H. and Graham, D. : 1982. *HortScience*, 17, 238–239.
63. Raison, J.K. and Lyons, J.M. : 1970. *Plant Physiol.* 45, 382–385.
64. Scott, K.J., Blake, J.R., Strachan, G., Tugwell, B.L. and McGlasson, W.B. : 1971. *Trop. Agric. (Trinidad)* 48, 245–254.
65. Scott, K.J. and Wills, R.B.H. : 1977. *HortScience* 12, 71–72.
66. Scott, K.J., Hardisty, S.E. and Stafford, I.A. : 1980. *CSIRO Food Res. Q.* 40, 29–32.
67. Chaplin, G.R. and Scott, K.J. : 1980. *HortScience* 15, 514–515.
68. Wade, N.L. : 1979. in reference (10), pp.81-96.
69. Wilson, J.M. : 1979. in reference (10), pp.47–65.
70. Scott, K.J. and Chaplin, G.R. : 1978. *Trop. Agric. (Trinidad)* 55, 87–90.

Wound Repair and Tumor Induction in Higher Plants

Günter Kahl

INTRODUCTION

Plant systems have evolved a comparably simple, but nevertheless efficient capability to react upon the various kinds of injuries to which they all are subject during their life cycle. Either they use the dead cells at a wound as a physical barrier, or they additionally accumulate bacteriostatic and fungistatic substances, more generally protectors, at the wound surface, or they form a protective tissue. Such protective tissues may consist of callus-like proliferations or a regular wound periderm. In any case, numerous new cell divisions at the wound site are induced together with massive modifications of the cell walls with lignin or suberin. These defense mechanisms altogether prevent extensive loss of water from the tissues as well as the invasion of the cells by pathogens such as viruses, bacteria and fungi.

Sometimes through the interference of physical, chemical or biological agents the normal wound-healing process is disturbed or does not come to an end but is indefinitely perpetuated. This is especially true for the infection of wounds with *Agrobacterium tumefaciens,* a ubiquitously occurring soil bacterium belonging to the family Rhizobiaceae. Agrobacteria, after the site-specific attachment to walls of wounded cells, transfer part of a plasmid (pTi, tumor inducing plasmid) into the wound cell where it is stably integrated into the nuclear genome. Transcription of the transferred DNA (T-DNA) by host cell RNA polymerase II yields a set of RNAs, one of which is also translated into a protein, an opine synthesizing enzyme. The action of this enzyme leads to the accumulation of an opine, an amino acid derivative, in the transformed cell. The functions of the other T-DNA transcripts are not yet clear, but may be necessary for the induction and maintenance of the tumorous state of the cell. The latter is characterized by continuous proliferations, mostly into unorganized tumor cell masses. In these tissues, the high levels of auxins and cyto-kinins of the wound-healing period are sustained and induce a never-ending cycle of mitoses.

It is only during a distinct period in the whole wound-healing process of a sus-ceptible plant that the transfer of the T-DNA and/or its integration is allowed. Thus, it is impossible to infect wounded tissues immediately after injury or after the healing of the wound. It seems, that only the activated state of the wounded cell can be exploited by the pathogen. A complete understanding of the induction of tumors by *A. tumefaciens* consequently presupposes a detailed knowledge of the wound-healing process.

WOUNDING OF PLANT TISSUES

Wounds may occur in the normal life cycle of a plant or may be induced by physical, chemical or biological agents of the environment. During rapid growth processes, *i.e.*, the elongation phase of stems, some cells are proliferating and elongating much faster than their neighbors. This causes tissue tensions, which in turn lead to internal wounds by tearing. Quite frequently rapid dilatation growth of plant organs (*i.e.*, fruits or stems) generates holes within the expanding tissues. In a process resembling normal wound-healing the annual abscission of leaves is prepared : a specialized tissue separates the leaf-stalk from its base and leads to the repulsion of the leaf. During the growth of roots the rhizodermal tissue is damaged and partly replaced by suberized cortical tissues in a wound repair process. Thus, intrinsic parameters quite frequently cause injuries of various dimensions within plant tissues.

The main physical causes of wounds in plants are of meteorological origin : hail, rain, snow, ice, frost, wind or lightning are often responsible for economically important damage to the plant, especially in the crop plant world. Falling rock and fire may also be included in this category. The chemical causes of wounds are less conspicuous, but nevertheless widespread in occurrence. The extensive usage in agri-culture of various salts, bacteriocides, fungicides, insecticides, and herbicides as well as other poisons for pest control, sprays and fertilizers of different compositions either alone or in combination damages or even destroys plants. Industrial or private pollution (*i.e.*, CO and sulfur compounds) especially pronounced in areas with a high density of population injures plant organs to various extents. Heavy metals are deposited on the surface of exposed plants and at high concentrations lead to ruptures of the epidermis. In contrast, the biological causes of wounds are generally well known : viruses, bacteria, fungi, and the various activities of lower and higher animals including man.

The injured tissues of plants react differently upon wounding depending on their morphogenetic capabilities. The most archaic type of wound-reaction is to use the damaged cells and their precipitated constituents as a barrier, which protects from the invasion of pathogens as well as extensive water efflux. This type is charac-

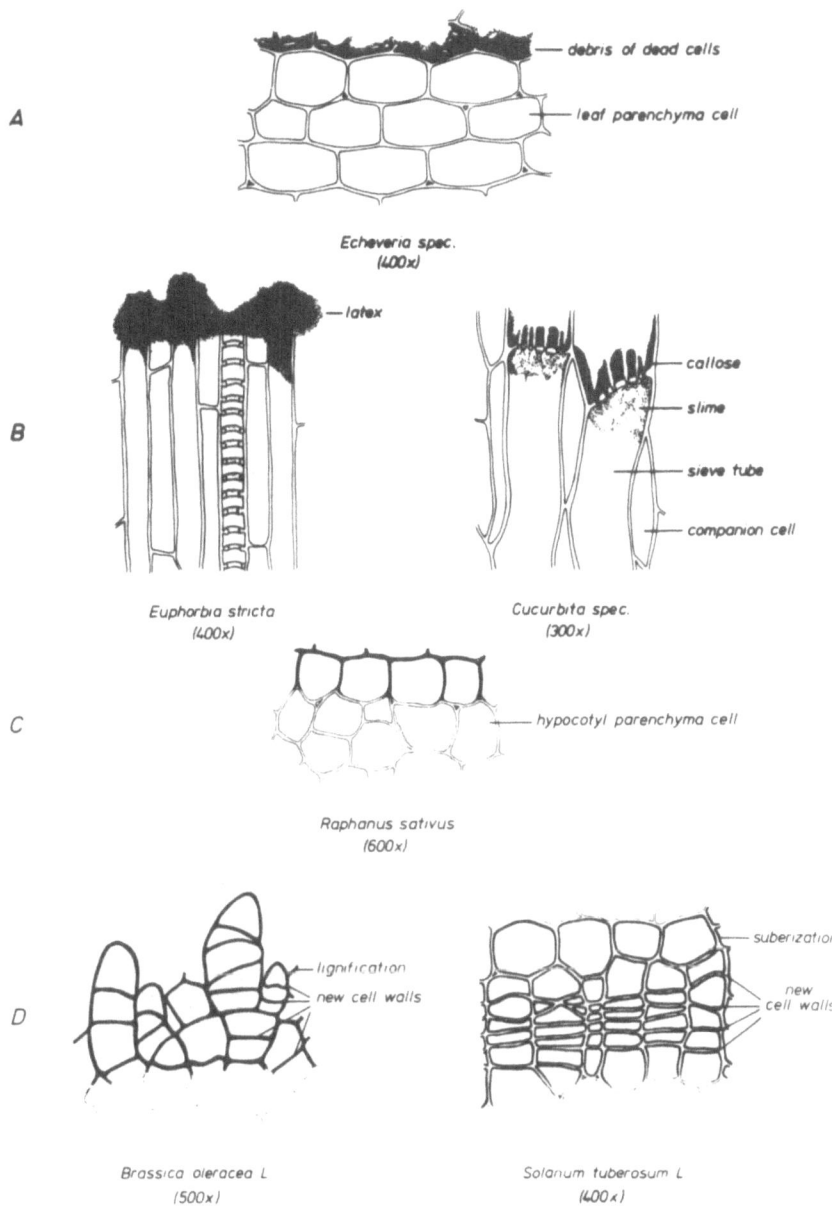

Fig. 1. Different types of wound-healing tissues. A , injured leaf tissues; B , secretion of latex (left) or callose (right); C, lignification of cell walls adjacent to a wound site (*Raphanus sativus*); D , wound-induced callus (*Brassica oleracea*, left) and periderm formation (*Solanum tuberosum*, right).

teristic for most leaf tissues (Fig. 1A). A more advanced type absorbs the residues of the killed cells at the wound site and starts to produce resins, gums, latex or callose in the wound area (Fig. 1B) or to modify the cell walls adjacent to the wound by silicic acid, lignin or suberin (Fig. 1C). Quite frequently the cells at a wound site are induced to undergo mitoses leading to a complex pattern of wound-healing: callus or periderm formation (Fig. 1D).

Although the cells of different plants react with different cytological measures upon wounding, their basic metabolic reactions are quite similar. A series of primary events after wounding leads to a transient disorganization of the membrane systems of the cell. Consequently some blocked metabolic pathways are unblocked and reactions are induced leading to the activation of genes and the synthesis of various enzymes. The ensuing metabolic sequences guarantee an adequate wound response of the cell. Most of the activities of the wound-healing cell are under hormonal control, as is the termination of wound-healing. Since most of the work on wound-healing in higher plants has been performed with white potato tuber tissue, the following report focusses on wound reactions of the potato tuber model system.

PRIMARY EVENTS AFTER WOUNDING

Wounding of whatever kind induces three classes of reactions in the cells at the wound site : immediate and fast reactions which are independent of gene activity, and slow reactions, which depend on previous transcription. Wounding instantane-

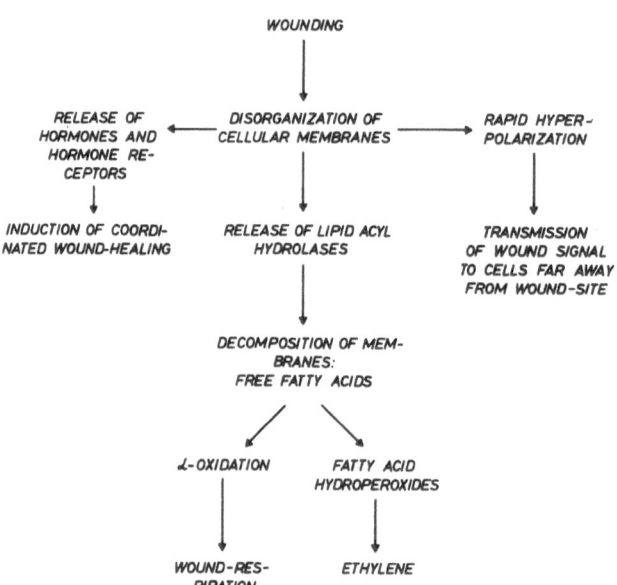

Fig. 2. Early events after the wounding of white potato tuber parenchyma.

ously tears the membranes of the ruptured cells, which leads to an immediate membrane disorder within the adjacent cells. The symplasmic continuum transmits the signal to remote cells, which are thus informed of the event of wounding. Membrane depolarization ensues and the membrane components are suddenly and extensively degraded by liberated enzymes, *i.e.*, lipolytic acyl hydrolases (*1–4*). In a matter of seconds or minutes after wounding, some 35% of the total

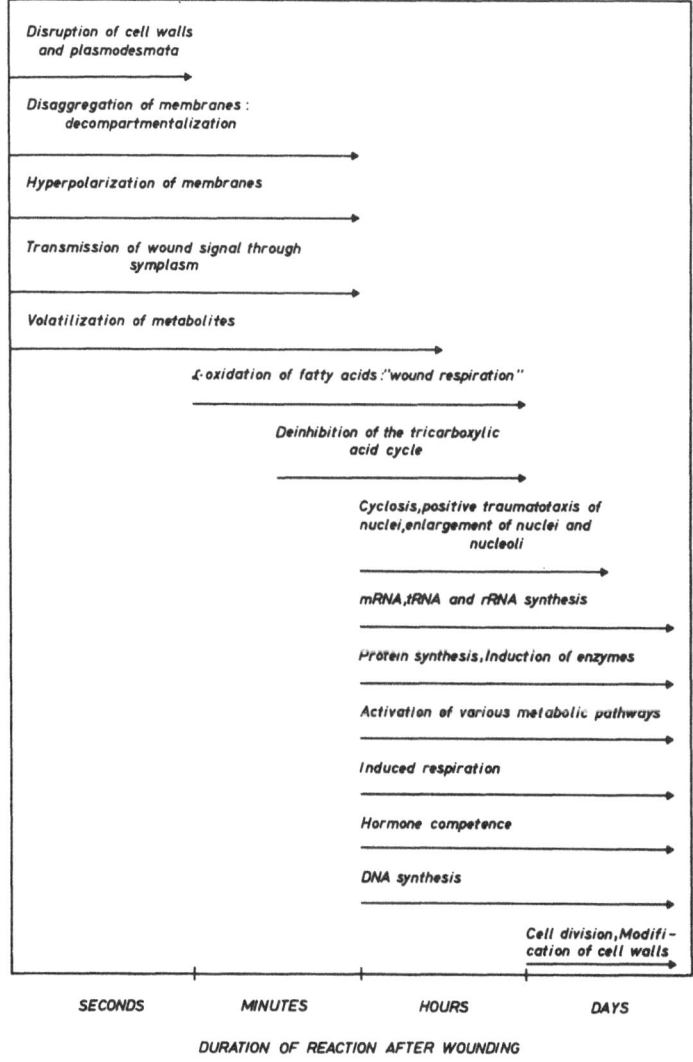

Fig. 3. Ultrafast, fast and slow reactions upon wounding of a white potato tuber parenchyma cell.

membrane phospholipids and more than 50% of the cellular galactolipids are thus destroyed (5). The liberated fatty acids in turn stimulate acyl hydrolases in an auto-catalytic cycle, feed an initial "wound respiration" via α-oxidation (see 6), penetrate the neighboring cells and aid the propagation of the wound stimulus, until they are in turn oxidized enzymatically to fatty acid hydroperoxides, carbonyl compounds and other oxygenated derivatives (Fig. 2).

The wound-induced changes of the membrane system release previously bound substances as volatiles, amoung them ethylene ("wound ethylene") (7) and ethane (8), carbon dioxide, and a series of aromatic compounds and aldehydes (9). Volatilization of i.e., acetaldehyde, which inhibits certain steps in the respiratory electron transport chain, may allow the onset of the previously blocked respiration (10). However, the tricarboxylic acid cycle (TCAC) is also blocked in intact organs, the inhibitor being γ-hydroxy-α-ketoglutarate (11,12). After wounding this compound is rapidly destroyed and an immediate rise in succinate oxidation follows (13).

The cell in a yet unknown way controls these degradative processes and rapidly restores the damaged membranes. This is also observable by electron microscopy. Immediately after wounding the integrity of the endoplasmic reticulum (ER) and mitochondrial membranes have been shown to be severely impaired. The lamellae of the ER are reduced to small vesicles. After prolonged wound-healing the lamellae are again reconstituted into unusually long ER strands and dictyosomes (14,15).

It is evident from these few comments that the architecture of cells adjacent to a wound site is muddled and severely impaired. The membranes temporarily lose their integrity, compartmentation no longer exists and the cytoplasm more closely resembles a homogenate in which enzymes and substrates, neatly separated from each other in an undisturbed cell, have access to each other. Unblocking processes are induced by the release of volatiles and other processes in turn are blocked. However, the cell at a wound soon regains control over these processes : restoration of membranes ensues, genetic activity is induced and the cell's synthetic machinery becomes active. This is certainly a very complex interplay between self-assembly processes, feed-back and feed-forward reactions triggering coordinate gene action and a whole series of different, mutually interfering processes at any level of the cell (Fig. 3).

ACTIVATION OF METABOLIC PATHWAYS

Once the cell gets over the severe irritations accompanying wounding, a coordinated activation of a series of metabolic activities ensues. The cell soon reaches a level of activity which is comparable to that of a highly active embryonic cell and which is documented by an enhanced cytoplasmic streaming, a positive traumatotaxis of the nuclei and an enlargement of both nuclei and nucleoli. All these reactions occur relatively early in the wound-healing cycle and precede the more slowly induced activations of various metabolic pathways, which in turn are dependent on translation of new messages into proteins. The development and control of these capacities of the wound cell will be exemplified by the induction of respiration ("induced respiration").

After the comparably short period of extreme perturbation of the exposed cell, which is accompanied by the transient "wound respiration" depending on fatty acids as substrates, the cell near a wound develops an increased respiration ("induced respiration"). This respiratory outburst reaches its maximum about 2–3 days after wounding and then levels off. Throughout the whole wound-healing period, this respiration if fed through an enhanced breakdown of starch, an accelerated catabolism of sugars *via* glycolysis as well as an activation of the Krebs cycle. This induced respiration provides the synthetic machinery of the regenerating cell with ATP. Some energy may also be lost as heat.

Wounding modifies the respiratory chain : the comparably low respiration shortly after wounding is extremely sensitive to cyanide (inhibition to some 70–95%), the induced respiration becomes more and more resistant ("cyanide-resistant respiration"). This increasing insensitivity of the respiration is due to the establishment of a non-phosphorylating alternate electron transport system by-passing the conventional cytochrome route. This alternate pathway branches out at the level of the flavoprotein-ubiquinone region and involves one or more iron-sulfur proteins which are sensitive to hydroxamic acid and other chelating compounds (*16*). The development of this system is dependent on one or several protein factors synthesized in the cytoplasm. These may include an autoxidizable cytochrome b_7, probably representing the alternate terminal oxidase (*17*). In fact, *b*-type cytochromes are

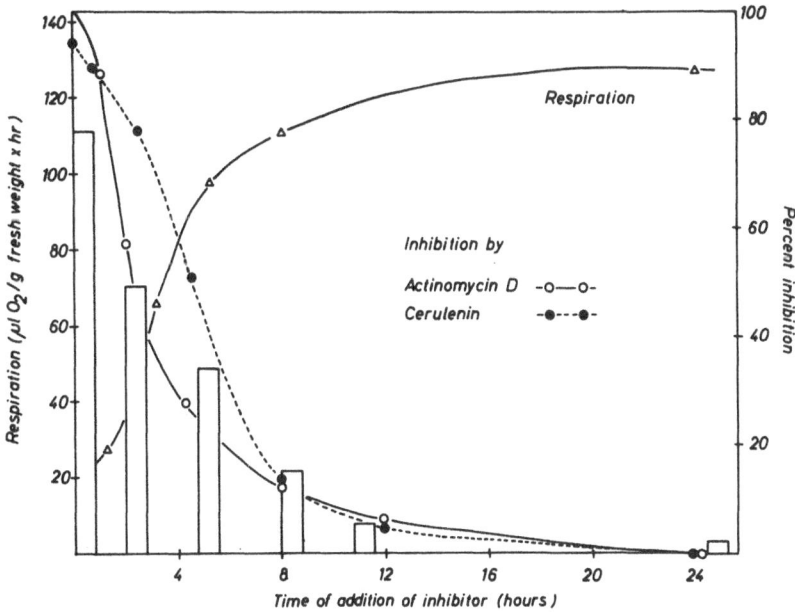

Fig. 4. The effect of time of addition of different inhibitors on the development of wound-induced respiration and choline incorporation in potato tuber tissue (bars). Modified after ref. 5.

actively sythesized after wounding (18). The function of this alternate pathway is still obscure ; it may be a source of superoxides and peroxides (19).

The development of the induced respiration is inhibited by actinomycin D and puromycin, if these inhibitors are added during the first 8–10 hr after wounding (20,21). Thus, RNA and protein synthesis are both necessary for this process to develop. Among the proteins newly synthesized certainly are membrane and enzyme proteins of mitochondria appearing de novo. Membrane lipid biosynthesis also seems necessary for the development of the induced respiration : the key enzymes in lecithine synthesis, phosphorylcholine cytidyltransferase, phosphoryl choline glyceridetransferase, and phosphatidyl phosphatase are all synthesized after wounding. Therefore the incorporation of choline into phospholipids of membranes is also inhibited by actinomycin D (22). Inhibition of fatty acid synthesis by cerulenin and choline incorporation into membranes by dimethylaminoethanol completely suppress the induced respiration, again indicating the importance of membrane biosynthesis (Fig. 4).

ENZYME INDUCTION AND GENE ACTIVATION AFTER WOUNDING

Within only a few hours after wounding, the cells at a wound site commence an active synthesis of proteins. Although ribosomes are present even in resting tissues, they obviously lack the messages to form polysomes. This can be proved in vitro. If virtually inactive ribosomes from intact tissues are supplied with poly(U) in vitro, they are capable of greater activity (23) suggesting that protein synthesis in resting cells is drastically restricted by the absence of messengers. However, shortly after wounding a faint protein synthesis driven by preexisting messengers on preexisting ribosomes occurs (24,25).

The dramatic increase in protein synthesis after an injury is also reflected by ribosome-polysome profiles and can be documented by electron micrographs of wounded cells. Whereas resting cells contain mostly monosomes in their cytoplasm, wounding induces a rapid polysome formation. The percentage of membrane-bound polysomes also increases (24). Taken together, wounding induces a rapid increase in ribosomal activity, which is highest about 8–12 hr after injury. During that time it manufactures nearly all proteins needed for the construction of new membranes, cell walls, ribosomes, plastids, mitochondria, and the cytoplasm and thereby supports the wound-healing process (26). For some of the proteins, including enzymes, a de novo synthesis has been demonstrated. Thus, the first enzyme of the shikimate pathway, phenylalanine ammonia lyase (PAL) is hardly detectable in intact tissues. After a lag phase of 4–6 hr following wounding, PAL activity appears, increases to a maximum some 24–48 hr after wounding and then levels off (27). The appearance of PAL after wounding is due to de novo synthesis of the enzyme protein and hence a consequence of induction. Uritani and coworkers, in a series of elegant experiments, have shown that the number of PAL molecules per cell in fact increases after wounding. PAL from wounded sweet potato tissues was purified to near electrophoretic homogeneity (28) and antibodies were raised against the PAL antigen through immunization of white rabbits. Then antigens were prepared from intact and wounded sweet potato root tissue. No precipitin band was formed with anti-serum and antigen from resting root tissue, using the Ouchterlony double diffusion test. This indicated that no PAL proteins are present in the intact root. Wounding, however, leads to the appearance of antigenic material, the concentration of which increases up to the second day of wound-healing and then decreases again (27). Labeling experiments also point to a *de novo* synthesis of PAL. Wounded tissues were incubated with ^{14}C-leucine and the label incorporated into PAL was monitored by radioactivity measurement in the corresponding immunoprecipitate and in the subunits of the enzyme after SDS-polyacrylamide gel electrophoresis. After a lag period of 4–6 hr label was introduced into the subunits, reaching a maximum about 8 hr after wounding. The incorporation of label is then maintained at a constant level, although PAL activity begins to decrease (29). This may be interpreted as the presence of a PAL degradation system which comes into action after the enzyme reaches maximal activity (30). Wound-induced synthesis of an enzyme, its accumu-lation, and the induction of a (specific?) enzyme degradative system with subsequent decrease in the number of active enzymes may be a characteristic feature of other systems as well (*i.e.,* glucose-6-phosphate dehydrogenase) (30). Wounding, then, induces two fundamentally different kinds of processes in the cells at a wound : rapidly occurring unblocking processes and comparably slow processes, which pre-suppose the induction of enzymes and consequently transcription. Wound-healing is therefore under gene control.

It has been amply demonstrated that wound-healing is accompanied by a tre-mendous increase in RNA content and that the inhibition of this process inhibits the healing (31). Among the different RNAs synthesized after an injury rRNA is domi-nant in concentration over tRNAs and mRNAs. During the first 2–3 hr after wound-ing, however, the synthesis of rRNA and tRNA is negligible, whereas mRNA for-mation proceeds intensely at the same time (32,33). Immediately after wounding, preexisting mRNAs are obviously used for translation but they code for the same proteins as those synthesized after wounding (34,35). The pattern of active genes in the intact cells is therefore retained shortly after wounding. mRNA synthesis con-tinues for another 10–15 hr.

Starting about 4–6 hr after wounding, rRNA synthesis comes into play with the synthesis of a 2.3×10^6 dalton precursor, which is cleaved into one 1.4×10^6 and one 0.9×10^6 dalton component. These fragments are further processed to 1.3×10^6 and 0.7×10^6 dalton rRNAs of the large and small ribosomal subunits respectively (36,37). tRNA synthesis is also a characteristic feature of wounded tissues (38,39). The tRNA of intact tissues seems to be undermethylated and inactive ; wounding induces the synthesis of tRNAs which are rapidly and extensively methy-lated (40,41) and concomitantly develop an increased amino acid acceptor activity.

All these events are in good agreement with electron microscopic and autoradio-graphic data. Thus, nucleoli of wounded cells increase considerably in volume. Distinct changes in structural organization accompany this process. Whereas the com-pact and relatively small nucleoli of quiescent cells possess only a thin peripheral granular zone embedded in masses of specialized chromatin together with few nucleolar vacuoles, wounding leads to the development of defined granular zones and nucleolar vacuoles. The latter increase in size after wounding and show rhythmic pulsations during the whole wound-healing period. At the same time the nucleolar chromatin (L-zone) disperses, indicative of active transcription of the ribosomal cistrons (42–47). The dynamics of rRNA synthesis can be revealed by pulse-chase experiments. Most label of administered ^{3}H-uridine appears in the chromatin region but not in the nucleolus immediately after wounding. About 1 hr later label also accumulates in the nucleolus. rRNA proceeds quite rapidly, but processing takes some time so that the ribosomal constituents do not appear in the cytoplasm until about 5–6 hr after wounding (33).

The activities of both DNA-dependent RNA polymerase I and II also reflect these dramatic changes. The action of these enzymes is comparably low in intact tissues but rises vigorously after wounding (45,48–50). Polymerase I usually increases its activity about 4-fold during 18 hr of wound-healing, whereas polymerase II only doubles its activity in the same period (45). The polymerases read out newly activated genes after wounding : nearest neighbor frequency analysis reveals changes in the overall base composition of the RNA synthesized *in vitro* on chromatin from resting as compared to wounded tissues.

THE HORMONAL REGULATION OF WOUND-HEALING

Very early in plant wound research hormones were suspected of regulating wound-healing. Haberlandt (51,52) in a series of classical experiments with wounded potato tuber tissue hypothesized about the induction of mitoses at a wound surface. He thought a so-called "wound hormone" or "nekro-hormone", generated by degradation of cytoplasmic constituents of wounded or killed cells was responsible for the wound-induced cell divisions. In the neighborhood of the phloem (then called "leptom") a so-called "lepto-hormone" was thought to aid the "wound-hormone", so that the frequency of mitoses around the phloem was distinctly higher than elsewhere. Although the concept of the "wound hormone" could not be maintained, we know of many hormones interfering with various processes of the wounded plant. Naphthalene acetic acid if supplied externally to wounded tissues accelerates mitotic frequency, as does gibberellic acid and kinetin. Combinations of various hormones were more effective than distinct hormones alone ; the hormonal balance seems to be the determining factor (53). Higher concentrations of kinetin as compared to auxin seem to advance callus formation, while higher concentrations of auxin promote periderm formation in wounded potato tuber tissue. A series of metabolic reactions is also drastically influenced by hormones. Many enzymes are more active in auxin- and gibberellic acid-treated wound tissues (*i.e.*, DNA-dependent RNA polymerases, histone and nonhistone protein kinases, DNA polymerases). Consequently, the synthesis of rRNA, tRNA, and mRNA (54) and the phosphorylation of various chromosomal (55) and ribosomal proteins (56) are much higher in hormone-treated wound tissues. As judged from nearest neighbor analyses of the RNAs synthesized in wounded or GA_3-treated tissues, respectively, the hormone induces the synthesis of qualitatively different RNAs. The simplest interpretation of this fact is that GA_3 induces the switching-on of specific genes not active in wounded tissues. Moreover it is apparent that after wounding the tissue develops a responsiveness towards the hormones. For example, freshly wounded tissue does not react upon any hormone, but after some hours of wound-healing it becomes competent (57). This is true for all hormones so far tested. Lastly, all the well-known phytohormones are definitely more "active" (i.e., exhibit more effectiveness in biological tests) or are present in higher quantities after wounding. This is true for auxins which are concentrated in the phloem (58,59), gibberellins (60,61), and cytokinins as well (62,63).

Although we do not understand the molecular mechanism by which all the different hormones act in the cell, it is evident that they function by regulating profound reactions after wounding (i.e., RNA synthesis). Even the termination of wound-healing seems to be under hormonal control, at least in the potato tuber model system. Some 2–3 days after wounding of these tissues suberin synthesis starts. This polymer is deposited onto the wound-exposed cell walls and leads to a tight closure of the wound, stopping the free exchange of gases like CO_2 and O_2. Depletion of oxygen for respiratory processes in turn leads to a decrease in cellular activity. Now, L-abscisic acid (ABA) appears in increasing concentrations about 24–36 hr after wounding. ABA triggers the synthesis of a "suberin-inducing factor" (SIF) which initiates transcription of the ω-hydroxyacid dehydrogenase gene. De novo synthesis of this enzyme is a prerequisite of suberin biosynthesis. After maximal suberin formation the enzyme is again degraded and suberin synthesis stops (64). The wound-healing process is terminated.

ABNORMAL WOUND-HEALING INDUCED BY A. TUMEFACIENS

Frequently a ubiquitously occurring gram-negative bacterium of the Rhizobiaceae, A. tumefaciens, settles at the wound site of a plant. The residues of the destroyed cells accumulating at the wound ("wound sap") create favorable conditions for the bacteria. Wound sap components may serve as nutrients (65,66) and may thereby influence bacterial metabolism and growth (67). Special sites of the cell walls of the destroyed cells allow an attachment of the bacteria. The existence of these sites was first suggested when it was found that tumor induction was delayed by the simultaneous inoculation of wounds with both avirulent and virulent bacteria (66). The competition between the two types of bacteria for specific receptor sites has been amply confirmed (68–70) and occurs only when avirulent cells precede or are inoculated together with the virulent bacteria. The attachment process is rapid : after only 15 min the site attachment is complete in wounds of bean leaves (66). The attachment is also specific : avirulent agrobacteria but neither closely related Rhizobium species nor E.coli cells compete for attachment sites (68). The specificity of this process is best explained by the assumption that complementary surface charges or configurations are present on both the bacterial and the host sites. The effective component of the bacterial envelope is a lipopolysaccharide, especially the polysaccharide moiety. The specific binding site of the plant cell probably originates in the middle lamella of the cell wall and consists at least partly of polygalacturo- nates (71). These sites are no constituent of the plant cell *a priori*, but are either synthesized or unmasked for recognition during maturation of the cell (72). Mono- cotyledons apparently do not possess these attachment sites nor do tumor cells. This may be due to extensive methylation of pectins in the walls of these cells which prevents such attachment (71).

Once the attachment has taken place, several changes in the bacterial cell enve- lope and the plant's cell wall occur. These changes have been documented by elec- tron micrographs (68). It is evident, then, that wounding of plants removes physical barriers (*i.e.*, cuticles) and allows the approach and settlement of virulent bacteria. Moreover, it exposes specific receptor sites. The complementarity of these receptors with the attachment sites of the bacterial cell envelope is a fundamental recognition element in the infection process.

After the site-specific attachment the bacteria in a conjugation-like process inject part of a large plasmid into the plant cell where it is stably integrated into the nuclear genome. As a consequence of this transfer of genetic material the wound cell does not return to a quiescent state again ; the normal wound-healing process is perpetuated. The wound-induced activation of metabolism, the extensive synthesis of various phytohormones and the induction of mitotic activity remain and lead to an increasing proliferation of the transformed cells : the crown gall tumor becomes manifest (Fig.5).

Some strains of *Agrobacterium* induce the formation of disorganized tumors

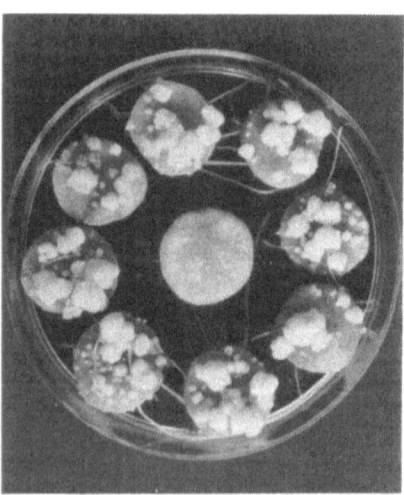

Fig. 5. Crown gall tumors on white potato tuber parenchyma induced by virulent strain C58 of *A. tumefaciens*. The control disk in the center represents the wounded, uninfected tissue.

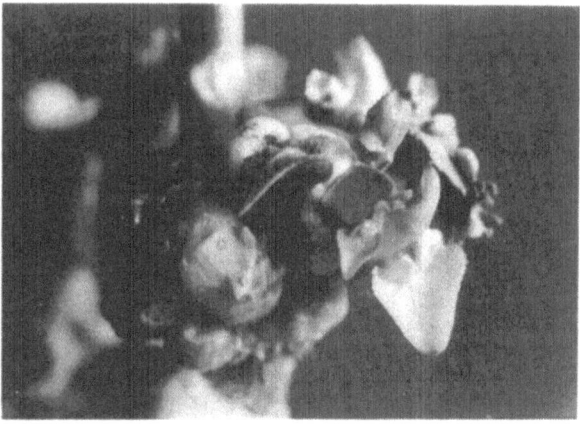

Fig. 6. Teratoma formation on a *Bryophyllum daigremontianum* host plant induced
by *A. tumefaciens* strain C58.

on susceptible plants, other strains the formation of teratomas in which aberrant
shoots and leaves develop in the tumor tissue (Fig. 6).

To summarize, it seems that the tumorous outgrowth of infected wound cells
represents a type of wound-healing which begins normally, but by an as yet unknown
mechanism the genes controlling wound-healing remain active indefinitely and
proliferation escapes control.

THE Ti-PLASMID OF *A. TUMEFACIENS*

The hypothetical tumor-inducing principle of *A. tumefaciens* (TIP) (*73*) has been
shown to be located on relatively large plasmids ("Ti-plasmids" for tumor-inducing
plasmids) by several criteria :

1) Agrobacteria, which lose their Ti-plasmid inevitably lose their oncogenicity
(*74,75*).

2) Transfer of a Ti-plasmid from an oncogenic to a non-oncogenic strain by
conjugation or transformation confers oncogenicity to the previous non-oncogenic
strain (*76,77*).

3) *Agrobacterium* strains harboring mutated Ti-plasmids are non-oncogenic
(*78*).

The Ti-plasmids with molecular weights ranging between 90 and 156 million can
be grouped into three categories :

1) "Octopine" plasmids, which code for N^2-(D-1-carboxyethyl) derivatives of
L-arginine (octopine), L-ornithine (octopinic acid), L-lysine (lysopine), and L-histidine
(histopine) as well as agropine, a bicyclic derivative of glutamic acid and a hexitol
sugar (Fig. 7).

2) "Nopaline" plasmids, which code for N^2-(1,3-dicarboxypropyl) derivatives

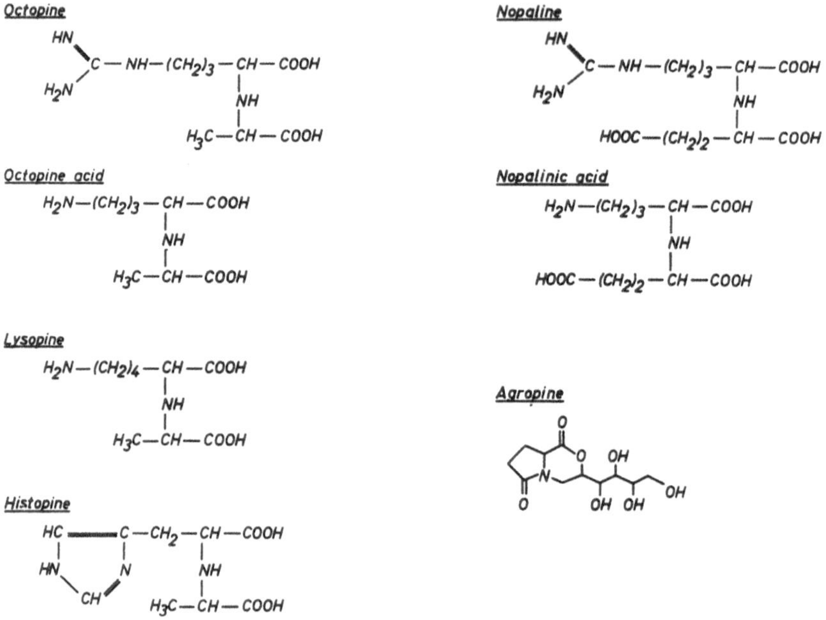

Fig. 7. Structural formula of the opines, synthesized in crown gall tumor cells and coded by T-DNA from nopaline (left), octopine (right, upper half) and agropine plasmids.

of L-arginine (nopaline), L-ornithine (nopalinic acid, ornaline) and so-called agrocinopines (phosphorylated opines) (79).

3) "Agropine" plasmids, which code for agropine (formerly called "null-type" plasmids) and also agrocinopines.

These opines are only synthesized in the transformed plant cells at the direction of Ti-plasmid-borne genes. Yet other information located on that plasmid confers on the bacterium the ability to catabolize specifically those opines whose synthesis it elicits in the tumor tissue. This represents the "biological rationale for the existence of the crown gall tumor" (80). This phenomenon will be dealt with on page 210.

Other plasmid-encoded functions have been defined by antibiotic resistance transposon insertion mutants as well as deletion mutants (78) and comprise transfer functions (Tra), incompatibility (Inc), and phage exclusion functions (Ape), genes for the catabolism of nopaline (Noc) or octopine (Occ), arginine (Arc), agropine (Agr), and agrocinopine catabolism (Psc). Moreover, regions of homology between nopaline and octopine plasmids must be directly or indirectly involved in neoplastic transformations (Onc regions ; oncogenicity). One of these Onc functions is transferred to the host cell together with other functions (T-region) (81) and maintained in the host cell's nucleus as T-DNA. The other Onc functions are distributed over one half of the pTi map, but what they code for is still unknown (Fig. 8).

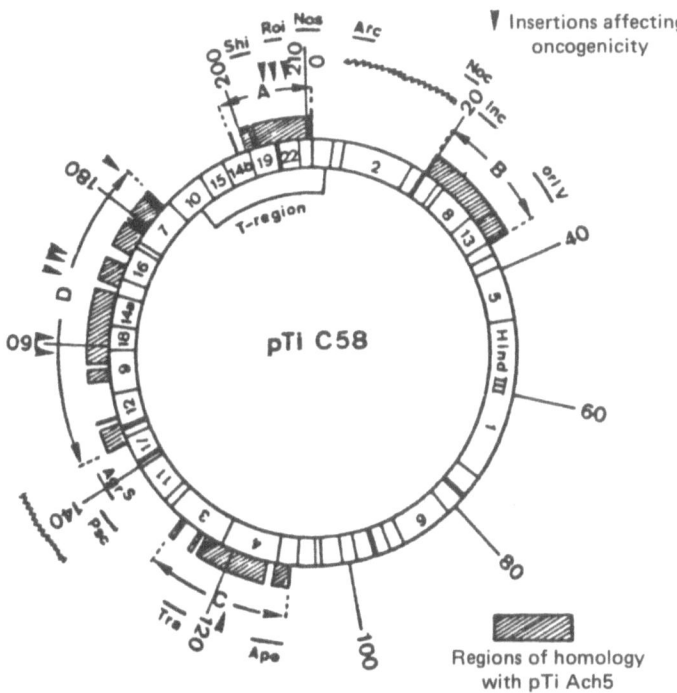

Fig. 8. Functional map of a nopaline Ti-plasmid (C58). *Hin*d III restriction fragments
are numbered in the order of their molecular weight. Hatched areas represent regions
of homology between nopaline and octopine plasmids. Arrow heads indicate positions
of regions (onc regions) which are necessary for tumor induction but not part of the
T-region (*101*).

Inc, incompatibility ; Nos, nopaline synthesis ; Shi, shoot induction ; Noc, nopaline
catabolism ; Roi, root induction ; Arc, arginine catabolism ; Tra, transfer genes ; Agrc,
agropine catabolism ; Ape, phage exclusion ; Psc, catabolism of phosphorylated sugars ;
oriV, origin of replication

The T-region itself, though differing in length in octopine and nopaline plasmids,
carries functions essential for tumor induction and maintenance ("core"), opine
synthesis (right part) and functions such as shoot induction, root induction or shoot
and root induction and host range. The host range-DNA determines the variety of
dicotyledonous plants on which tumors are induced. With the aid of all this information functional maps of the T-regions of the various Ti-plasmids can be constructed
(Fig. 9). The most important conclusion of this genetic analysis is that this T-region
must contain genes which in an unknown way transform plant cells, maybe by
direct or indirect control of the synthesis of phytohormones. Direct control would
mean that genes of the core T-DNA code for auxins and cytokinins, indirect control implies that these genes regulate the action of the plant's auxin and cytokinin
genes.

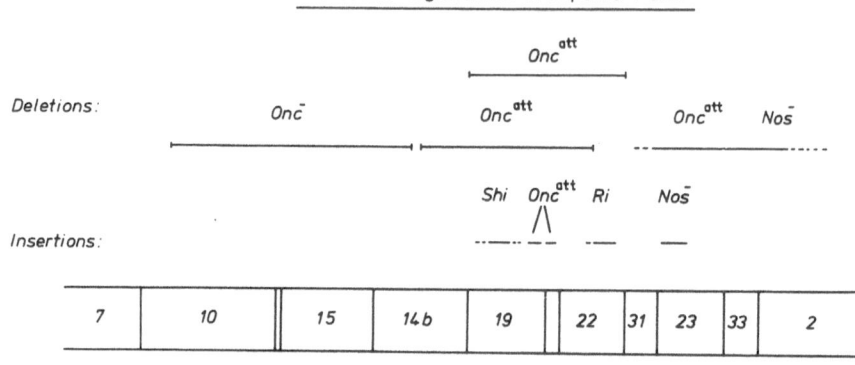

Fig. 9. Functional organization of the T-DNA in crown gall tumor cells producing nopaline (as revealed by deletion or insertion mutants of pTi C58 ; 78).

TRANSFER OF THE T-REGION OF THE Ti-PLASMIDS INTO THE HOST CELL, ITS INTEGRATION AND TRANSCRIPTION

The transfer of the T-region of Ti-plasmids from the bacterium to the plant cell has been proved by a series of experiments using DNA/DNA hybridizations. DNA from crown gall tissues was cut with various restriction endonucleases and radioactively labelled fragments covering the whole Ti-plasmid were used as probes in the Southern blotting analysis (81–84). In essence, the results of these analyses showed that all crown gall tumors contain a DNA segment (T-DNA) which is homologous to and colinear with the T-region of pTi. Thus, no major rearrangements occur during the transfer and integration process. The T-DNA found in tumors induced on tobacco by the nopaline strains C58 and T37 comprises about 23kbp, whereas that of octopine tobacco crown gall tumors is smaller (14–15 kbp) and variable in size (84). It seems that the T-DNA is present in multiple copies in the plant cell, the upper limit being 20 copies. These copies are organized as tandem repeats which are inserted into repetitive sequences of the plant cell's DNA (81). Nucleotide sequences bordering the T-DNA have also been shown to be arranged in repetitive elements suggesting an involvement of these border sequences in integration, very much like bacterial transposable elements (85).

The T-DNA is integrated exclusively into the nuclear DNA. DNA preparations from both purified chloroplasts and mitochondria of crown gall tumors contain no sequences homologous to Ti-plasmid DNA. On the contrary, purified nuclei of tumors harbor DNA which hybridizes to the T-region DNA (86,87).

There is sound evidence that T-DNA sequences are also transcribed. The first indication of expression of plasmid genes was the demonstration of opines in crown gall tumors, substances which have never been found in healthy plant tissues. Direct proof for transcripts deriving from pTi-sequences came from experiments in which total RNA from crown gall tumors was hybridized to a specific fragment of the Ti-plasmid (88). Using a similar approach it was demonstrated that in three different octopine tumors the right part of the T-DNA was most actively expressed, whereas only faint transcription occurred in the left part (89). Other studies have confirmed and detailed these results since then (90–92). Both strands of the T-DNA code for RNA. The Tl-DNA in octopine crown gall tissues is transcribed into seven polyadeny-lated RNAs, which range in size from 670 to 2,700 bases. At least two, possibly four transcripts are common to both octopine and nopaline tumors and originate from the core T-DNA region (91). All the different RNAs are expressed from individual promoters. The functions of several transcripts have been determined : two out of the seven RNAs are sufficient to allow tumor growth ; four, maybe five RNAs sup-press tumor organ development ; shoot and root formation are controlled by dif-ferent transcripts as is T-DNA transfer and oncogenicity (92).

Transcription of T-DNA genes is exerted through host DNA-dependent RNA polymerase II, since low concentrations of α-amanitin (0.7 μg/ml) inhibited the ap-pearance of transcripts homologous to T-DNA sequences in isolated tumor cell nuclei (91).

Taken together, these studies revealed that the T-DNA is transcribed into seven transcripts by RNA polymerase II of the host cell nucleus. The pattern of trans-cription is characteristic : some regions (i.e., the opine and tumor growth genes) are more actively transcribed than others. No extensive differences exist between nuclear and polysomal transcripts, which means that processing of the transcripts is of minor importance, if at all (90). As is already suggested by the appearance of opines in the tumors, transcription of the opine genes is also followed by the translation of the RNAs. As a consequence, T-DNA specific proteins are constituents of crown gall tumor cells (92 ; i.e., lysopine dehydrogenase, 93).

PHYTOHORMONES AND TUMOROUS PROLIFERATION

It has been known for some time that crown gall tumors no longer require outside sources of both auxin and cytokinin for their growth in culture (94,95). They synthesize and contain higher levels of both regulators than do normal tissues and therefore gain autonomy. The persistent presence of both hormones in crown gall cells keeps these cells dividing indefinitely. Thus, tumorous proliferation is thought to be a consequence of the immense production of these phytohormones. Among these are indole acetic acid (IAA), ribosyl-trans-zeatin, ribosylzeatin and the side chain glucosides of the latter compounds (96). All these components are also synthe-sized and secreted by virulent strains of A. tumefaciens (97,98) and may be a de-termining factor in initiation of tumorous growth. However, early log-phase cultures of the bacterium do not secrete significant amounts of auxins or cytokinins into media although they are most infectious (99). Thus we are left with the conclusion that bacteria-borne phytohormones of whatever kind do not induce tumorous proliferation. On the contrary, it seems that the T-DNA carries genes directly or indirectly involved in hormone biosynthesis. A direct involvement of T-DNA may be brought about by a gene coding for a tRNA, from which a base with cytokinin-like activity is excised. Whereas such a specific excision enzyme has been found in Ehrlich ascites tumors which cuts the Q base out of the tumor tRNA and replaces it with guanine, no evidence has been reported in plant tumors. An indirect involvement may be envisaged as interference with the activity of phytohormone genes of the host by positional or other regulatory effects.

Whatever the function of T-DNA genes in hormone metabolism, the wound-activated synthesis of both auxins and cytokinins does not come to an end but is perpetuated in the presence of *Agrobacterium*. Actually, a biphasic response to wounding and infection can be detected : wound-healing is accompanied by a rise in concentration of isopentenyladenosin and, to a lesser extent, *trans*-zeatin and auxin, which decreases again at the end of the healing period. Infection with agrobacteria, however, again leads to a dramatic increase in all hormones (*98*). In crown gall cell cultures the dynamics of the phytohormones suggests a complicated interplay of the different classes of hormones and abscisic acid (*100*). Thus, mitotic frequency seems to be directly correlated with increased levels of cytokinins and decreased concentrations of abscisic acid.

To sum up, our understanding of the mechanisms of T-DNA-induced transformation presupposes a considerable extension of our knowledge of how hormones work on a molecular level.

"GENETIC COLONIZATION" : A NOVEL TYPE OF PARASITISM

While it is not yet completely evident what the benefit of tumor formation for the plant may be, it is doubtless of importance for the bacteria. These agrobacteria, and probably other bacteria, have evolved a unique mechanism for the transfer of specific genetic information to plants. The information is stably inserted into the plant's genome and, at least partly, expressed to new phenotypes such as proliferation and the synthesis of opines. These opines are synthesized only by transformed cells which increase in number during tumor growth. Therefore more and more opines are synthesized and released into the immediate environment of the tumor. The agrobacteria present in the tumor surroundings are able to take up the opines and to use them as an energy, C and/or N source. In other words, they benefit from these opines for their own growth and proliferation. Since only agrobacteria possess plasmids with genes specifying opine catabolism and utilization, they have an important selective advantage over the millions of other competing bacteria. Therefore *Agrobacterium* has created by way of natural genetic engineering an ecological niche for itself.

Thus, in a special type of parasitism the parasite manages to introduce plasmid genes into a host cell's genome. As a consequence it forces the host to synthesize opines which can only be catabolized by the parasite. By simultaneously inducing the proliferation of the host cell, the bacterium secures an ever increasing source of opines for its growth. In essence, the parasite broaches the photosynthetic machinery for its own purposes. This kind of specialized parasitism has been termed "genetic colonization" by J. Schell (101).

CONDITIONING OF THE WOUND CELL AND TRANSFORMATION

It has been known from the earliest experiments on the induction of crown gall tumors that wounding of a potential host is essential for the transformation process (102). All attempts to induce primary tumors without any wound have thus far been unsuccessful. The various hypotheses put forward to explain the necessity of wounds for tumor induction can be grouped into two broad categories. First, wounding permits the penetration of the bacteria into the host plant ("portal of entry"). Second, it leads to a transient activation of the host cell's metabolism, which in turn makes the cell competent to react upon the T-DNA with tumorous proliferation ("conditioning").

Wounding doubtless functions by providing a portal of entry for the bacteria. The removal of mechanical barriers (i.e., cuticles), the exposure of plasmodesmata and specific receptor sites of middle lamella origin, and perhaps activating or nutritional compounds of the "wound sap" altogether influence the interaction between bacterium and plant.

The site-specific attachment of agrobacteria to complementary address sites of wound-exposed cell walls has been elucidated in some detail. The existence of such sites was first postulated when it was found that tumor induction was delayed by the simultaneous inoculation of wounds with both avirulent and virulent bacteria (103). If the avirulent cells are added after the inoculation of a wound by virulent bacteria, then no competition takes place (104). The attachment process is rapid ; after only 15 min the attachment is complete (103). It is also very specific. Competition is only exerted by avirulent agrobacteria but not by closely related Rhizobium species or by E.coli cells (105). The specificity of this process is best explained by the assumption that complementary surface charges or configurations are present on both the bacterial and the host sites. These allow some kind of conjugative mechanism between Agrobacterium and the dicot host cell. The receptors of the host are only exposed by the drastic act of wounding.

The bacterial attachment site seems to consist of lipopolysaccharides, more concretely the polysaccharide moiety (106). The specific binding site of the plant probably originates in the middle lamella of the cell wall and consists partly if not entirely of polygalacturonic acid. This compound is highly methylated in the cell walls of monocotyledons, which are nonsusceptible to Agrobacterium and hence tumor formation. Demethylation, however, converts the pectins of monokots to address sites for the bacterium (106).

Another important consequence of wounding is an activation of metabolism in the wound cell. Transformation can only take place, if the host cell is in this activated state, if it is "conditioned". When a plant is wounded, then the great variety of reactions discussed in the preceding chapters are set into motion in the living cell adjacent to the wound, which altogether result in the healing of that wound. Normally upon repair of the wound, the cells again return to the quiescent stage they had been before. The wound response then is transient in nature. Virulent agrobacteria, if present at the wound-site, interfere with this wound-healing process. The T-DNA quite obviously fixes the activated wound response indefinitely, the return to normality is blocked. Doubtless the permanent switching on of genes coding for the synthesis of certain plant hormones plays a crucial role in this process, since growth of plant tissues is regulated by the interaction of at least two such regulators (auxins and cytokinins).

Wounded cells are not susceptible to tumor formation throughout the entire wound-healing period. Only during a relatively short time in the wound-healing cycle can the cells be transformed (107). Shortly after wounding the cells are resistant to an infection but become gradually more and more susceptible to transformation, reaching a maximum of vulnerability after about 2–3 days of wound-healing, which afterwards again declines (108,109). Tumors initiated before or after this period develop more slowly and remain smaller in size. Thus, if the host cells have not yet acquired maximum susceptibility ("competence") after wounding, transformation will not occur despite the presence of virulent bacteria in intimate contact with the plant cells. The process by which the cells at a wound-site first become sensitive and then refractory to the tumorigenic stimulus of A. tumefaciens has been termed "conditioning" (110). If potential host cells are not adequately conditioned (i.e., in the earliest and latest stages of wound-healing or in most normal cells not influenced by a wound), then the transformation will not occur, even though virulent bacteria attach to the corresponding plant cell receptors. In some plants the minimum conditioning is complete before the first wound-induced cell divisions occur (111). These facts signal that transformation of a normal cell to a tumor cell presupposes a state of active wound response. Indicator for such a state doubtless may be active cell divisions, although in many wound systems no new mitoses are necessary for tumor formation (112,113).

Conditioning, then, is a transient process and represents an important part of the normal wound response. However, it is still obscure what conditioning means in molecular terms. A more detailed knowledge of the wound processes in general is necessary to allow a more precise and, perhaps a new definition of this process.

PERSPECTIVE

The various wound-healing processes in lower and higher plants as well as crown gall tumor induction by A. tumefaciens in wounded plants will remain areas of intensive and profitable research. What we urgently need is a more detailed knowledge of how the activity of different genes is induced after wounding. How are they controlled at the end of wound-healing? What triggers cell divisions in wounded tissues? How do the main controlling hormones, auxins and cytokinins, work at the molecular level? What is the function of abscisic acid in wound-healing? What signals the end of wound-healing?

Even more questions remain to be answered as to the interaction of Agro-

bacterium with the wounded plant cell. What is the mechanism of recognition, what is the chemical nature of the receptors? How is the transfer of the T-region of the Ti-plasmid achieved? What are the functions of the Onc-regions? Where and how are the copies of the T-DNA made? What enzymes catalyze the replication of the T-DNA? The mechanism and location of integration of the T-DNA is not known. It remains to be solved whether the insertion locus is always the same or is variable. How many transcripts are coded by the T-DNA ; what are their functions?

Since highly active groups are working on all these questions, new insights into the mechanisms of wound-healing and crown gall tumor formation are to be expected in the near future. It is also emerging that crown gall research in the time ahead will focus on the capability of the Ti-plasmid to transfer genes with desirable information into higher plants. The use of the pTi as a vector for introducing new properties into crop plants may very soon prove to be as important to mankind as the old art of breeding and selection.

Acknowledgements
    The following persons contributed to the work of the author and should be considered as coauthors : C.Gräf, E.Hadjiloucas, S.Kost, K.Oba (Nagoya University, Japan), S.Ranostaj, W.Schäfer, R.K.Tripathi (G.B.Pant University, India), M.Wechselberger, F.Weigand, K.Weising, B.Wielgat (Polish Academy of Sciences, Warsaw, Poland), and H.J.Zimmermann. The experiments of this group were supported by the Thyssen-Foundation (Germany) and the Alexander von Humboldt-Foundation (Germany).

REFERENCES

1.   Hirayama, O., Matsuda, H., Takeda, H., Maenaka, K. and Takatsuka, A. : 1975. *Biochim. Biophys.Acta* **384**, 127–137.
2.   Shephard, D.V. and Pitt, D. : 1976. *Phytochemistry* **15**, 1471–1474.
3.   Hasson, E.P. and Laties, G.G. : 1976. *Plant Physiol.* 57, 142–147.
4.   Galliard, T. : 1978. *In* Biochemistry of Wounded Plant Tissues (Kahl, G., ed.), pp.155–201, Walter de Gruyter, Berlin.
5.   Laties, G.G. : 1978. *In* Biochemistry of Wounded Plant Tissues (Kahl, G., ed.), pp.421–466, Walter de Gruyter, Berlin.
6.   Kahl, G. : 1974. *Bot. Rev.* 40, 263–314.
7.   Yang, S.F. and Pratt, H.K. : 1978. *In* Biochemistry of Wounded Plant Tissues (Kahl, G., ed.), pp.595–622, Walter de Gruyter, Berlin.
8.   Elstner, E.F. and Konze, J.R. : 1976. *Nature* 263, 351–352.
9.   Kahl, G. : 1982. *In* Molecular Biology of Wound Healing : The Conditioning Phenomenon (Kahl, G. and Schell, J., eds.), pp.211–267, Academic Press, New York.
10.   Laties, G.G. : 1963. *In* Control Mechanisms in Respiration and Fermentation (Wright, B., ed.), pp.129–155, Ronald Press, New York.
11.   Payes, B. and Laties, G.G. : 1963. *Biochem.Biophys.Res.Commun.* 10, 460–466.
12.   Laties, G.G. : 1967. *Phytochemistry* 6, 181–185.
13.   Laties, G.G. : 1978. *In* Biochemisty of Wounded Plant Tissues (Kahl, G., ed.), pp.421–466, Walter de Gruyter, Berlin.
14.   Jackman, M.E. and van Steveninck, R.F.M. : 1967. *Aust. J. Biol. Sci.* 20, 1063–1068.

214   G. KAHL

15. van Steveninck, R.F.M. and Jackman, M.E. : 1967. *Aust. J. Biol. Sci.* **20**, 749–760.
16. Lance, C. and Dizengremel, P. : 1978. *In* Biochemistry of Wounded Plant Tissues (Kahl, G., ed.), pp.467–501, Walter de Gruyter, Berlin.
17. Hackett, D.P., Haas, D.W., Griffiths, S.K. and Niederpruem, D.J. : 1960. *Plant Physiol.* **35**, 8–19.
18. Kanazawa,Y., Asahi, T. and Uritani, I. : 1967. *Plant Cell Physiol.* **8**, 249–262.
19. Rich, P.R., Boveris, A., Bonner, W.D. and Moore, A.L. : 1976. *Biochem. Biophys. Res. Commun.* **3**, 695–703.
20. Click, R.E. and Hackett, D.P. : 1963. *Proc. Natl. Acad. Sci. U.S.* **50**, 243–250.
21. Laties, G.G. : 1965. *Plant Physiol.* **40**, 1237–1241.
22. Waring, A.J. and Laties, G.G. : 1977. *Plant Physiol.* **59**(6).
23. Kahl, G. : 1971. *Z. Naturforsch.* **26B**, 1058–1064.
24. Sparkuhl, J., Gare, R.L. and Setterfield, G. : 1976. *Planta* **129**, 97–104.
25. Byrne, H. and Setterfield, G. : 1977. *Planta* **143**, 75–83.
26. Kahl, G. : 1973. *Bot. Rev.* **39**, 274–299.
27. Tanaka, Y. and Uritani, I. : 1976. *J.Biochem. (Tokyo)* **79**, 217–219.
28. Tanaka, Y. and Uritani, I. : 1977. *J.Biochem. (Tokyo)* **81**, 963–970.
29. Tanaka, Y. and Uritani, I. : 1977. *Eur. J. Biochem.* **73**, 255–260.
30. Kahl, G. : 1978. *In* Biochemistry of Wounded Plant Tissues (Kahl, G., ed.), pp.347–390, Walter de Gruyter, Berlin.
31. Kahl, G. : 1973. *Bot.Rev.* **39**, 274–299.
32. Sato, T., Watanabe, A. and Imaseki, H. : 1978. *Plant Cell Physiol.* **19**, 609–616.
33. Byrne, H. and Setterfield, G. : 1977. *Planta* **136**, 203–210.
34. Byrne, H. and Setterfield, G. : 1978. *Planta* **143**, 75–83.
35. Sato, T., Ishizuka, M., Watanabe, A. and Imaseki, H. : 1980. *Plant Cell Physiol.* **21**, 137–147.
36. Rogers, M.E., Loening, U.E. and Fraser, R.S.S. : 1970. *J. Mol. Biol.* **49**, 681–692.
37. Leaver, C.J. and Key, J.L. : 1970. *J. Mol. Biol.* **49**, 671–680.
38. Kahl, G. : 1971. *Z. Naturforsch.* **26B**, 1058–1064.
39. Frazer, R.S.S. : 1975. *Eur. J. Biochem.* **50**, 529–537.
40. Stone, B.P. and Cherry, J.H. : 1972. *Planta* **102**, 179–189.
41. King, B. and Chapman, J.M. : 1973. *Planta* **114**, 227–238.
42. Rose, R.J., Setterfield, G. and Fowke, L.C. : 1972. *Exp. Cell Res.* **71**,1–16.
43. Jordan, E.G. and Chapman, J.M. : 1971. *J. Exp. Bot.* **22**, 627–634.
44. Barckhausen, R. : 1978. *In* Biochemistry of Wounded Plant Tissues (Kahl, G., ed.), pp.1–42, Walter de Gruyter, Berlin.
45. Kahl, G. and Wielgat, B. : 1976. *Physiol. Vég.* **14**, 725–738.
46. Jordan, E.G. and Chapman, J.M. : 1973. *J. Exp. Bot.* **24**, 197–209.
47. Setterfield, G., Sparkuhl, J. and Byrne, H. : 1978. *In* Biochemistry of Wounded Plant Tissues (Kahl, G., ed.), pp.571–594, Walter de Gruyter, Berlin.
48. Duda, C.T. and Cherry, J.H. : 1971. *Plant Physiol.* **47**, 262–268.
49. Kahl, G. and Wechselberger, M. : 1977. *Z.Naturforschung* **32C**, 229–235.
50. Kamisaka, S. and Masuda, Y. : 1971. *Plant Cell Physiol.* **12**, 201–209.
51. Haberlandt, G. : 1921. *Beitr. Allg. Bot.* **2**, 1–53.
52. Haberlandt, G. : 1922. *Biol. Zentralbl.* **42**, 145.
53. Rosenstock, G. and Kahl, G. : 1978. *In* Biochemistry of Wounded Plant Tissues (Kahl, G., ed.), pp.623–671, Walter de Gruyter, Berlin.
54. Wielgat, B. and Kahl, G. : 1979. *Plant Physiol.* **64**, 867–871.
55. Schäfer, W. and Kahl, G. : 1981. *Plant Mol. Biol.* **1**, 5–17.
56. Schäfer, W. and Kahl, G. : unpublished.
57. Wielgat, B. and Kahl, G. : 1979. *Plant Physiol.* **64**, 863–866.
58. Hemberg, T. : 1943. *Ark. Bot.* **308**, 1–8.

59. Hemberg, T. : 1944. *Sven. Bot. Tidskr.* 38, 428.
60. Rappaport, L. and Sachs, M. : 1967. *Nature* 214, 1149–1150.
61. Edelman, J. and Bradshaw, J.M. : 1969. *Planta* 84, 94–96.
62. Conrad, K. and Köhn, B. : 1975. *Phytochemistry* 14, 325–328.
63. Weiler, E.W. and Spanier, K. : 1981. *Planta* 153, 326–337.
64. Kollattukudy, P.E. : 1978. *In* Biochemistry of Wounded Plant Tissues (Kahl, G., ed.), pp. 43–84, Walter de Gruyter, Berlin.
65. Lippincott, B.B. and Lippincott, J.A. : 1966. *J.Bacteriol.* 92, 937–945.
66. Lippincott, B.B. and Lippincott, J.A. : 1969. *J.Bacteriol.* 97, 620–628.
67. Pohjakallio, O. : 1963. *Z. Pflanzenkr. Pflanzenschutz* 71, 99–107.
68. Schilperoort, R.A. : 1969. Ph.D.Thesis, University of Leiden, Leiden.
69. Kerr, A. : 1969. *Aust. J. Biol. Sci.* 22, 111–116.
70. Manigault, P. : 1970. *Ann. Inst. Pasteur Paris* 119, 347–359.
71. Lippincott, J.A. and Lippincott, B.B. : 1977. *Annu. Rev. Microbiol.* 29, 377–405.
72. Lippincott, B.B., Whatley, M.H. and Lippincott, J.A. : 1977. *Plant Physiol.* 59, 388–390.
73. Braun, A.C. and White, P.R. : 1943. *Phytopathology* 33, 85.
74. Van Larebeke, N., Engler, G., Holsters, M., Van Den Elsacker, S., Zaenen, J., Schilperoort, R.A. and Schell, J. : 1974. *Nature* 252, 169–170.
75. Watson, B., Currier, T.C., Gordon, M.P., Chilton, M.D. and Nester, E.W. : 1975. *J.Bacteriol.* 123, 255–264.
76. Kerr, A. : 1969. *Nature* 223, 1175–1176.
77. Van Larebeke, N., Genetello, C., Schell, J., Schilperoort, R.A., Hermans, A.K., Hernalsteens, J.P. and Van Montagu, M. : 1975. *Nature* 255, 742–743.
78. Holsters, M., Silva, B., Van Vliet, F., Genetello, C., De Block, M., Dhaese, P., Depicker, A., Inzé, D., Engler, G., Villaroel, R., Van Montagu, M. and Schell, J. : 1980. *Plasmid* 3, 212–230.
79. Ellis, J.G. and Murphy, P.J. : 1981. *Mol. Gen. Genet.* 181, 36–43.
80. Guyon, P., Chilton, M.D., Petit, A. and Tempé, J. : 1980. *Proc. Natl. Acad. Sci. U.S.* 77, 2693–2697.
81. Zambryski, P., Holsters, M., Kruger, K., Depicker, A., Schell, J., Van Montagu, M. and Goodman, H.M. : 1980. *Science* 209, 1385–1391.
82. Thomashow, M.F., Nutter, R., Montoya, A.L., Gordon, M.P. and Nester, E.W. : 1980. *Cell* 19, 729–739.
83. Yadav, N.S., Postle, K., Saiki, R.K., Thomashow, M.F. and Chilton, M.D. : 1980. *Nature* 287, 458–461.
84. Lemmers, M., De Beuckeleer, M., Holsters, M., Zambryski, P., Depicker, A., Hernalsteens, J.P., Van Montagu, M. and Schell, J. : 1980. *J. Mol. Biol.* 144, 355–378.
85. Calos, M.P. and Miller, J.H. : 1980. *Cell* 20, 579–595.
86. Willmitzer, L., De Beuckeleer, M., Lemmers, M., Van Montagu, M. and Schell, J. : 1980. *Nature* 287, 359–361.
87. Chilton, M.D., Saiki, R.K., Yadav, N., Gordon, M.P. and Quetier, F. : 1980. *Proc. Natl. Acad. Sci. U.S.* 77, 4060–4064.
88. Drummond, N.H., Grodon, M.P., Nester, E.W. and Chilton, M.D. : 1977. *Nature* 269, 535–536.
89. Gurley, W.B., Kemp, J.D., Albert, M.J., Sutton, D.W. and Callis, J. : 1979. *Proc. Natl. Acad. Sci. U.S.* 76, 2828–2832.
90. Willmitzer, L., Otten L., Simons, G., Schmalenbach, W., Schröder, J., Schröder, G., Van Montagu, M., de Vos, G. and Schell, J. : 1981. *Mol. Gen. Genet.* 182, 255–262.
91. Willmitzer, L., Simons, G. and Schell, J. : 1982. *EMBO J.* 1, 139–146.
92. Leemans, J., Deblaere, R., Willmitzer, L., De Greve, H., Hernalsteens, J.P., Van Montagu, M. and Schell, J. : 1982. *EMBO J.* 1, 147–152.
93. Willmitzer, L., Schmalenbach, W. and Schell, J. : 1981. *Nucleic Acids Res.* 9, 4801–4812.

94. McPherson, J.C., Nester, E.W. and Gordon, M.P. : 1980. *Proc. Natl. Acad. Sci. U.S.* 77, 2666–2670.
95. Schröder, J., Schröder, G., Huisman, H., Schilperoort, R.A. and Schell, J. : 1981. *FEBS Lett.* 129, 166–168.
96. Braun, A.C. : 1978. *Biochim. Biophys. Acta* 516, 167–191.
97. Morris, R.O. : 1977. *Plant Physiol.* 59, 1029–1033.
98. Weiler, E.W. and Spanier, K. : 1981. *Planta* 153, 326–337
99. Lippincott, J.A. and Heberlein, G.T. : 1965. *Am. J. Bot.* 52, 396–403.
100. Weiler, E.W. : 1981. *Naturwissenshaften* 67, 377.
101. Schell, J., Van Montagu, M., De Beuckeleer, M., De Block, M., Depicker, A., De Wilde, M., Engler, G., Genetello, C., Hernalsteens, J.P., Holsters, M., Seurinck, J., Silva, B., Van Vliet, F. and Villarroel, R. : 1979. *Proc. R. Soc. Lond. B* 204, 251–266.
102. Smith, E.F., Brown, N.A. and Townsend, C.O. : 1911. *U.S.Dep. Agric. Bur. Plant Ind. Bull.* 213, 1–201.
103. Lippincott, B.B. and Lippincott, J.A. : 1969. *J. Bacteriol.* 97, 620–628.
104. Lippincott, J.A. and Lippincott, B.B. : 1975. *Annu. Rev. Microbiol.* 29, 377–405.
105. Schilperoort, R.A. : 1969. Ph. D. Thesis. University of Leiden, Leiden.
106. Lippincott, B.B., Whatley, M.H. and Lippincott, J.A. : 1977. *Plant Physiol.* 59, 388–390.
107. Braun, A.C. : 1962. *Annu. Rev. Plant Physiol.* 13, 533–558.
108. Klein, R.M. : 1954. *Brookhaven Symp. Biol.* 6, 97–114.
109. Braun, A.C. : 1954. *Brookhaven Symp. Biol.* 6, 115–127.
110. Braun, A.C. and Stonier, T. : 1958. *Protoplasmatologia* 10, Part 5a.
111. Lipetz, J. : 1966. *Cancer Res.* 26, 1597–1604.
112. Lippincott, J.A. and Heberlein, G.T. : 1965. *Am. J. Bot.* 52, 396–403.
113. Kurkdjian, A., Manigault, P. and Beardsley, R. : 1969. *Can. J. Bot.* 47, 803–808.

The Role of Secondary Metabolism in Plant-Pathogen Interactions

Tsune Kosuge and Luca Comai

INTRODUCTION

Both plants and microorganisms produce many compounds which have no recognized role in the maintenance of fundamental biological processes (1). Yet it is becoming evident that some of these compounds, known as secondary compounds, fulfill important functions in host-pathogen interactions (1). Certain secondary metabolites produced by pathogens act as factors conferring virulence on plants (2,3) ; plants on the other hand produce secondary metabolites that function as disease resistance factors (4). The production of a secondary metabolite therefore can help determine if the outcome of a plant-pathogen association is a compatible or an incompatible interation. Because of space limitations, the discussion below will be devoted mostly to the relationship of expression of virulence to secondary metabolism in a plant pathogen and will use the interaction between *Pseudomonas savastanoi* and its host, oleander, as a model.

VIRULENCE AND PATHOGENICITY IN PLANT-PATHOGEN INTERACTIONS

Microorganisms possess many attributes that help them function as pathogens on plants. The general term pathogenicity is defined here as the capacity of an organism to cause disease on a plant ; virulence is one component of pathogenicity and concerns those attributes of a pathogen directly involved with formation and severity of symptoms caused on plants. It is important to emphasize, however, that in many interactions expression of virulence represents one of the later steps in the overall

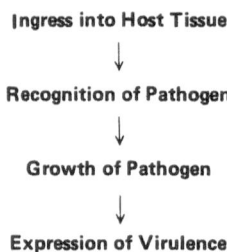

Fig. 1. Diagram of possible processes concerned with pathogenesis. The steps indicated involve attributes concerned with pathogenicity of a microorganism that causes a disease on plants. Some of the steps shown may overlap since a process such as ingress into plant tissue may be associated with an enzymic virulence factor such as pectate lyase (5). An incompatible interaction will result if one or more of the processes shown is inhibited or arrested by factors produced by the plant.

process of pathogenesis, and can involve production of more than one factor and more than one type of a factor (Fig. 1).

Furthermore expression of virulence can only be detected by the reaction of the plant to infection ; that primary symptoms of a disease are the reactions of the plant to the interaction between the pathogen's virulence factors and their target(s) in the plant. The response of a plant to a virulence factor can be affected by its physiological state as conditioned by the environment. Likewise a virulence factor could affect more than one site in the plant and some of the processes concerned with expression of virulence may overlap or may be difficult to distinguish from other steps of pathogenesis.

A virulence factor can be an enzyme such as pectate lyase (5), or a secondary metabolite such as a host selective toxin (6), or a growth hormone such as indole-acetic acid (2,3). Since virulence usually can be lost without lethal effects on the organism, many virulence factors may be produced as secondary metabolites. If virulence can be associated with production of a secondary metabolite, concepts of metabolic regulation can be used to explain how expression of virulence is controlled in a pathogen.

An understanding of virulence and its expression is not complete without in depth information of the virulence factor itself namely, its structure, route of synthesis, and mechanisms controlling its metabolism.

SIGNIFICANCE OF SECONDARY METABOLISM IN EXPRESSION OF VIRULENCE

It is commonly observed that virulence in a pathogen population varies widely and can be lost without loss of viability in culture. This observation is compatible with the notion that secondary metabolism is not essential for maintenance of fundamental processes in the pathogen. By relating expression of virulence to a known pathway of metabolism in the pathogen it is possible to prove that virulence can be modified by altered metabolism ; that factors regulating metabolism are directly concerned with expression of virulence. Thus if a virulence factor is identified and its biosynthetic pathway defined, enzymes concerned with its synthesis can be studied. Genes coding for synthesis of the enzymes then can be identified and by appropriate genetic techniques the gene(s) and gene products for virulence confirmed. It is also certain that mechanisms controlling production and turnover of virulence factors are influenced by the environment of host tissue. Hence, it will be necessary to determine what these controls are and how they respond to the conditions found in host tissue. In the following sections we will describe evidence which supports notion that indoleacetic acid (IAA) is a the secondary metabolite and functions as a virulence factor.

THE ROLE OF INDOLEACETIC ACID AS A VIRULENCE FACTOR

In the case of *Pseudomonas syringae* p.v. *savastanoi* (*P. savastanoi*) production of IAA is necessary for expression of virulence on its host oleander and olive (*7*). In this bacterium, indoleacetic acid is produced from L-tryptophan by the reactions (*8*): a) L-tryptophan → indoleacetamide, b) indoleacetamide → indoleacetic acid.

The enzyme and genetic determinants are respectively, a) tryptophan 2-monooxygenase, *iaaM*, b) indoleacetamide hydrolase, *iaaH*. Mutants deficient in indoleacetic acid production are weakly virulent or avirulent (Fig. 2) (*7*). In certain isolates of *P.*

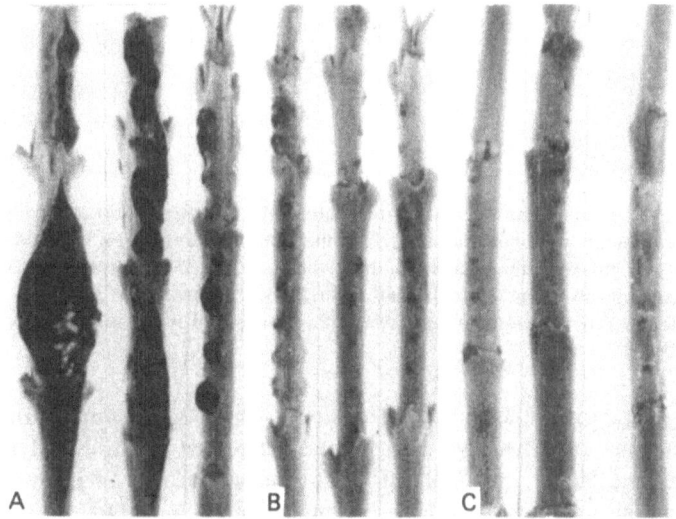

Fig. 2.   Oleander plants inoculated with different isolates of *P. savastanoi*. Stem of a plant inoculated (center) 2009−3, an Iaa⁻ mutant ; with 2009, an Iaa⁺ parental strain (left) ; stem of a plant injected with sterile culture medium (right). Plants were inoculated 6 weeks prior to photographing.

220   T. KOSUGE AND L. COMAI

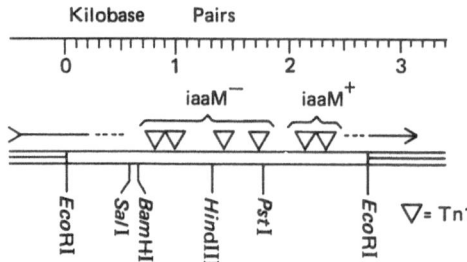

Fig. 3. Physical map of *Eco*RI-M, a 2.7 kb fragment of *pIAA1* which has been cloned
in *E. coli* (2). The approximate location of *iaaM* in the fragment was determined by Tn1
insertions which inactivate the monooxygenase gene and yield the iaaM⁻ phenotype.
Certain restricition endonuclease sites are also indicated.

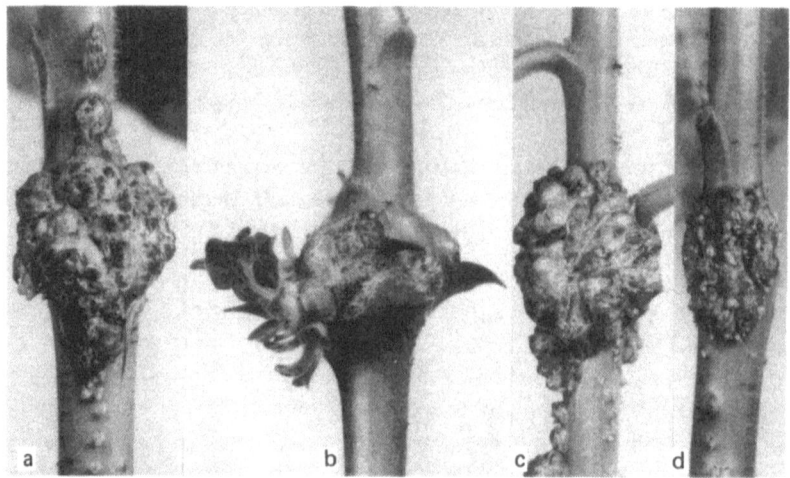

Fig. 4. Tumors on tobacco plants inoculated with *Agrobacterium tumefaciens* mutants.
The mutants were constructed by Garfinkel and Nester (*11*) and Garfinkel *et al.* (*12*)
by Tn5 insertion mutagenesis of the T-region of the Ti plasmids. Plants were inoculated with a) parental strain A348 (pPh1j1) ; b) mutant tms-328::Tn5 ; c) mutant
tml-358::Tn5 ; d) mutant tmr-149::Tn5. Photographs were taken 6 weeks after inoculation.

savastanoi, genes for indoleacetic acid synthesis occur on a plasmid pIAA (Fig. 3)
(*9*). Two versions of pIAA have been found so far : the 52 kb pIAA1 in oleander
strains 2009 and 2015 ; the 72 kb pIAA2 in Italian oleander strains (*10*). Loss of
pIAA1 by curing attenuates virulence in mutants such as 2009-3 ; virulence is restored by transformation with pIAA1 (*9*). It is interesting that the virulence factor,
IAA, is a natural growth hormone of the plant. However, in the normal plant, there
are mechanisms which maintain IAA at levels appropriate for the particular type of
tissue and its stage of development. In the association between *P. savastanoi* and its

host, IAA is introduced by the pathogen in amounts causing an imbalance in the host phytohormone levels and ultimately promotes unregulated or abnormal growth responses to occur in the host. Another phytohormone might also function as a virulence factor since Iaa⁻ mutants of *P. savastanoi* stimulate limited amount of tissue proliferation when inoculated into host tissue.

Indoleacetic acid together with cytokinins are strongly implicated as virulence factors in crown gall incited by *Agrobacterium tumefaciens* (6). *A. tumefaciens* mutants created by transposon mutagenesis of the T-region (T-DNA) of the Ti plasmid cause 'production of tumors of different morphologies, shoot forming and root-forming tumors *(11–15)* (Fig. 4).

The *Agrobacterium*-plant interaction appears to represent a highly specialized case since a segment of DNA (the T-region) carried in the Ti-plasmid of the pathogen is integrated into the host DNA where it is expressed and confers tumorgenicity *(16)*. Since tumor tissue shows cytokinin and auxin independence for growth in culture in contrast to normal plant callus tissue, production of the two phytohormones by host tissue may be regulated by genetic determinants on the T-DNA *(12,14–16)*.

T-DNA functions as a vehicle for delivery of virulence factors from pathogen to host and may also find wide application as a gene delivery system in plant genetic engineering *(17)*.

FACTORS ATTENUATING VIRULENCE AND LOSS OF SECONDARY METABOLISM

Loss of pIAA1 by curing attenuates virulence in mutants of this bacterium and virulence is restored when mutants are transformed with pIAA1. Virulence also is attenuated by insertion of IS::PS1, a 1.3 kb IS-element in the monooxygenase gene *(18)*. A number of copies of the insertion element occur on plasmids and the chromosome of all strains of *P. savastanoi* examined so far *(18)*. The insertion has polar effects since the hydrolase gene is also not expressed in these mutants.

The insertion in the monooxygenase gene effectively switches off flow of tryptophan into production of indoleacetic acid ; it is noteworthy that insertion mutants are commonly obtained by selection for resistance to α-methyltryptophan, a false feedback inhibitor of anthranilate synthetase *(7)*. Mutations by insertion elements are responsible for altering virulence and pathogenicity in other plant pathogenic bacteria. Indeed such an event had been demonstrated earlier by Garfinkel and Nester *(11)* who found that Tn5 insertions in Ti plasmid of *A. tumefaciens* created avirulent mutants as well as those which cause formation of morphologically distinct tumors on host plants. Insertional inactivation may be responsible for the spontaneous loss of virulence in *P. syringae* pv. *solanacearum* due to altered colony type. Although in the latter case the basis for loss of virulence is not understood, it may be due to an alteration in the bacterium's extracellular polysaccharide perhaps responsible for recognition between host and pathogen. Insertional mutations may represent a mechanism responsible for loss of virulence in plant pathogens in general. Insertion sequences which cause mutations were first found in antibiotic resistance and their prevalence represents a general phenomenon for altering gene expression in both eukaryotes and prokaryotes (19).

Although it is possible reversion to virulence can occur in insertion mutants by excision of the IS element, such excisions are not always precise and can leave behind deletions (19). Such a phenomenon could have given rise to the Iaa⁻ avirulent mutant of *P. savastanoi* which has a deletion in the monooxygenase gene (L. Comai, unpublished observations). Attempts to demonstrate spontaneous reversions to virulence in several avirulent Iaa⁻ insertion mutants of *P. savastanoi* have not been successful (L. Comai, M.L. Smidt, and T. Kosuge, unpublished observations).

ECOLOGY

In oleander strains of *P. savastanoi* the Iaa genes, *iaaM* and *iaaH*, are borne on a plasmid, *pIAA* (9). Thus certain analogies can be drawn between pIAA and R-plasmids. Since the monooxygenase effectively oxidizes and therefore detoxifies 5-methyltryptophan which otherwise would be inhibitory to the bacterium, pIAA functions both as an R-plasmid and a virulence plasmid.

It is significant that genes associated with extending metabolic capabilities of bacteria are commonly plasmid borne. For example, genes confering antibiotic resistance, hydrocarbon degradation, and pesticide detoxification are commonly plasmid borne (20). These metabolic activities can be classified as secondary metabolism and can extend the ability of the concerned organism to occupy ecologic niches unavailable to other organisms not possessing similar capabilities. Although production of IAA appears to offer no selective advantage to *P. savastanoi* under greenhouse conditions, it could under field conditions in California and in the Mediterranean area where both oleander and olive plants are grown. These plants are adapted to surviving in hot dry summers characteristic of their habitats and pathogens of these plants should possess similar capabilities. Perhaps galls or knots formed in the early spring provide protection for the bacterium during adverse summer months.

RELATIONSHIP OF PRIMARY METABOLISM TO EXPRESSION
OF VIRULENCE

Since production of indoleacetic acid depends upon the availability of the primary metabolite, tryptophan, mechanisms controlling production and utilization of the amino acid could in turn affect expression of virulence. The priority for typtophan utilization is protein synthesis and secondarily IAA production ; hence there are metabolic controls which assure these requirements are met. Tryptophan synthesis also is regulated in a fashion designed to make an organism efficient in its utilization of nutrients and energy (21).

Mutations to relaxed control of tryptophan synthesis will result in increased IAA production and increased virulence (3,7). However, relaxed tryptophan synthesis reduces efficiency of carbon, nitrogen and energy utilization in primary metabolism and reduces fitness of the pathogen to survive or compete under nutrient stress. Hence there will be a trade off between increased virulence due to increased IAA production and decreased fitness due to less efficient tryptophan metabolism. The same concept can be applied to other pathogens in which secondary metabolites serve as virulence determinants. Indeed it is common that highly virulent strains of a pathogen are less fit then their less virulent counterparts. Altered metabolic regulation therefore relates virulence with fitness of a pathogen.

THE ACTION OF SECONDARY METABOLITES AS TOXINS

There are a number of other secondary metabolites that can function as virulence determinants. These include toxins, both host-selective and non-selective. Among the most interesting ones of this group are toxins which inhibit enzymes concerned with primary metabolism. Phaseolotoxin produced by *P. syringae* pv. *phaseolicola* inhibits ornithine transcarbamylase of the host plant (22). Tabtoxin produced by *P. syringae tabaci* is particularly interesting because it must undergo host modification to yield the active toxin, tabtoxinine β-lactam, an inhibitor of glutamine synthetase (22,23). The inhibitory activity of the 2 toxins on plant systems is sufficient to disrupt plant metabolism and cause disease. For lack of conclusive evidence it is not certain if the antimetabolite activity alone of these secondary metabolites is sufficient to explain their toxic effects on plants. In both cases the toxins are not the sole virulence factor produced by the pathogens. Mutants incapable of toxin production nevertheless still cause disease symptoms, albeit at a reduced level.

THE INTERACTION OF PATHOGEN AND PLANT SECONDARY METABOLISM

While virulence factors of a pathogen disrupt metabolism and development of the plant, the plant itself may produce secondary compounds which inhibit the invading organism. Such plant metabolites, known as phytoalexins, include isoflavonoids, terpenes, and acetylenic compounds (1,24). The role of these compounds in disease resistance awaits conclusive evidence from experiments with molecular biology approaches. Nonetheless one of the most interesting aspects of the phytoalexin response is that contact with a microorganism or components of the microorganism switches on secondary metabolism for production of phytoalexins. This implies there is some form of "recognition" that occurs between the pathogen and the plant (25). This interesting phenomenon is being actively pursued at the molecular level as a plant response to stress (26,27), since it may reveal an important mechanism in plant disease resistance (4).

CONCLUDING REMARKS

The area of secondary metabolism in host-pathogen interactions offers much potential for exciting findings in the future. The new approaches in plant and microbial molecular biology offers opportunities for revealing mechanisms and concepts on expression of virulence ; it will be possible to generate new information leading to the understanding of how a microorganism functions as a pathogen.

The role of secondary metabolism in disease resistance in plants likewise will be understood and mechanisms triggering responses of plants to pathogen invasion will be revealed. At last, products of disease resistance genes may be identified and the corresponding fundamental basis for disease resistance may be delineated.

Perhaps the most useful findings that will emerge from these discussions is that we can now equate expression of virulence to control of metabolism ; we can identify products of genes for virulence, and determine how they act as virulence factors. If we understand the pathway of synthesis of the virulence factor, we can identify key gene products that regulate production of the virulence factor and so control expression of virulence. We can understand how virulence can be lost without affecting viability of the organism in culture.

REFERENCES

1. Bell, E.A. and Charlwood, B.V. : 1980. Secondary Plant Products, Springer-Verlag, Berlin, Heidelberg, New York.
2. Comai, L. and Kosuge, T. : 1982. *J. Bacteriol.* **149**, 40–46.
3. Kosuge, T. and Comai, L. : 1982. *In* Plant Infection (Asada, Y. *et al.*, eds.), pp.175–186, Japan Scientific Societies Press, Tokyo.
4. Bell, A.A. : 1981. *Annu. Rev. Plant Physiol.* **32**, 21–81.
5. Chatterjee, A.K. and Starr, M.P. : 1977. *J. Bacteriol.* **132**, 862–869.
6. Strobel, G.A. : 1982. *Annu. Rev. Biochem.* **51**, 309–333.
7. Smidt, M.L. and Kosuge, T. : 1978. *Physiol. Plant Pathol.* **13**, 203–214.
8. Kosuge, T., Heskett, M.G. and Wilson, E.E. : 1966. *J. Biol. Chem.* **241**, 3738–3744.
9. Comai, L. and Kosuge, T. : 1980. *J. Bacteriol.* **143**, 950–957.
10. Comai, L., Surico, G. and Kosuge, T. : 1982. *J. Gen. Microbiol.* **128**, 2157–2163.
11. Garfinkel, D.J. and Nester, E.W. : 1980. *J. Bacteriol.* **144**, 732–743.
12. Garfinkel, D.J., Simpson, R.B., Ream, L.W., White, F.F., Gordon, M.P. and Nester, E.W. : 1981. *Cell* **27**, 143–153.
13. Amasino, R.M. and Miller, C.O. : 1982. *Plant Physiol.* **69**, 389–392.
14. Leemans, J., Deblaere, R., Willmitzer, L., DeGreve, H., Hernalsteens, J.P., Van Montagu, M. and Schell, J. : 1982. *EMBO J.* **1**, 147–152.
15. Ooms, G., Hooykas, P.J.J., Moolenaar, G. and Schilperoort, R.A. : 1981. *Gene* **14**, 33–50.
16. Nester, E.W. and Kosuge, T. : 1981. *Annu. Rev. Microbiol.* **35**, 531–565.
17. DePicker, A. : 1983. *In* Genetic Engineering of Plant (Kosuge, T., Meredith, C.P. and Hollaender, A., eds.), Plenum Press, Inc., in press.
18. Comai, L. and Kosuge, T. : 1982. submitted.
19. Starlinger, P. : 1980. *Plasmid* **3**, 241–259.
20. Chakrabarty, A.M. : 1980. *In* Plasmids and Transposons – Environmental Affects and Maintenance Mechanisms (Stuttard, C. and Rozee, K.R., eds.), pp. 21–30, Academic Press, New York.
21. Crawford, I. and Stauffer, G.V. : 1980. *Annu. Rev. Biochem.* **49**, 163–195.
22. Mitchell, R.W. : 1981. *In* Toxins in Plant Disease, pp. 259–293, Academic Press, New York.
23. Durbin, R.D. : 1981. *In* Toxins in Plant Disease, Academic Press, New York.
24. Conn, E.E. : 1981. *In* The Biochemistry of Plants – Secondary Plant Products, Academic Press, New York.

25. Keen, N.T. : 1982. *In* Advances in Plant Pathology, Vol. I (Ingram, D.S. and Williams, P.H., eds.), pp.35–82, Academic Press, New York.
26. Hahlbrock, K., Lamb, C.J., Purwin, C., Ebel, J., Fautz, E. and Schafter, E. : 1981. *Plant Physiol.* 67, 768–773.
27. Hahlbrock, K. and Grisebach, H. : 1979. *Annu. Rev. Plant Physiol.* 30, 105–136.

Approaches for the Detection of the Products of Genes for Resistance and Avirulence in the Wheat and Stem Rust System

W.K.Kim, R.Rohringer, N.K.Howes and J.Chong

INTRODUCTION

There is considerable information on the inheritance of resistance of wheat to stem rust and the genetics of pathogenicity. Single genes for resistance have been identified and transferred into susceptible wheat varieties by backcrossing or by chromosome substitution and backcrossing. These so-called single-gene lines are widely used as differential hosts for studies of virulence in the rust population and in studies on the biochemical basis of resistance.

Most studies of the biochemistry of host/parasite interactions have been limited to comparing various biochemical changes between resistant reacting and susceptible reacting single-gene lines after infection (1–4). Such biochemical studies undoubtedly contributed to a knowledge of biochemical symptomatology, yet in most cases the observed changes probably were the result of the expression of the resistant reaction, and no chemical changes have been correlated with the action of specific genes for resistance.

Host/rust systems involve gene-for-gene interactions, in which a specific gene for resistance in the host is matched by a corresponding gene for avirulence in the parasite (5). Therefore, any proposed biochemical mechanism of resistance in the wheat/stem rust system must be consistent with this gene-for-gene complementarity. It is hypothesized that the informational molecules specifying this recognition between host and parasite must be present at the interaction site to establish such host/parasite specificity. Identification of such molecules is essential for the elucidation of the biochemical basis of resistance of wheat to stem rust.

There are three critical questions which have not been resolved until the present time : (1) At what stage of fungal development does the gene-specific interaction occur ; (2) What physical structures of the host and parasite are involved in the interaction ; (3) What are the chemical properties of the gene products.

In this chapter, our earlier studies are reviewed and new approaches are proposed for the detection of specific gene products.

BACKGROUND FOR NEW APPROACHES

In histological and ultrastructural studies of resistance of wheat to stem rust we concentrated our effort on the *Sr6* gene for resistance. This gene is temperature sensitive, being expressed in plants grown at temperature below 20°C but inactive at temperature above 25°C. Furthermore, the gene is only expressed in the presence of light and in cells containing chloroplasts. In studies aimed at identifying possible products of a gene for resistance, the *Sr6* gene has a number of advantages; near-isogenic lines differing in resistance by the *Sr6* gene of the cultivars Chinese Spring and Prelude are available, the location of the *Sr6* gene on the 2D^S chromosome arm is known, several flanking markers closely linked to the *Sr6* have been discovered, and ditelosomic stocks of Chinese Spring lacking the 2D^S chromosome arm are available. We are exploiting these advantages in the present studies.

Histological Study (Stage of Fungal Development Where the Gene-Specific Interaction Occurs) (6)

Seedlings of resistant *(Sr6)* and susceptible *(sr6)* near-isogenic lines of wheat were inoculated with a race of stem rust *(Puccinia graminis* f. sp. *tritici)* that was avirulent on the line containing the *Sr6* gene, and were kept at 19,25,26 and 27°C. In leaves containing the *Sr6* gene a smaller percentage of rust colonies at 25°C had necrotic cells associated with them than those were grown at 19°C. When infected plants containing the *Sr6* gene were kept for varying times at 25°C and then were transferred to 19°C there was significantly less fungal growth and more necrosis than in plants kept continuously at 25°C. This necrosis occurred largely in those cells that were invaded after the transfer to 19°C where the *Sr6* gene was activated. Conclusion – the gene-specific interaction occurred at the interface between host cells and the newly formed haustorium.

Ultrastructural Study (Physical Structures Involved) (7)

The interaction between avirulent wheat stem rust and wheat mesophyll cells containing the temperature-sensitive *Sr6* gene for stem rust resistance was studied by electron microscopy, using the plant material described above. Mesophyll cells that were invaded at 26°C did not develop any sign of incompatibility after they were transferred to 19°C at which temperature incompatibility is normally expressed. In host tissue that appeared to be invaded after the change from 26 to 19°C, the earliest ultrastructural symptoms of incompatibility were ; a more electron dense and often perforated invaginated host plasmalemma, disruptions of the non-invaginated host plasmalemma ; and at a later stage, disorganization of the haustorial cytoplasm and extrahaustorial sheath, and eventually both host cell and haustorium die (8). Conclusion − one of the earliest symptoms of incompatibility is a lesion in the host plasmalemma after having made contact with the haustorial wall.

Biochemical Study (Chemical Properties of Gene Products) (9)
Seedlings of resistant wheat leaves were inoculated with an avirulent race of stem rust and kept for 2 days at 26°C, treated with blasticidin S (an inhibitor of plant protein synthesis, but at a concentration which had no effect on fungal growth), polyoxin D (at concentration which inhibited rust development, but had no phytotoxicity on the host) or buffer (15 mM NaCl, 1.5 mM sodium citrate, pH 7.0), and transferred to 19°C. In leaves treated with either inhibitor there were fewer necrotic cells associated with rust colonies than in leaves treated with buffer. Conclusion − in order to have host necrosis, both fungal growth and host protein synthesis are required.

NEW APPROACHES

Three areas of research are being pursued for the isolation and the identification of the gene products : (A) Studies directed towards isolation of products of genes for avirulence, (B) Studies directed towards isolation of products of genes for resistance and (C) Development of bioassays for detection of gene products.

Studies Directed towards Isolation of Products of Genes for Avirulence
1) Polypeptides of Genetically Different Races of Stem Rust
Dormant uredospores of five physiologic races of wheat stem rust, which can be distinguished by the differential expression of resistance of different host cultivars, were extracted with a buffer containing the detergent Nonidet P 40. Polypeptides were separated and analyzed by two-dimensional isoelectric focusing-polyacrylamide gel electrophoresis (IEF-PAGE) (10). Comparing the polypeptide patterns of all 5 races, more than 270 polypeptides in each extract consistently had similar relative positions and intensities. An additional 18 polypeptides were detected which were not common to all 5 races. Figure 1 is a composite diagram based on the analysis of extracts from 5 races, showing the position of both common polypeptides and of polypeptides not common between races. Races C17(56) and C45(56A) had only one polypeptide "spot" difference (protein g). The other four race combinations differed by 6 to 14 polypeptides. No two races had identical protein patterns (Table I). Some of these differences correlated with differences in virulence of the races (11). More races are under investigation to determine if this correlation holds.
2) Macromolecular Wall Components of Uredosporelings
A considerable amount of information is available on the nature of polysaccharides and glycoproteins from cell walls of several fungi in relation to the structure

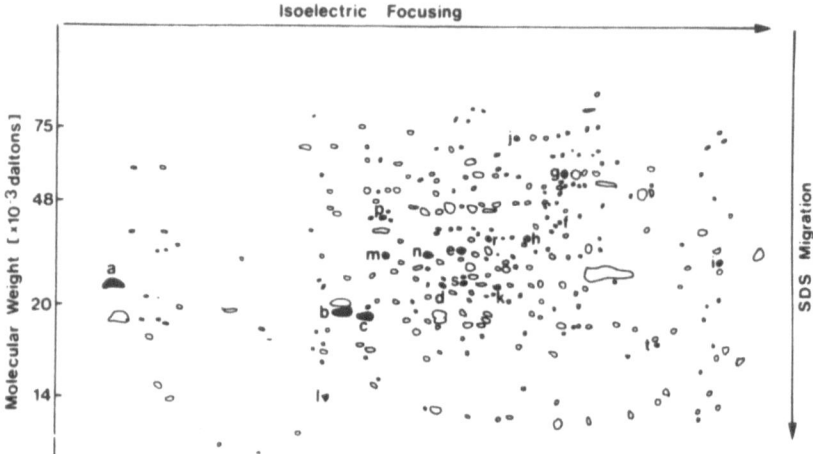

Fig. 1. Composite diagram of two-dimentional IEF-SDS PAGE of polypeptides of uredospores to show variable polypeptides (dark spots) between races of wheat stem rust.

TABLE I
Avirulence Genes and Uredospores Polypeptides that Differ between Races

Races	Avirulent to host Sr single gene lines														
	Sr5	6	7a	8	9a	9b	9d	9e	10	11	13	17	Tt1	Tt2	GB
C1 (17)	+	+	+	−	+	+	−	+	+	+	+	+	−	+	−
C10 (15B-1)	−	+	+	+	−	−	−	−	−	−	−	−	−	−	+
C17 (56)	−	+	−	+	+	+	+	+	−	+	+	−	+	+	−
C45 (56A)	−	−	−	+	+	+	+	+	−	+	+	−	+	+	−
C57 (32)	−	−	−	−	+	−	−	+	−	−	−	−	+	+	−

Races	Uredospore polypeptides										
Polypeptide	a	c	d	(e,h)	f	g	(b,i,j,k,l,t)	m	n	p	(r,s)
C1 (17)	−	+	+	+	+	−	+	−	+	+	+
C10 (15B-1)	+	+	−	+	−	−	+	+	*	−	+
C17 (56)	+	+	+	−	−	+	−	+	*	−	+
C45 (56A)	+	+	+	−	−	−	−	+	*	−	+
C57 (32)	−	−	+	+	+	−	−	+	+	+	−

*Faint spot.

and the function of the cell surface, and in particular to differentiation and morphogenesis (12).

Polysaccharides or glycoproteins from fungal cell walls have been implicated as possible determinants of specificity in the gene-for-gene recognition by resistant hosts of avirulent races of pathogens (13), or as elicitors of phytoalexin accumulation

TABLE II
Bound Sugars in Each Extract and Concanavalin A-Sepharose 4B Column Eluates of Extracts
from Walls of Uredosporelings of Wheat Stem Rust (% of total extracted)

	Walls of sporelings extracted with									
Bound sugars	Imidazole/ EDTA		NH$_4$OH		NaOH		Tris- trichloro- acetate		Deoxycholate	
	A	B	A	B	A	B	A	A	A	B
Mannose	74.2	66.7	40.0	76.2	39.0	70.2	29.2	56.0	21.5	68.1
Glucose	12.9	33.3	9.8	23.8	13.7	29.8	15.8	44.0	4.0	26.6
Galactose	7.8	0	17.4	0	17.8	0	13.0	0	57.8	5.3
Fucose	0	0	12.3	0	9.6	0	32.4	0.	5.8	0
Ribose	5.1	0	11.5	0	12.7	0	9.5	0	6.3	0
Rhamnose	0	0	0	0	0	0	0	0	4.3	0
N-Acetyl- glucosamine	0	0	8.9	0	7.2	0	0	0	0	0

A, extractives ; B, eluates from affinity columns.
Imidazole/EDTA, 50mM imidazole/HCl/10mM EDTA, pH 7.2 (2 hr, 100°C); NH$_4$OH, 25%
NH$_4$OH (3 hr, 37°C); NaOH, 0.2 M NaOH (24 hr, 0°C); Tris/TCA, 50mM Tris/HCl/500 mM
Na-trichloroacetate, pH 7.4 (1.5 hr, 4°C); Deoxycholate, 1% Na-deoxycholate, pH 7.4 (1.5 hr,
20°C).

in a number of other host/parasite interactions (*13*). Although no such relationship
has yet been established for rust/host interactions, walls of germinated uredosporelings were purified and extracted by several relatively mild methods (see Table II) to
release macromolecular wall components (*14*). These procedures released between
0.5 and 2.5% of the wall material. Bound sugars and bound amino acids of the extractives are shown in Tables II and III, respectively.

The five most abundant bound sugars in each of the 5 extracts were mannose,
glucose, galactose, fucose and ribose. Approximately 1/3 of the protein and bound
carbohydrate in each extract was retained by, and eluted from, columns of Concanavalin A-Sepharose 4B. Bound mannose predominated in all column eluates, constituting between 56 and 76% of total carbohydrate in the eluates. The remainder con-
sisted of glucose, except in the column eluate of the deoxycholate extract where a
small amount of galactose was also detected. The bound amino acid composition was
similar in each extract, with aspartic acid, serine, glycine, and glutamic acid predomi-
nating. The Concanavalin A-Sepharose 4B column eluates contained less bound
glutamic acid, glycine, and alanine, but more aspartic acid, threonine, and serine
than the corresponding extracts before column fractionation.

TABLE III

Amino Acid Compositions in Each Extractive and Eluates from Concanavalin A-Sepharose 4B Affinity Column (Values expressed as a percentage of total amino acid residues (nmoles) per g spore equiv. present in each sample)

Amino acids	Imidazole/ EDTA		NH$_4$OH		NaOH		Tris/TCA		Deoxycholate	
	A	B	A	B	A	B	A	B	A	B
Asp + Asn	9.3	10.5	8.4	9.3	9.5	10.8	11.4	9.4	11.4	12.2
Thr	4.7	5.9	4.3	5.2	5.0	6.4	5.1	6.1	5.9	7.1
Ser	7.5	10.5	6.9	7.7	6.7	8.6	7.3	10.1	7.7	10.6
Glu + Gln	10.0	5.6	8.4	11.6	8.5	7.9	9.1	6.6	9.2	6.8
Pro	4.9	3.8	5.7	8.0	5.9	5.8	5.3	4.4	5.8	5.5
Gly	9.8	8.2	10.6	7.5	9.6	7.7	7.4	7.3	7.8	6.9
Ala	9.3	7.9	8.6	7.1	8.5	8.1	7.9	7.9	8.3	8.1
Cys	–	0.6	–	–	–	–	–	–	–	–
Val	4.2	5.1	4.5	4.8	5.0	5.4	4.7	5.4	4.9	5.5
Met	1.8	2.0	1.4	0.7	1.3	0.3	0.7	1.0	0.8	0.3
Ile	2.7	4.3	2.9	3.4	3.5	4.0	3.2	4.2	3.7	4.1
Leu	4.9	6.6	5.9	6.2	6.7	7.4	6.8	7.2	7.8	7.3
Tyr	1.6	1.5	3.4	2.6	3.3	2.4	2.0	1.2	2.5	1.9
Phe	2.0	3.3	2.7	3.3	2.9	4.1	2.7	3.5	3.2	4.1
His	2.0	2.0	1.9	1.9	1.6	2.1	1.7	2.1	1.8	2.1
Lys	5.5	4.8	4.8	3.9	5.5	5.4	5.4	5.8	5.3	4.9
NH$_3$	17.3	15.3	15.6	14.3	13.4	10.3	10.5	15.9	10.2	10.1
Arg	2.7	2.0	3.2	2.4	3.2	3.2	3.4	1.9	3.7	2.7
Lipophilic	13.8	19.3	16.1	17.7	18.1	20.9	17.4	20.3	19.6	21.0

A, extractives; B, eluates from affinity columns.
Lipophilic amino acids : valine, leucine, isoleucine, and phenylalanine

In each extract, the relative abundance of bound lipophilic amino acids increased after binding to, and elution from, Concanavalin A-Sepharose 4B, indicating that proteins relatively rich in lipophilic amino acids also glycosylated with mannose and/or glucose. These properties are characteristic of membrane-bound proteins.

3) Changes in Polypeptide Patterns during Germination of Uredospores

Polypeptides of dormant uredospores and of non-differentiated and differentiated uredosporelings (15) were separated and analyzed by two-dimensional IEF-PAGE. During germination (germ tube development-non-differentiating) and differentiation (development of infection structures), changes in polypeptide occurred (16), however, these were not the same polypeptides which differed between different physiologic races of stem rust. This suggests that non-differentiated uredosporelings could be used in a study to determine in which cellular compartment (cytoplasmic or membrane-bound) the polypeptides of interest are located.

Studies Directed towards Isolation of Products of Genes for Resistance

Evidence was presented that a structural component of the host plasmalemma might represent the product of the gene for resistance. Therefore, considerable emphasis was placed on attempts to isolate constituents of the host plasmalemma without destroying their biological activity. Since no adequate methods of obtaining plasmalemma-enriched fractions from wheat leaves are available, several methods of protein extraction were compared and the proteins extracted were separated by two-dimensional IEF-PAGE.

1) Isolation of Plasma Membrane

A plasmalemma-enriched fraction was prepared from the following sources — mesophyll protoplasts, mesophyll cells, green and etiolated leaves.

a) Isolation of mesophyll protoplasts : Protoplasts were isolated from primary leaves of 7-day old seedlings of wheat by the method of Earle *et al.* (*17*) except for the following modifications ; the concentration of Cellulysine (purified) (*18*) and Driselase (purified) (*18*) in osmoticum, OS-I (*19*) were 0.5% and 0.25%, respectively. Viability of isolated protoplasts was estimated by counting them after staining with fluorescein diacetate (*20*).

b) Isolation of mesophyll cells : The isolation method was essentially the same as that for protoplasts, except that pectolyase Y-23 concentration was 0.5%. Viability was checked with the fluorescence microscope using the autofluorescence of chloroplasts as the criterion.

c) Isolation of plasma membrane fraction : Two grams of primary leaves of 7-day old wheat seedlings (green or etiolated leaves) were ground in liquid nitrogen. They were incubated in a mixture of cell wall degrading enzymes (0.5% Cellulysine, 0.25% Driselase in OS-I osmoticum) for 2 hr at 21°C. The suspension was further homogenized in OS-I osmoticum containing 0.1g PVP 40 per 2 g fresh meterial at 4°C. The homogenates were centrifuged at 1,000×*g* for 15 min to precipitate cell organells. The second supernatant fraction was centrifuged at 40,000×*g* for 30 min to yield the crude membrane fraction, which was further washed 3 times with fresh osmoticum (40,000×*g*).

A plasma membrane fraction was also isolated from homogenates of isolated mesophyll cells or from isolated protoplasts as described above.

2) Extraction of Membrane Proteins

Method-1. Primary leaves of 7-day old seedlings of wheat were ground in liquid nitrogen, and total soluble proteins were extracted with a buffer containing SDS and Nonidet P 40 (50 mM Tris/HCl, pH 6.8, 0.04% SDS, 1 mM DTT, 0.05 mM MgCl$_2$, 1 mM EDTA, 2% Nonidet P 40) (*21*). The homogenate was centrifuged at 100,000×*g* for 90 min at 4°C. The supernatant was precipitated with 9 vol of cold acetone.

Method-2. Membrane proteins were extracted from green or from etiolated leaves, which had been ground in liquid nitrogen, with a buffer containing sodium deoxycholate (50 mM Tris/HCl, pH 7.0 and 1% sodium deoxycholate). The homogenate was treated as in Method-1.

3) Polypeptide Patterns

a) Differences between extraction methods : There were approximately 200 polypeptides in the extract of green leaves prepared by Method-1 whereas approximately 60 polypeptides were identified in extracts of green leaves prepared by Method-2. Membrane proteins prepared from green leaves and a plasma membrane-enriched fraction prepared by using Method-2 yielded almost the identical polypeptide patterns, suggesting that green leaves extracted by Method-1 could be a good source of membrane proteins. This is more convenient than having to prepare a plasma membrane fraction which normally results in poor recovery. It also minimized the possibility of protein degradation that may occur during preparation of the plasma membrane fraction.

b) Genetically different plants (isogenic and ditelosomic lines) : Using Method-1, total soluble polypeptides were extracted from isogenic lines of Chinese Spring (*Sr6* and *sr6*) and from the $2D^L$ ditelosomic line of Chinese Spring. The polypeptides prepared were then separated by two-dimensional IEF-PAGE. One polypeptide (polypeptide α) was absent in the extract prepared from the $2D^L$ ditelosomic line, but present in extracts prepared from the other lines. Two polypeptides (β and ε) were present in Chinese Spring *Sr6* but absent in other lines. This indicates that the gene controlling the synthesis of these polypeptides are located on the $2D^S$ chromosome arm. None of these polypeptides were detected in extracts of the membrane-enriched fraction prepared by Method-2, indicating that they were not membrane proteins or were readily solubilized during the preparation of the membrane fraction.

Further investigations are required to determine the location of these polypeptides in the cell and to determine if any genes controlling their synthesis are located at the same locus as the *Sr6* gene for resistance.

c) Development of bioassays for detection of gene products : It is hypothesized that the incompatible interaction between wheat and stem rust might result from the interaction of a structural component of the invaginated host plasmalemma with a structural component of the haustorial body wall. If the gene-for-gene interaction is analogous to a surface to surface interaction in other biological systems, there should be receptor molecules in the host plasmalemma which bind to components of the parasite wall, perhaps analogous to an affinity binding system.

Two types of affinity binding assays were experimentally tested ; (1) *an assay using the electron microscope* and (2) *an assay using protoplasts*.

An assay using the electron microscope : The method employing gold markers for visualizing lectin binding sites (gold granules with an average diameter of 5 nm coated with lectins) (22,23) may be useful to detect the binding of fungal components containing the product of the gene for avirulence to the host plasmalemma if the former can be conjugated with gold particles. Between 80 to 90% of our fungal extractives (14) can be conjugated with gold particles (22,24). In a possible binding assay, the gold conjugates can be brought in contact with ultra-thin sections of normally incompatible wheat leaves to test for the presence of fungal wall specific receptors in the host plasmalemma. Fungal wall components also can be conjugated with fluorescein isothiocyanate (FITC), and injected into rust-infected wheat leaves trace the binding sites in the host plasmalemma.

Alternatively, the host membrane components which may contain the product of the gene for resistance, conjugated with gold particles, or with FITC, could be used to test for binding to the newly formed haustorial wall.

Applied to the *Sr6/P6* system, this would have 2 advantages : previous studies have shown that the gene for resistance is already expressed in non-infected leaves, and the assay system contains all structures of the fungus, including the haustorial neck wall that is most likely to be involved in the gene-specific interaction. It is not yet known whether the product of the gene for resistance can, in fact, be extracted and conjugated to gold in a biologically active form and whether possible "receptors" in the fungus survive the histochemical processing prior to the assay.

An assay using protoplasts : Radioactively labeled fungal cell wall components isolated from the avirulent race can be incubated with mesophyll protoplasts isolated from susceptible and resistant lines of wheat to determine if fungal wall components bind to their surface in a gene-specific manner. If our working hypothesis is correct, we expect more radioactivity to be retained on the surface of protoplasts isolated from the resistant line of wheat than on that of susceptible protoplasts. We found that 25 to 30% of the activity bound to protoplasts of both lines of wheat (*24*), indicating a considerable amount of non-specific binding. Evidently, the extracts require extensive purification before such an assay can become useful. Also, there is a great deal of uncertainty as to whether possible gene-specific "receptors" survived the method of protoplast preparation involving macerating enzymes.

CONCLUDING REMARKS

Our approach has been to develop methods for fractionating wheat leaf and stem rust macromolecules in order to detect the compounds involved in the initial expression of resistance. Using these methods we have shown that some rust polypeptides differ between physiologic races and are possible candidates for specifying avirulence in this pathogen. The cell location and nature of these polypeptides has not been determined. Genetic studies will be carried out on stem rust to determine if the genes specifying these polypeptides are located in the same position on the chromosome as the genes for avirulence.

We have tentatively identified polypeptides present in resistant wheat leaves where genes controlling their synthesis are located on the same chromosome arm as the *Sr6* gene for resistance. There is no evidence to suggest that these polypeptides are present in the host plasmalemma.

REFERENCES

1. Rohringer, R. and Samborski, D. J. : 1967. *Annu. Rev. Phytopathol.* 5, 77–86.
2. Kim, W. K. and Rohringer, R. : 1969. *Can. J. Bot.* 47, 1425–1433.
3. Samborski, D. J. and Rohringer, R. : 1970. *Phytochemistry* 9, 1939–1945.

4. Samborski, D. J., Rohringer, R. and Kim, W. K. : 1978. *In* Plant Disease, Vol. 3, pp.375–390, Academic Press, New York.
5. Ellingboe, A. H. : 1976. *In* Physiological Plant Pathology, Encyclopedia of Plant Physiology, Vol. 4 (Heitefuss, R. and Williams, P. H., eds.), pp.760–778, Springer-Verlag, New York.
6. Samborski, D. J., Kim, W. K., Rohringer, R., Howes, N. K. and Baker, R. J. : 1977. *Can. J. Bot.* 55, 1445–1452
7. Harder, D. E., Samborski, D. J., Rohringer, R., Rimmer, S. R., Kim, W. K. and Chong, J. : 1979. *Can. J. Bot.* 57, 2626–2634.
8. Skipp, R. A., Harder, D. E. and Samborski, D. J. : 1974. *Can. J. Bot.* 52, 2615–2620.
9. Kim, W. K., Rohringer, R., Samborski, D. J. and Howes, N. K. : 1977. *Can. J. Bot.* 55, 568–573.
10. Howes, N. K., Kim, W. K. and Rohringer, R. : 1982. *Physiol. Plant Pathol.* 21, in press.
11. Green, G. J. : 1978. *Can. Plant Dis. Survey* 58, 44–48.
12. Bartnicki-Garcia, S. : 1968. *Annu. Rev. Microbiol.* 22, 87–108.
13. Callow, J. A. : 1977. *In* Advances in Botanical Research, Vol. 4 (Preston, R.D. and Woolhouse, H. W., eds.), pp.1–49, Academic Press, New York.
14. Kim, W. K. and Rohringer, R. : 1981. Proc. 13th Int'nl. Bot. Congress, Sydney, Australia. pp.157.
15. Kim, W. K. : 1971. *Can. J. Bot.* 49, 1119–1122.
16. Kim, W. K., Howes, N. K. and Rohringer, R. : 1983. *Can. J. Plant Pathol.* 5, in press.
17. Earle, E. D., Gracen, V. E., Yoder, O. C. and Gemmill, K. P. : 1978. *Plant Physiol.* 61, 420–424.
18. Constabel, F. : 1975. *In* Plant Tissue Culture Methods (Gamborg, O. L. and Wetter, L. R., eds.), National Research Council (14383), pp.11–21.
19. Larkin, P. J. : 1977. *J. Cell Sci.* 26, 31–46.
20. Larkin, P. J. : 1976. *Plants* 128, 213–216.
21. Gabriel, D. W. and Ellingboe, A. H. : 1982. *Phytopathology* 72, 1496–1499.
22. Rohringer, R., Chong, J., Kim, W. K. and Harder, D. E. : 1981. *In* Active Defense Mechanisms in Plants, (Wood, R. K. S., eds.), NATO Advanced Inst., pp.349–350.
23. Horisberger, M. and Rosset, J. : 1977. *J. Histochem. Cytochem.* 25, 295–305.
24. Kim, W. K. and Rohringer, R. : 1982. unpublished.

Biochemistry of Oxidase Enzymes as Related to the Etiology of and Resistance to Plant Diseases, the Hemorrhagic Sweet Clover Disease, Farmer's Lung Disease, and Feed or Food Production

Mark A. Stahmann

INTRODUCTION

It is a special pleasure for me to participate in this symposium honoring my good friend Professor I. Uritani. I consider it an extraordinary privilege to have had the opportunity to work with Uritani both in Madison and Nagoya. He is among the few really great plant biochemists I have known.

My interest in the biochemistry of plant disease began in the thirties when I was shown field plots for study of disease resistance. Susceptible plant varities were often all killed ; adjacent resistant varieties remained healthy. In some cases, this resistance was due to a single gene. It seemed to me that a better understanding of the biochemical role of resistant genes and how they block disease development could lead to better ways to control disease.

Since genes control protein synthesis, I undertook a study of protein changes in resistant and susceptible plants following infection. We tried in the fifties to use microelectrophoresis of aqueous plant extracts but the resolving power was not enough, so we shifted to gel electrophoresis. By combining the high resolving power of gel electrophoresis with the sharp specificity of antigen-antibody or enzyme-substrate interaction, we were able to show marked changes in specific proteins or enzymes following inoculation with various plant pathogens.

THE BLACK ROT OF SWEET POTATO

In the early sixties, I was attracted by techniques developed by Uritani in his studies of the black rot disease of sweet potato. Sweet potato root tissue was cut into pieces and inoculated on one side with spores from pathogenic or non-pathogenic isolates of *Ceratocystis fimbriata.* The pieces were incubated in a closed moist chamber. Both isolates grew on the surface ; however, only the pathogen could penetrate susceptible varieties, but not resistant varieties, and turn the piece coal black. Hence, the name, black rot. After a short incubation period, a cylinder was cut and sections taken at known distances below the infected surface. These sections were homogenized in cold buffers and the soluble proteins separated by gel electrophoreses. Protein bands in gels were made visible by staining or by reaction with antisera against the extracts, or with substrates that gave colored bands for specific enzymes.

Uritani applied starch gel immunoelectrophoresis to extracts of resistant and susceptible diseased and healthy sliced tissue (*1*). Proteins diffusing from the starch gel were reacted with antisera against either diseased or sliced extracts. In response to the infection, only the resistant sweet potato tissue produced two new bands. These were shown as arcs by the reaction of antibodies with antigens from tissues. The antigens which formed these two new bands were found only in extracts from resistant diseased tissues, but not in extracts of sliced resistant or in extracts of diseased or healthy susceptible tissue. These new antigens, which were seen only after inoculation in extracts from resistant tissue adjacent to the infected surface but not in susceptible tissue, were identified as a new peroxidase and a new polyphenol oxidase. They were found in tissue 2 to 3 mm below the infected surface layer containing the fungus and were formed in response to the fungal growth on the surface of only the resistant tissue.

A clue to the nature of the inducer or hormone that moved from the infected surface layer and induced the formation of new oxidases in the underlying cells came from an unexpected observation that uninoculated pieces of sweet potato tissue incubated in closed chambers with infected pieces showed increased peroxidase activity. This indicated a volatile substance was involved. I wondered if this volatile substance could be ethylene and could ethylene induce resistance to the spread of fungus and the synthesis of peroxidase and polyphenol oxidase in the underlying tissue.

Since we had no ethylene in the laboratory at that time, I asked D. Weber to incubate pieces of tissue for 2 days above healthy or diseased tissue and above a banana, orange, or apple before inoculating. The results clearly indicated that resistance or immunity was induced in susceptible sweet potato tissue when the tissue was incubated for 2 days above diseased sweet potato tissue or above an apple (*2*). I thought the apple would produce more ethylene than the orange or banana so we then ordered a tank of ethylene.

Resistance or immunity in susceptible sweet potato tissue was induced by inoculation with non-pathogenic isolates of *C. fimbriata* from pine or almond (*3*). This is somewhat similar to the immunity in animals that follows vaccination. When susceptible tissue was inoculated with the non-pathogenic isolate 2 days before challenge by inoculation with a pathogenic isolate, the infection of susceptible tissue was confined to the surface just as with resistant tissue. Only the piece of susceptible tissue not first inoculated with the non-pathogen developed typical black rot symptoms.

We concluded that like vaccination, the prior inoculation with the related non-pathogen had induced immunity or resistance to subsequent infection by the pathogen.

Gas analyses showed that ethylene was produced when sweet potato tissue was inoculated with the pathogenic isolate or with certain non-pathogenic isolates of *C. fimbriata* which induced this resistance. Exposure of susceptible sweet potato tissue to air containing 8 parts per million of ethylene induced a similar resistance to subsequent infection (*4*). Following exposure to ethylene the blackening and growth of the fungus on the susceptible tissue was confined to the surface ; a reaction like that seen in tissue from resistant varieties. Peroxidase and polyphenol oxidase activity increased in tissue 2 mm below the surface of susceptible sweet potato tissue that was exposed to 8 or 24 parts per million of ethylene. There was a large increase in oxidase activity, about 10-fold, and new isozymes of peroxidase and polyphenol oxidase were formed following exposure to 8 or 24 parts per million of ethylene. These data suggest that ethylene was the stimulus that diffused from the inoculated surface into adjoining tissues and induced metabolic changes including synthesis of oxidase enzymes that led to a local resistance or immunity that blocked spread of the disease.

PEROXIDASE EFFECTS ON SWEET POTATO ROOT TISSUE

In Uritani's laboratory we studied the effect of vacuum infiltration of solutions of commercial peroxidase into slices of sweet potato root tissue. Slices of root tissue (4×3 mm) were vacuum infiltered with 3.0 ml of an aqueous solution containing 1 mg/ml of commercial peroxidase. The slices were incubated for 24 hr under oxygen in closed 30 ml tubes containing 2.0 ml of water while supported on a thin glass rod so that only an edge of each slice touched the water. The amount of ethylene formed within 24 hr was estimated by gas chromatography and the amount of water taken up by the slices was measured by the increase in weight.

As shown in Table I, the slices of sweet potato or carrot tissue that were infiltered with the peroxidase formed much more ethylene than the controls infiltered with water. Water uptake as measured by the increase in weight was about twice that of the controls.

The increased water uptake by the peroxidase treated slices caused them to curl and split. This reminded us of the effect of auxin. Figure 1 shows turgid round sweet potato cells from the peroxidase treated tissue that were filled with many vacuoles. The tissue was so soft that such cells could be easily separated. This may be like the cell separation that occurs in abcission layer formation and was induced in sweet potato tissue by infiltration with commercial peroxidase which also increased ethylene formation.

We did not publish these results because we thought further study was needed, for the commercial peroxidase may have contained other enzymes like pectinase which could have softened the tissue. Inasmuch as ethylene induced the synthesis of

TABLE I
Ethylene Formation Following Treatment of Root Tissues with a Commercial Peroxidase

| Tissue | Incubation time (hr) | Ethylene formed after infiltration with | | Increase in ethylene (%) |
		Water (mml/g)	Peroxidase (mml/g)	
Sweet potato	7	0.1	0.3	300
Sweet potato	23	1.8	3.8	110
Carrot	5	0.8	1.7	143
Carrot	22	2.8	22.6	700

Seven slices (4.0×13 mm) were infiltered with 3.0 ml of water or a solution (1 mg/ml) of a commercial peroxidase preparation and incubated under oxygen in closed 30 ml tubes containing 2.0 ml of water. Ethylene concentration was measured by gas chromatography.

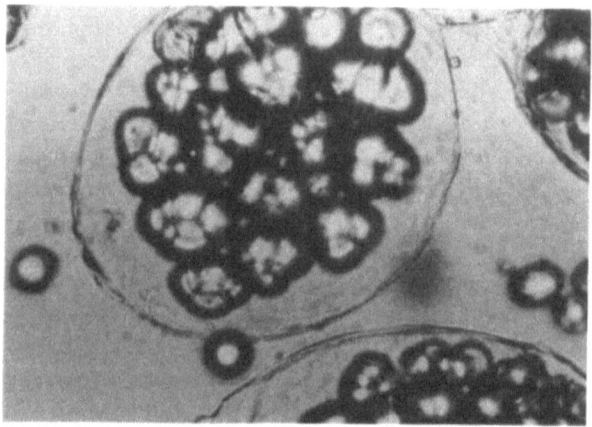

Fig. 1. Round turgid cells from infiltration of peroxidase into sweet potato root tissue.

peroxidase in sweet potato tissue and peroxidase stimulated ethylene formation, it suggests that this may be the basis for a biochemical amplification system that some plants may use to alter their metabolism in times of stress from disease or injury.

The products from the oxidation of indole-3-acetic acid by peroxidase, probably 3-methylene oxindole, bind to histones. We observed a 10-fold increase in the binding of radioactivity to calf thymus histone after oxidation by peroxidase of labeled indole-acetic acid (5). Such reactions may be involved in inducing the changes in enzyme synthesis or the auxin-like effects which have been observed in many disease plant tissues or on treating sweet potato tissue with ethylene or peroxidase. These observations suggest that peroxidase may be involved in activation of plant genomes during injury or disease and suggests a regulatory role which merits further study (6).

CROSSLINKING OF PROTEINS BY OXIDASE

In all our studies of enzyme changes in plants following inoculation with bacterial, fungal or viral pathogens, we observed that the activity of peroxidase and polyphenol oxidases would always increase and was greatest in resistant varieties. Crosslinks between polypeptide chains of elastin or fibrin are formed by oxidases. Free radicals from a lipid oxidizing system polymerized proteins. Because I thought that an oxidative crosslinking might also be involved in the formation of the infection barriers about infection sites in plants, we studied the action of peroxidase on various proteins.

The reaction of peroxidase and catechol with peptide bound lysine was studied in model *in vitro* systems using polylysine or proteins. When polylysine, casein, lactoglobin, or other proteins were reacted with peroxidase, catechol and hydrogen peroxide, more than half of the lysine was sometimes lost. Inclusion of benzene sulfinic acid, a quinone trapping agent, prevented this loss. Oxidation of the reaction mixtures with performic acid before acid hydrolysis produced a new ninhydrin positive peak which we identified by co-chromatography and mass spectroscopy as aminoadipic acid. Reduction before acid hydrolysis gave a new peak at the position reported for chloro-norleucine (7).

This oxidative deamination of lysyl residues may be involved in the crosslinking of proteins by peroxidase. Gel electrophoresis showed that new protein bands were formed by reaction of ovalbumin and other proteins with peroxidase, hydrogen peroxide, and a phenol. The position of the peaks corresponded almost exactly to that expected for dimers, trimers, and higher polymers of ovalbumin. The molecular weights of the protein polymers were determined by using gels of varying polyacrylamide concentrations. The observed molecular weights of polymers of ovalbumin produced by reaction with peroxidase were those expected if covalent crosslinks were formed between protein molecules. The bands corresponding to the di, tri, and higher polymers remained after treatment with dodecyl sulfate, urea and mercaptoethanol which breaks all non-covalent bonds in proteins (8).

Many phenolic compounds which are widely distributed in plant tissue formed protein polymers when reacted at physiological concentrations with peroxidase and hydrogen peroxide (9). Proposed mechanisms for this crosslinking or polymerization of protein molecules by peroxidase and such phenols include :

1) Oxidation of epsilon-amino groups of lysine to lysyl aldehyde which condenses with a lysyl residue of a second protein molecule to form dimers linked through Shiff's base formation.

2) Formation of free radicals on tyrosyl residues in proteins which then react to form dityrosyl links between two proteins molecules.

3) Oxidation of phenols to quinones (or to free radicals) followed by a nucleophilic reaction of the quinone (or free radical) with amino or sulfhydryl groups of one protein molecule and then a second oxidation and nucleophilic reaction of the quinones (or free radicals) with second protein molecule to form crosslinks between molecules.

All of these results suggest to me that the oxidative crosslinking of proteins by peroxidase or polyphenol oxidase may be a nonspecific defensive reaction in plants. This is somewhat similar to the crosslinking, or agglutination of protein antigens in animals by antibodies, but without its high specificity. However, in plants, this high specificity of antibodies is not needed for plants can survive even though such an oxidative defense reaction may inactivate enzymes and kill many cells of the host plant as well as the pathogen or some leaves or parts initially infected. This theory requires that oxidase enzymes should increase in plants following infection and that this increase be greater or faster in resistant varieties. In all our studies, we observed increases in oxidase activity following inoculation of resistant plants with bacteria, viral or fungal pathogens which would support this theory. A few examples from our studies will illustrate this increase in oxidase activity after inoculation with typical plant pathogens.

INCREASE IN OXIDASES FOLLOWING INOCULATION OF RESISTANT AND SUSCEPTIBLE PLANTS

Bacterial Disease

The injection into tobacco half leaves of heat killed cells of *Pseudomonas tabaci* induced a large increase in peroxidase activity. It also induced the formation of new isozymes of peroxidase and an increased resistance to subsequent infection (*10*).

This increased resistance in tobacco leaves was shown by using the left half of each leaf as a control injected with water. The right half of the tobacco leaves were injected with boiled peroxidase. The right half of other leaves were injected with 50 mg/ml of commercial peroxidase. Both halves were inoculated with *P. tabaci* cells on the next day and severe disease symptoms developed on the left half leaves 4 days later. The injection of commercial peroxidase blocked disease development in the right half leaves. Boiled peroxidase did not. Injection of heat killed *P. tabaci* cells or material extracted from killed cells also protected the right half leaves from subsequent infection. This protection against bacterial infection by prior injection of killed cells has some features common to vaccination.

Virus Disease

We observed an increase in peroxidase activity in pinto bean leaves following infection by southern bean mosaic virus. This increase in peroxidase following viral infection was greatest in the lesions and the surrounding area. Accompanying this increase was the formation of two new isozymes of peroxidase. Experiments with protein synthesis inhibitors indicated that the formation of new peroxidases in the infected tissues involved protein synthesis (*11*).

Fungal Diseases

Filtrates obtained from 3-day-old cultures of *Fusarium oxysporium* f. sp. *conglutinans* on living cabbage showed a 2,000% increase in peroxidase over uninoculated tissue. When the fungus was grown on autoclaved cabbage extracts, no peroxidase was detected indicating that the peroxidase was formed by the plant (*12*).

The peroxidase and polyphenol oxidase activities from culture filtrates of *F. oxysporium* f. *pisi* grown on resistant and susceptible tomato stem tissue was measured. The activities of both enzymes from cultures on stem tissue from resistant tissue was much higher than from susceptible tissue. Similar results were obtained with pea and cabbage tissues. Thus the results obtained with tomato, pea, and cabbage tissue showed that following inoculation with the *Fusarium* fungus, the level of peroxidase and polyphenol oxidase activity in these tissue greatly increased and that this increase was much greater in tissue from resistant varieties than from susceptible varieties. The enzymatic oxidation of catechol or caffeic acid produced an inhibition of pectic enzymes not shown by the unoxidized phenol. This oxidation also produced an inhibition of fungal growth also not produced by the unoxidized phenols (*13*).

The peroxidase activity increased about 900% in resistant isogenic wheat leaves following inoculation with urediospores of *Puccinia graminis*. There was no increase after inoculation of the susceptible line or in uninoculated leaves. Rust urediospores germinated on nutrient media supplemented with peroxidase showed slow germ tube elongation, suppressed aerial growth and short branched mycelia that resembles the fleck reaction. The increase in peroxidase in the resistant host and the suppression of mycelial development by peroxidase indicate that oxidases may be important in rust resistance (*14*).

The inoculation of isogenic lines of barley with the ectoparasite *Erysiphe graminis* caused a 3 to 5-fold increase in peroxidase within 24 hr. One new major isozyme of peroxidase that was not seen in conidia or in healthy barley leaves was formed. Since this increase occurred in the same period when resistance was expressed and was greater in the immune line, was suggest it plays a role in resistance and merits further study (*15*).

Oxidases as a Biochemical Defense against Infection

I conclude that our studies of protein changes in plants infected with fungal, bacterial or viral pathogens and of the polymerization of proteins by oxidases would support a hypothesis that the increase in peroxidase or polyphenol oxidase following infection may be a nonspecific defense reaction of plants. I suggest that this defense reaction involves a nonspecific oxidative crosslinking of proteins which may inactivate enzymes and kill cells of both the pathogen and host. Such induced oxidases also are involved in the formation of the lignin-like polymers which often form a barrier to penetration of plant pathogens. Lignification as a mechanism of disease resistance has been reviewed (*16*). In some respects this biochemical crosslinking by oxidases of proteins, phenolic compounds and probably also of carbohydrates and nucleic acids reminds me of the agglutination of antigens by antibodies, but without the high specificity of antibodies. However, plants may not require a specific defense system, for plants have the ability to carry on all their metabolic functions even though many of the leaves, roots or stems are killed by the infection or by the plant's biochemical defense. In contrast, animals require a specific defense against pathogens so that the specialized cells or tissues that carry out essential metabolic functions survive the infection. I suggest that this hypothesis merits further thought and study.

At this point in my preparation of this paper, I received the letter from the organizers suggesting that we write an essay-type review describing our research and scientific philosophy, so I decided to include short sections on three studies in plant biochemistry that had world wide impact on agriculture and medicine. The first is an untold story of the development of anticoagulant coumarins that are now used in many countries as rodenticides and as anticoagulant drugs. The second describes how the etiology of a crippling and sometimes fatal disease called Farmer's Lung was solved. The third summarizes our study on a way to increase feed and food production by the preparation and anaerobic fermentation of juices and fibrous residues from forage plants.

THE HEMORRHAGIC SWEET CLOVER DISEASE AND THE DEVELOPMENT OF ANTICOAGULANT COUMARINS

When I came to Wisconsin in 1936 I first worked with W. Roberts in K.P. Link's laboratory. I helped Willard in his development of a quantitative method to determine coumarins in plant tissue that was used by the Genetics Department in breeding for low coumarin clover strains. Willard undertook the isolation of the hemorrhagic principle of spoiled sweet clover hay that had caused fatal bleeding in cattle by using rabbits for prothrombin assays and reported in his thesis that the toxic principle had been concentrated a 1,000-fold (17). H. Campbell continued work on the isolation and reported in his thesis that he had concentrated the toxic principle about 50,000-fold and described only a single isolation of 5–6 mg of an active substance with "a sharp melting point at 253–255°C" which had lowered the prothrombin time of a single rabbit (18).

Campbell left the University soon after filing his thesis, and for several months. Huebner tried without success to repeat the isolation of the active crystals following the procedure described in Campbell's thesis which I understood was accomplished only once. I was then asked by Link to try to isolate the hemorrhagic agent. I postponed my final Ph. D. examination and worked for a few months, but I too could not isolate crystals following the procedure described by Campbell. However, by extracting large amounts of spoiled clover hay in wooden beer barrels instead of beakers and by following only part of the isolation schemes described by Roberts and Campbell and by developing additional steps for the isolation, I was able to obtain 1,800 mg of beautiful crystals with high prothrombin lowering activity and a very sharp melting point of 288–289°C. This melting point was 34 degrees higher than the sharp melting point that was reported by Campbell which indicates that my crystals were not the same as those isolated by Campbell. Campbell and Link then published the isolation scheme outlined in the Campbell thesis but added additional recrystallization steps not previously described and the melting point, microphotograph, molecular weight, electrometric titration, elementary analysis, and biological activity of the crystals isolated by me (*19*). Link first wanted to send my crystals to an out-standing organic chemist in Switzerland for characterization; I asked that he let us try for 6 months to identify them. Within that time, we completed the identification and synthesis of the hemorrhagic agent as 3,3″-methylenebis(4-hydroxycoumarin), Dicumarol (*20*).

When I first suggested to Link that this compound might be used as an anti-coagulant drug, he said that our job was to isolate the toxin and that it was so toxic that it would never be used in medicine and that our work was finished. This greatly disappointed me, so I went back to the laboratory and began the synthesis of about a pound of very pure 3,3″-methylenebis (4-hydroxycoumarin). After several week's work and recrystallization seven times, I took a large bottle of it into Link's office and asked him to please take some of this very pure anticoagulant to the University Medical School to test on dogs. These tests were so promising that additional amounts were sent to the University Medical School, to the Mayo Clinic, to clinics in New York and to the Abbott and Lilly Laboratories for further evaluation and the first tests on human patients. We talked with officials of these drug companies and the Wisconsin Alumni Research Foundation ; then I worked with Schley, the attorney hired by the Wisconsin Alumni Research Foundation, to prepare the patents on the anticoagulant drug, Dicumarol.

As soon as we had identified the hemorrhagic agent, we realized that many analogs could be synthesized and assayed for prothrombin lowering activity. Link then asked me to be in charge of the entire project for I had finished my Ph.D. degree and could spend full time in the laboratory. I knew this synthetic program would require large amounts of 4-hydroxycoumarin which was then not available. Because there was no large equipment in the laboratory suitable for its synthesis, I went to a hardware store and purchased a cast iron, three leg camping kettle. Then I drilled holes in the lid for a stirrer, thermometer and reflux condenser and bolted it down. I poured in acetylmethyl salicylate that I made by acylation of methyl salicylate and mineral oil purchased from the corner drug store that was usually sold as a laxative. I heated the mixture with bunsen burners and carefully dropped in pieces of metallic sodium. There was a small explosion as each piece hit the hot reaction mixture and I was told I would kill myself doing that. Nevertheless, I alone repeated this synthesis until I had enough 4-hydroxycoumarin for the synthetic research by the entire labo-ratory and to synthesize enough Dicumarol for the first clinical tests. About a hun-dred compounds were synthesized by me and others and assayed for prothrombin lowering activity.

A few weeks before I left Madison to take a position with Max Bergman at the Rockefeller Institute, I read Michael's original papers in German. I decided that 4-hydroxycoumarin should undergo a Michael condensation with unsaturated ketones and so I carried out the initial experiments on this Michael condensation in Madison. However, I completed the final synthesis and the proof of structure of the anticoagu-lant compound now known as Warfarin or Coumadin by working nights at the Rocke-feller Institute. A report of this work was mailed to Link with the recommendation that we apply for patent protection. Link wrote me that, "...my report would suffice to establish that I was sole inventor." However, he advised against a patent application and asked M. Ikawa, a beginning graduate who had worked with me, to repeat my studies. Later we published on the condensation of a,b-unsaturated ketones with 4-hydroxycoumarins (21).

The patent office allows patent applications on published material only if authors file within 1 year of publication. About 2 weeks before this year expired, the Foundation followed my recommendations and asked me to work again with Schley to write the basic patent on Warfarin. Schley said it was clear that I could file in my name alone, but since three authors were on our publication, a joint application would be stronger. I agreed.

I was asked to return to Madison in the fall of 1945 to take charge of Link's laboratory while he was in a sanitorium recovering from tuberculosis. I wanted to show that the anticoagulant coumarins were vitamin K antagonists ; so, with the help of graduate students in both biochemistry and plant pathology, I started a study of the effect of these anticoagulants on bacteria. We discovered that some anticoagulant coumarins would stop bacterial growth at levels as low as one part in 2 million. I discussed this finding with E. McCoy of the Bacteriology Department who advised that we determine their mouse toxicity. These mouse assays showed that the compound now known as Warfarin caused fatal hemorrhage at a level as low as 0.001% and was about 100 to 200 times more toxic to mice than Dicumarol. These findings were included in my annual reports for 1946–1948 and a summary was published by the College of Agriculture. This was the first published report of the rodenticidal action of phenylacetyl-4-hydroxycoumarin (Warfarin) (22).

The first field use of anticoagulants for rodent control was reported about the same time by O'Conner who found Dicumarol effective as a rodenticide. My work had clearly shown that Warfarin would be a much better rodenticide than Dicumarol. Knowing this, Link then left the sanitorium against the doctor's advice and told me and our chairman that he alone wanted to continue our studies on anticoagulant coumarins, forced me out of his laboratory and said that I should find other work. At this point in my career I accepted another position and almost left the University of Wisconsin. However, the chairman and others convinced me that I should not leave and that I should stop my anticoagulant research and return to research on the plant biochemistry of disease.

In this way our studies on the biochemistry of sweet clover hay led to an understanding of the etiology of the hemorrhagic sweet clover disease of cattle and to worldwide application in medicine of a new class of anticoagulant drugs and in agriculture of a new type of rodenticide. I suggest tht oxidase enzymes in the spoiled clover hay had oxidized coumarin in the clover 4-hydroxycoumarin which reacted with formaldehyde to form the anticoagulant 3,3'methylenebis(4-hydroxycoumarin) which I had isolated.

ANTIGENS IN MOLDY HAY AS THE CAUSE OF FARMER'S LUNG DISEASE

Although Farmer's lung disease was described in 1932, the exact cause was not known. It is a serious and crippling occupational disease associated with inhalation of dusts, most frequently from hay. The symptoms are shortness of breath, cough, fever, chills and serious changes in the lung tissues. It was suggested that Farmer's lung disease might arise from mechanical irritation, fungal infection or a hypersensitivity to molds or mold products.

In collaboration with the Medical School, we obtained sera from Farmer's lung patients and moldy hay from their farms. Trichloroacetic acid extracts of the hay gave specific precipitation lines with the sera from patients with acute symptoms and with two-thirds of the sera from recovered patients, but not with sera of healthy farmers with no history of Farmer's lung disease (23).

The antigens in the extracts were purified and shown to be composed of peptides linked to a carbohydrate. Trichloroacetic acid soluble antigens from the fungus *Thermopolyspora polyspora* cross-reacted with antisera against the antigens from moldy hays. Pronase digestion did not destroy the antigens, which contained a peptide of lysine, glutamic acid, alanine, and *allo*-isoleucine and a carbohydrate composed of galactose, arabinose, and glucosamine. When patients with a history of Farmer's lung disease were given aerosols of our purified antigens, typical acute symptoms developed after several hours. Our second study of the biochemistry of hay showed that Farmer's lung disease was caused by a delayed hypersensitivity to fungal antigens in moldy hays to which patients had been exposed (24).

In this way, a combined application of plant biochemistry and immunology led to an understanding of the etiology of Farmer's lung disease and provided antigens useful in diagnosis.

AN APPLICATION OF PLANT BIOCHEMISTRY TO INCREASE FEED AND FOOD PRODUCTION

In closing, I will briefly discuss our studies that were directed to a simple way to increase feed and food production from green plants. This involves separating from the fiber a part of the abundant protein and other nutrients that are produced by forage plants. It is done by maceration of fresh green tissue and squeezing out a green juice that contains proteins, carbohydrates, vitamins, and minerals. The green juice or its coagulum formed by heating or fermenting can then be used as a supplement to feeds or foods for monogastric animals. The fibrous residue is a good feed for herbivores and equal on a weight basis to the original forage and can be preserved as silage.

As populations grow it will be necessary to increase the efficiency of agriculture. This requires better utilization of our most efficient forage crops. Our study of the average yields of the essential amino acids (or protein) per acre for 25 common crops grown in the United States from 1953 to 1962 showed that forage crops produced many times more protein than seed crops. Thus, average protein yields from alfalfa were five times that of wheat and about four times that of rice. Forage crops can be grown on poor land and multiple harvests are possible in regions where now only one seed crop can be harvested (25).

In 1961 I obtained samples from England of leaf protein concentrates prepared by heat coagulation of juice from various crops. Our amino acid analyses showed a good amino acid composition, about like that of milk, and little variation between species (26). I then talked with our Dean of Agriculture, a dairy economist, about engineering support. He said, "I would hate to hurt the dairy industry," for he thought it might be competitive with milk.

I thought that if only good silage could be made from the fibrous residue, it would help the dairy industry. But in 1961, I had been told that good silage could not be made from the fibrous residues and was referred to a paper by an international authority on silage production who wrote that, "Silage made from the fibrous residue were unpalatable to all classes of ruminants." Nevertheless, I started experiments on silage making. After many failures, we learned how to make good silage from the fibrous residues of alfalfa and other plants and from various plant wastes like pea vines and beet tops. Cows readily ate all of this silage and good milk production was maintained (27).

Our analyses of alfalfa juice showed a very high content of thiamine, riboflavin, carotene, and xanthophyll (28). The mineral content was also high. In my estimation, plant juice concentrates should be used as a supplement to existing feeds and foods as a good source of vitamins and minerals as well as protein and carbohydrates and should not be considered as a source of only protein.

The energy costs to heat coagulate the protein in alfalfa juice is greater than the total energy for fertilizer, tillage, harvesting plus maceration, and juice expression. Hoping to reduce this energy cost, I undertook studies on the coagulation and preservation of the protein in plant juices by anaerobic fermentation. After some failures, I was able to make good liquid silage from juices of 12 forage plants or plant wastes produced in temperate climates and from an equal number of forage plants grown in tropical regions. Acid forming bacteria in the juice lowered the pH from an initial pH of about 6 to around pH 4, usually within 1 or 2 days. Protein in the juice was coagulated and settled to the bottom. Good liquid silage was made from alfalfa juice in quantities ranging from about 50 ml to 1,000 gallons (29).

The saponins in alfalfa or its juice are toxic to monogastric animals and limit the amount that can be safely fed. During the anaerobic fermentation, saponins were destroyed ; this was an unexpected advantage.

During any anaerobic silage fermentation, there is the possibility of clostridial toxin formation. So we tested for toxins by injecting 0.5 ml of the supernatant from fresh, heated or fermented alfalfa juice into mice. All of the mice injected with 6 samples of the fresh or 6 samples of heated alfalfa juice died ; apparently from saponin toxicity. None of the mice injected with 42 samples of fermented alfalfa juice died. Saponin assays by red cell hemolysis showed no saponins in the fermented juice or its coagulum.

In my opinion, the application of this new energy saving harvesting system to the utilization of forage crops and plant wastes could significantly increase feed and food production. It could reduce the cost of feeds and make forage harvesting almost independent of weather. It would enable preservation of the abundant green plant material that grows well in many regions only during the rainy season. It would facilitate forage harvest in northern regions where it is now difficult to dry forage and could make possible forage harvest during the winter in regions like the southeast coastal plains of the United States. As populations grow and the need for more food increases, it may be necessary to raise $1-\frac{1}{2}$ or $1-\frac{3}{4}$ crops in regions where the climate now allows only one crop to be raised to maturity. The nutrients in the coagulum from juice of such immature crops can be used as feed or food for monogastric animals and the fibrous residue fed to ruminants. The minerals in the supernatant can be recycled as fertilizer.

Plant oxidases are very stable proteins and may retain activity after heating or drying. The interaction of oxidases with phenols and proteins during the drying of forages or a heating of their juice may lower the digestibility and destroy lysine, cysteine and methionine in the proteins and so reduce the nutritive quality of the hay or silage or the protein coagulum. This oxidative lowering of protein quality is enhanced when the forage is wet by rain during the drying period and so poor quality hay or silage is formed. A mechanical dewatering and the exclusion of oxygen in the anaerobic fermentation of the juice or fibrous residues blocks these oxidative reactions.

It is my hope that continued further applications of plant biochemistry to agronomy, engineering, nutrition, and animal science will allow us to realize more of the great potential to increase the production of feeds and foods by utilizing the entire plant in the form of the nutrients from plant juices for monogastric animals, including man, and by feeding the fibrous residues to ruminants.

It is my experience and my scientific philosophy that the application of plant biochemistry to problems of pathology, medicine, nutrition, and agriculture can give important solutions to some of the major problems that now face our society.

REFERENCES

1. Uritani, I. and Stahmann, M.A. : 1961. *Plant Physiol.* 36, 707–782.
2. Clare, B., Weber, D.J. and Stahmann, M.A. : 1966. *Science* 153, 62–63.
3. Weber, D.J. and Stahmann, M.A. : 1966. *Phytopathology* 56, 1066–1070.
4. Stahmann, M.A., Clare, B.C. and Woodbury, W. : 1966. *Plant Physiol.* 41, 1505–1512.
5. Demorest, D.M. and Stahmann, M.A. : 1972. *Biochem. Biophys. Res. Commun.* 47, 227–233.
6. Stahmann, M.A. and Demorest, D.M. : 1972. *Symp. Biol. Hung.* 13, 355–356.
7. Stahmann, M.A. and Spencer, A.K. : 1977. *Biopolymers* 16, 1299–1306.
8. Stahmann, M.A., Spencer, A.K. and Honold, G.R. : 1977. *Biopolymers* 16, 1307–1318.
9. Leatham, G.F., King, V. and Stahmann, M.A. : 1980. *Phytopathology* 70, 1134–1140.
10. Lovrekovich, L., Lovrekovich, H. and Stahmann, M.A. : 1968. *Phytophathology* 58, 193–198.
11. Farkas, G.L. and Stahmann, M.A. : 1966. *Phytopathology* 56, 669–677.
12. Heitfuss, R., Stahmann, M.A. and Walker, J.C. : 1960. *Phytopathology* 50, 372–375.

13. Stahmann, M.A. : 1965. *Tagensberichte* 74, 9–49.
14. Macko, V., Woodbury, W. and Stahmann, M.A. : 1968. *Phytopathology* 58, 1250–1254.
15. Hislop, E.C. and Stahmann, M.A. : 1971. *Physiol. Plant Pathol.* 1, 297–312.
16. Vance, C.P., Kirk, T.K. and Sherwood, R.T. : 1980. *Annu. Rev. Phytophathol.* 18, 259–288.
17. Roberts, W. : 1937. Ph.D. Thesis, pp.60–70. University of Wisconsin, Madison, Wisconsin.
18. Campbell, H.A. : 1939. Ph.D. Thesis, pp.5–45, University of Wisconsin, Madison, Wisconsin.
19. Campbell, H.A. and Link, K.P. : 1941. *J. Biol. Chem.* 138, 21–33.
20. Stahmann, M.A., Huebner, C.F. and Link, K.P. : 1941. *J. Biol. Chem.* 138, 513–526.
21. Ikawa, M., Stahmann, M.A. and Link, K.P. : 1944. *J. Am. Chem. Soc.* 66, 902–906.
22. Agriculture Experiment Station Annual Report : 1949. Bulletin 487, Part 1. pp.79–80. College of Agriculture, University of Wisconsin, Madison, Wisconsin.
23. Kobayashi, M., Stahmann, M.A., Rankin, J. and Dickie, H. : 1963. *Proc. Soc. Exp. Biol. Med.* 113, 472–476.
24. Laberge, D.E. and Stahmann, M.A. : 1966. *Proc. Soc. Exp. Biol. Med.* 121, 458–462.
25. Akeson, W.R. and Stahmann, M.A. : 1966. *Economic Bot.* 20, 244–250.
26. Gerloff, E.D., Lima, I.H. and Stahmann, M.A. : 1965. *Agric. Food Chem.* 13, 139–145.
27. Oelschlegel, F.J., Schroeder, J.R. and Stahmann, M.A. : 1969. *Agric. Food Chem.* 17, 791–795, 796–798.
28. Hartmann, G.H., Akeson, W.R. and Stahmann, M.A. : 1967. *Agric. Food Chem.* 15, 74–79.
29. Stahmann, M.A. : 1978. *In* Proceedings of the Second International Green Crop Drying Congress (Howarth, R.E., ed.), pp.42–49, University of Saskatchewan.

SUBJECT INDEX